2D Materials as Electrocatalysts

Online at: https://doi.org/10.1088/978-0-7503-5291-8

2D Materials as Electrocatalysts

Muhammad Ikram
Solar Cell Applications Research Lab, Department of Physics, Government College University Lahore, 54000 Punjab, Pakistan[†]

Ali Raza
Department of Physics 'Ettore Pancini', University of Naples 'Federico II', Piazzale Tecchio, 80, 80125 Naples, Italy[†]

Jahan Zeb Hassan
Solar Cell Applications Research Lab, Department of Physics, Government College University Lahore, 54000 Punjab, Pakistan[†]

Salamat Ali
Department of Physics, The University of Lahore, 54000 Punjab, Pakistan

IOP Publishing, Bristol, UK

[†] M Ikram, A Raza. and J Z Hassan contributed equally.

ISBN 978-0-7503-5291-8 (ebook)
ISBN 978-0-7503-5289-5 (print)
ISBN 978-0-7503-5292-5 (myPrint)
ISBN 978-0-7503-5290-1 (mobi)

DOI 10.1088/978-0-7503-5291-8

Version: 20241201

IOP ebooks

British Library Cataloguing-in-Publication Data: A catalogue record for this book is available from the British Library.

Published by IOP Publishing, wholly owned by The Institute of Physics, London

IOP Publishing, No.2 The Distillery, Glassfields, Avon Street, Bristol, BS2 0GR, UK

US Office: IOP Publishing, Inc., 190 North Independence Mall West, Suite 601, Philadelphia, PA 19106, USA

Contents

Preface viii

Acknowledgments x

Author biographies xi

1 Introduction to 2D electrocatalysts **1-1**

1.1 Pre-eminence of 2D nanocatalysts 1-2
1.2 Categories of 2D electrocatalysts 1-3
1.2.1 Graphene 1-3
1.2.2 MXenes 1-5
1.2.3 Transition metal dichalcogenides 1-7
1.2.4 Hexagonal boron nitride 1-9
1.2.5 Black phosphorus 1-10
1.2.6 Transition metal oxides 1-12
1.2.7 Graphitic carbon nitride 1-13
1.3 Conclusion, challenges and perspectives 1-14
Bibliography 1-16

2 Development protocols and catalyst design **2-1**

2.1 Development protocols 2-1
2.1.1 Chemical vapor deposition 2-1
2.1.2 Exfoliation 2-3
2.1.3 Hydro/solvothermal method 2-6
2.1.4 Other methods 2-7
2.2 Catalyst design 2-7
2.2.1 Controlling the electronic structure of electrocatalysts 2-7
2.2.2 Creating dual active sites 2-13
2.2.3 Adjusting the surface geometry of electrocatalysts 2-15
2.2.4 Lateral size and thickness regulation 2-18
2.3 Comparative analysis via experiments and computations 2-18
2.4 Conclusion, challenges and perspectives 2-20
Bibliography 2-21

3 Characteristics and performance of 2D electrocatalysts **3-1**

3.1 Electrochemical reactions at catalyst surfaces 3-1
3.2 Electrochemical stability of 2D materials as electrodes 3-2

3.3 Electron transfer at 2D materials in electrochemical reactions 3-4
3.4 Parameters for evaluating the performance of catalytic activity 3-8
3.4.1 Gibbs free energy and overpotential 3-8
3.4.2 Tafel slope and exchange current density 3-9
3.4.3 Turnover frequency 3-10
3.4.4 Stability 3-10
3.5 Conclusion, challenges and perspectives 3-11
Bibliography 3-12

4 The electrochemical hydrogen evolution reaction 4-1

4.1 Principles and mechanisms of the electrochemical HER 4-1
4.1.1 Acidic media 4-2
4.1.2 Alkaline media 4-2
4.2 Graphene-based electrocatalysts 4-3
4.2.1 Single-atom catalyst 4-3
4.2.2 Heteroatom-doped graphene 4-4
4.3 Transition metal dichalcogenides 4-7
4.3.1 TMD–metal electrocatalysts 4-8
4.3.2 TMD–carbon electrocatalysts 4-11
4.3.3 TMD–metal-oxide electrocatalysts 4-13
4.4 2D MXenes for the HER 4-16
4.5 Heterostructure schemes 4-23
4.5.1 Dimension-based schemes 4-24
4.5.2 Materials-based schemes 4-30
4.6 Conclusions, challenges and perspectives 4-34
Bibliography 4-35

5 The electrochemical oxygen reduction reaction 5-1

5.1 Reaction kinetics and mechanisms 5-1
5.2 Graphene-based electrocatalysts 5-3
5.2.1 Pt-supported graphene catalysts 5-4
5.2.2 Pt-supported heteroatom-doped graphene catalysts 5-8
5.2.3 Non-Pt-based precious metal catalysts on graphene 5-13
5.2.4 Non-precious metal catalysts on graphene 5-16
5.3 2D-MXenes for ORR 5-16
5.4 Transition metal dichalcogenides 5-20
5.5 Transition metal oxides 5-21

5.6 Transition metal hydroxides 5-21
5.7 Conclusion, challenges and perspectives 5-22
Bibliography 5-23

6 Electrochemical CO_2 reduction and beyond 6-1

6.1 Reaction kinetics and mechanisms 6-1
6.1.1 Measurement criteria 6-2
6.2 Graphene-based electrocatalysts 6-3
6.2.1 Alteration of graphene to improve CO_2 reduction 6-4
6.3 Transition metals 6-7
6.4 Transition metal-oxides 6-7
6.5 Transition metal dichalcogenides 6-10
6.6 2D-MXenes for CO_2 reduction 6-10
6.7 Other reactions 6-13
6.7.1 Nitrogen reduction reaction 6-13
6.7.2 Oxidation of carbon fuels 6-16
6.8 Industrial significance of electrocatalysis for clean energy 6-18
6.8.1 Activity at the edges 6-20
6.8.2 Mass transport 6-23
6.8.3 Introducing impurities and defects 6-23
6.9 Conclusion, challenges, and perspectives 6-28
Bibliography 6-30

7 Strategies to improve electrocatalytic activity 7-1

7.1 The hydrogen evolution reaction 7-1
7.1.1 Engineering protocols 7-1
7.1.2 Heteroatom doping 7-12
7.1.3 Formation of heterostructures 7-15
7.2 Oxygen reduction reactions 7-17
7.2.1 Defect engineering 7-17
7.2.2 Nonmetal doping 7-19
7.2.3 Metal doping 7-22
7.3 CO_2 reduction reactions 7-23
7.3.1 Surface modification 7-23
7.3.2 Surface-structure tuning 7-24
7.3.3 Optimization of the electrolyte and electrolyzer 7-25
7.4 Conclusion, challenges and perspectives 7-26
Bibliography 7-26

Preface

The field of electrocatalysis stands at the forefront of scientific research and technological innovation, driving the quest for sustainable energy solutions and environmental remediation. This book, *2D Materials as Electrocatalysts*, delves into the transformative potential of two-dimensional (2D) materials in catalyzing key electrochemical reactions, offering a comprehensive exploration of their properties, synthesis, and applications. The critical need for the development of efficient and affordable electrocatalysts is growing as the world struggles to shift from fossil fuels to cleaner energy sources.

An overview of 2D electrocatalysts is given at the start of the journey, laying the stage for realizing their unmatched benefits. The importance of 2D nanocatalysts is examined in depth, and then several 2D materials including graphene, MXenes, transition metal dichalcogenides, and black phosphorus are described in detail. The distinct structural and electronic behaviors of each material are explored, highlighting their potential for use in electrocatalytic processes. The design and production of 2D electrocatalysts are discussed thoroughly, ranging from various bottom-top to top-down approaches. The significance of catalyst design is also emphasized in this book, with methods for enhancing catalytic performance including controlling the electronic structure, generating active sites, and modifying surface areas highlighted. The performance characteristics of 2D electrocatalysts are scrutinized through a lens of electrochemical reactions and stability. A thorough grasp of their catalytic efficiency is provided by explaining the significant factors, including turnover frequency, Tafel slope, overpotential, and Gibbs free energy. This section is vital for beginners and researchers aiming to benchmark and enhance the performance of 2D materials in various electrochemical applications.

The book provides a thorough analysis of the electrochemical oxygen evolution and reduction reactions, along with the carbon dioxide reduction reaction, delving deeper into certain processes. We address the kinetics, electrochemistry mechanisms, and material-specific developments, providing insights into the difficulties and breakthroughs in these critical areas. Graphene-based electrocatalysts, MXenes, and transition metal compounds are among the key materials evaluated for their oxygen reduction reaction and carbon dioxide reduction reaction capabilities. The story then moves on to discuss the industrial importance of electrocatalysis, in particular in relation to clean energy. Comprehensive coverage is also given to the strategies for improving electrocatalytic activity through heteroatom doping, engineering techniques, heterostructure formation, and so on. Every chapter ends with a discussion of issues, viewpoints, and potential future perspectives, which invites readers to reflect on the field's continuous development and possible innovations.

In conclusion, *2D Materials as Electrocatalysts* is an indispensable resource for researchers, academicians, and industry professionals. It bridges the gap between fundamental science and practical applications, offering a rich tapestry of knowledge that underscores the pivotal role of 2D materials in shaping the future of electrocatalysis. This book aims to be a beacon for those dedicated to pioneering sustainable technologies and advancing the frontier of material science.

The Authors

Acknowledgments

We extend our heartfelt gratitude to Esha Shehzadi, sister of one of the authors, Ali Raza, for her invaluable assistance in securing permissions from publishers for the reuse of figures. Her meticulous efforts in figure editing have greatly enhanced the quality of this manuscript.

We also thank Laiba Jahan Zeb for her dedicated support in formatting and proofreading the manuscript. Her attention to detail and professionalism ensured that the content was presented with clarity and precision.

Finally, we acknowledge the unwavering support of our families and colleagues whose encouragement provided the foundation for this achievement. This book is a testament to their belief in our vision.

Author biographies

Muhammad Ikram

Dr Muhammad Ikram obtained his PhD degree in physics from the Department of Physics, Government College University (GCU) Lahore through the Pak–US joint project between the Department of Physics, GCU Lahore, Pakistan, and the University of Delaware, USA in 2015. He served as Deputy Director of Manuscript Science at the Punjab Textbook Board, Pakistan. Later on (2017–), Ikram joined the Department of Physics, GC University Lahore as an Assistant Professor of Physics and Principal of the Solar Cell Applications Lab and Advance Nanomaterials Research Lab. Ikram received the Seal of Excellence Marie Skłodowska Curie Actions Individual Fellowship in 2017 and 2020. Ikram is author of 16 international books chapters, six books and two US patents. His research interests involve the synthesis and characterization of inorganic semiconductor nanomaterials, and 2D materials for water treatment optoelectronic, energy storage and electrocatalytic applications.

Ali Raza

Ali Raza obtained his master's degree in material physics from Riphah International University, Islamabad, Pakistan, in 2019. He conducted his MPhil research at the Solar Cell Applications Lab at GCU Lahore, Pakistan, where he also worked as a Research Associate focusing on the catalytic and energy harvesting applications of pure and engineered 2D materials, including TMDCs, graphene, h-BN, and MXenes. Currently, Raza is a doctoral scholar at the Department of Physics 'Ettore Pancini', University of Naples Federico II, Naples, Italy. His research is centered on the fabrication of functional 2D materials for organic electronic and sensing applications.

Jahan Zeb Hassan

Jahan Zeb Hassan obtained his MPhil degree in physics from Riphah International University, Islamabad, Pakistan, in 2020. He completed his MPhil research in the Solar Cell Applications Lab at Government College University Lahore, Pakistan. Following this, he worked as a Research Assistant in the same laboratory, focusing on the synthesis and catalytic applications of nanomaterials and 2D materials for energy harvesting applications. Currently, Hassan serves as the Head of the Physics

Department at Punjab Group of Colleges, Kharian Campus, Gujrat, Pakistan. His current research involves designing and fabricating 2D materials for electrochemical analyses, energy harvesting, and storage applications.

Salamat Ali

Professor Dr Salamat Ali is working as a Professor of Physics at The University of Lahore (Main Campus) and retired as a Full Professor of Physics from GC University Lahore, Pakistan. He completed his PhD in 1996 at the University of Durham, UK, in superconductivity and magnetism, and his post-doc in 2006 at the K F University of Graz, Austria, with a specialization in nanotechnology. He also completed training in solar cell technology at the University of Delaware, USA, in 2012. His current research interests include 2D materials, renewable energy, water purification, sensor technology, and the development of medicines for cancer treatment. He has 38 years of experience in physics and material science.

IOP Publishing

Muhammad Ikram, Ali Raza, Jahan Zeb Hassan and Salamat Ali

Chapter 1

Introduction to 2D electrocatalysts

Graphene was the first material to be classified as a two-dimensional (2D) material. It is an atomically thin carbon material that is formed through peeling layers off graphite. Graphene displays extraordinary anisotropic electrical and physical characteristics. Other types of ultrathin 2D materials, such as single- or few-layered transition metal dichalcogenides (TMDs), metal oxides, layered double hydroxides (LDHs), hexagonal boron nitride (*h*-BN), graphitic carbon nitride (g-C_3N_4), as well as metal carbides and nitrides (known as MXenes), have also been investigated as a result of the precedent set by graphene. The atomic thickness of these materials ranges from one to several atoms, and one of their distinguishing characteristics is the way in which they are layered with distinct bonding interactions. Generally, anisotropy causes these ultrathin sheets to exhibit characteristics distinct from those of the parent bulk material. Because of their layered structure, early applications of layered 2D materials included their use as lubricants. Other types of 2D materials, such as TMDs, LDHs, metal oxides, and MXenes, all come with a variable bandgap and provide a wide variety of potential applications. It is now necessary to analyse the performance of numerous 2D materials regarding their use as catalysts to support recent breakthroughs in sustainable energy that are gathering attention on a world-wide scale. These categories include the hydrogen evolution reaction (HER), oxygen reduction reaction (ORR), oxygen evolution reaction (OER), and the carbon dioxide reduction reaction (CO_2RR); electrocatalysis is at the center of the process of converting clean energy using processes such as CO_2RRs in the technologies of the future. 2D materials are being researched for these reactions as potential cost-effective replacements for the costlier platinum (Pt)-based catalysts.

The importance of 2D materials in catalytic applications can be attributed to their mechanical properties, specific surface area, and electrical and thermal conductivity. Moreover, the 2D materials have the highest possible surface-to-bulk ratios that yield higher density for active sites and they are hence more suitable for surface-active applications. In addition, their outstanding mechanical qualities allow the life

doi:10.1088/978-0-7503-5291-8ch1

of the catalyst to be prolonged, and their higher thermal conductivity makes it easier for the catalyst to dissipate the heat generated by exothermic processes. In addition, the electrical characteristics of 2D materials are adjustable, which allows for more control over the catalytic activity of the material. When compared with bulk materials, the advantages of greater surface area, strengthened mechanical assemblies, and the extraordinary active site density of 2D materials in terms of the activity and stability of catalysts can be seen in the fact that, when manufacturing hierarchical composite catalysts, the utilization of 2D materials is the preferred choice for the building blocks, in contrast to their bulk counterparts. This chapter examines the differences and similarities between the electrochemical and electrocatalytic characteristics of several 2D materials. The individual components that make up each structure are described. In the second part of this chapter, we will address the inherent electrochemistry of the materials, which is caused by the materials' inherent activity or redox processes and affects the materials' stability within a predetermined potential window. Third, various 2D materials for strategic applications in electrocatalysis are also studied with a particular emphasis on the functions that edges and surface properties play in the process.

1.1 Pre-eminence of 2D nanocatalysts

The structural formations of 2D nanocatalysts demonstrate their essential effect in catalytic activity through their geometric arrangements that account for the distinctive properties of 2D catalysts, as described below.

1. **Surface-active sites**: The common geometric arrangements of 2D catalysts provide a well defined surface area that allows the exposure of surface-active sites with greater density and lattice planes. The greater density of surface-active sites can boost catalytic reactions on the surface of the material. An alternative strategy to increase the exposure of surface-active sites is to reduce the lateral size of 2D nanocatalysts [1]. For example, ultrathin MoS_2 exhibits superior HER activity compared to its bulk counterpart, which can be attributed to the enrichment of active sulfur edges [2].
2. **Carrier mobility:** Greater electron mobility has been comprehensively perceived in numerous 2D ultrathin materials, for example graphene, TMDs, and black phosphorus (BP) nanosheets [3]. Carrier mobilities for MoS_2 and graphene have been reported as around 10^1 cm^2 V^{-1} s^{-1} [4, 5] and 10^2–10^4 cm^2 V^{-1} s^{-1} [6, 7], respectively. The distinctive ultrathin assembly of catalysts offers comparatively higher charge migration because of the short transportation path and low value of primary resistance. An observable decrease in the HER activity of MoS_2 attached to extra atomic layers was described by Yu *et al*, which can be deduced to be due to the constraint of electron mobility along the vertical direction in a layered material [8].
3. **Energy band structures:** Varying the number of layers in a 2D crystal causes the bandgap energy to change, and the tunable band structure determines the light absorption characteristics for photocatalytic activity. The bandgap energy for Bi-based 2D layered materials can be tuned to 0.3–3.6 eV through

incorporating numerous anions and cations into an intrinsic assembly, whose resultant light reaction covers the ultraviolet to near-infrared range [9]. Further, the bandgap energy of MoS_2 can be modified by means of varying cumulate layers, and is found to be 1.8–1.9 eV for mono-layered and 1–1.2 eV for few-layered MoS_2 [10]. Additionally, the bandgap energy for g-C_3N_4 can be modified in the range 1.6–1.1 eV against a normal hydrogen electrode (NHE) [11]. Moreover, the interfacial desorption/adsorption-free energy between reactants and catalysts can also be controlled using a varying electron distribution along with adjustable band structure [12].

4. **Electronic properties:** The electronic characteristics can be controlled by adjusting the thickness of 2D catalysts [13]. The 2D electronic structures can regulate the bond strength between catalytic active sites and reactants and decrease the desorption kinetic barrier [14]. In this regard, Fang's group described a 2D mesoporous covalent organic framework based battery with increased areal capacitance (5.46 mF cm^{-2}), gravimetric power (55 kW kg^{-1}) and high power density (4.1–5.4 W cm^{-3}). These outcomes improved by two orders compared to conventional lithium thin-film batteries [15]. In addition, the 2D thin-film catalysts demonstrated numerous in-plane defects, which are conducive to electronic characteristics such as electrical conductivity [16] that improve the energy of electrical conduction produced via catalytic activity. Likewise, the plentiful defects of mono-layered WS_2 nanosheets are advantageous for the HER, which is associated with induced lattice distortions, as reported by Voiry *et al* [17].
5. **Mechanical properties:** 2D materials tend to possess remarkable mechanical characteristics [18] that produce higher catalyst durability. Moreover, the robustness of 2D materials suggests an opportunity for the development of hybrid nanocatalysts for advanced catalytic activity.

The characteristics mentioned above show the superiority of 2D nanocatalysts, and can yield significant advancements for electrocatalysis using these materials. However, their electrocatalytic activity requires further refinement, and a series of strategies is emerging gradually, which will be examined in the following chapters.

1.2 Categories of 2D electrocatalysts

1.2.1 Graphene

Electrocatalysts can significantly reduce the energy barrier and achieve the most effective water-splitting [19]. The catalyst performance standard for electrochemical hydrogen splitting is based on noble metals, i.e. catalysts based on noble metals provide the benchmark efficiency for electrochemical water-splitting [20]. However, large-scale H_2 evolution via noble metals is impractical because of their scarcity and exorbitant cost. Recently, studies have concentrated on (earth-abundant) water-splitting electrocatalysts which allow excellent activity and long-term durability. Because of their excellent catalytic performance and plentiful reserves, transition metal compounds (TMCs), e.g. transition metal oxides (TMOs), carbides, sulfides, selenides, nitrides, and phosphides, have been identified as potential replacements for noble-metal-based catalysts. Despite much research and enormous effort, the activity of TMCs in hydrogen splitting is relatively lower than for noble-metal-based catalysts. As a result of this challenge, numerous initiatives have been carried out to improve the effectiveness of TMCs in water-splitting. Current research on TMC catalyst phase structural tuning exhibits promising improvements for HER/OER catalytic performance [21]. Improving the water-splitting proficiency of TMCs through chemical doping using heteroatoms has also been investigated by measuring the electrical assembly of active surface edges precisely [22]. In addition to addressing the above-mentioned challenges, the graphene-based hybrids provide yet another method for highly effective water-splitting [23].

Graphene is a monoatomic crystal composed of carbon atoms (sp^2-bonded) in a honeycomb-like assembly in the form of a hexagonal lattice, as shown in figure 1.1. Consistent with fundamental carbon structures, graphene is recognized as an essential building block for a variety of different dimensional carbon nanomaterials, such as bulk graphite and other 3D carbon structures, 1D carbon nanotubes (CNTs) and 0D fullerenes. The great interest in graphene is due to its numerous distinctive features, including high (1.0 TPa) elastic modulus, enormous (2630 $m^2\ g^{-1}$) theoretical surface area [18], and high (5000 $W\ m^{-1}\ K^{-1}$) thermal [24] as well as

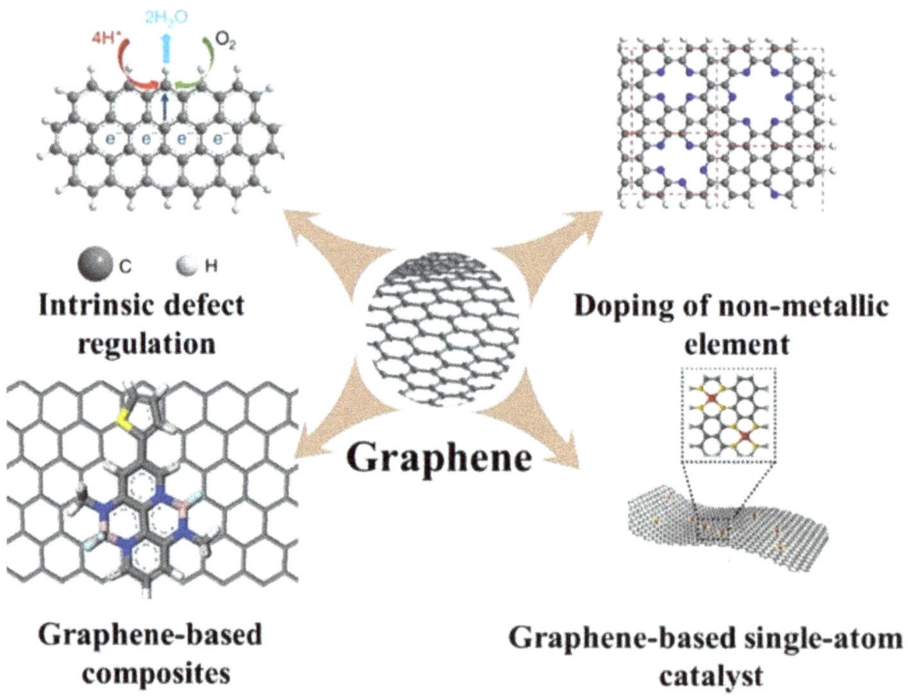

Figure 1.1. Electrocatalysts based on graphene are tuned by incorporating defects, non-metallic compounds, and monoatomic and additional active elements into the graphene framework. (Reproduced with permission from [31]. Copyright 2020 John Wiley and Sons. Data are modified from [32].)

electrical (10^6 S cm^{-1}) conductivities [25]. Graphene presents both challenges and opportunities in electrocatalytic applications. The adsorption of chemical intermediates is usually reported as being endothermic and causes poor intrinsic conductivity due to π bonding linkage (delocalized) of graphene. However, the basal plane of pristine graphene is generally regarded as inert for electrocatalytic activities. Therefore, it is considered desirable as a substrate for electrocatalytic applications, which is attributable to its excellent surface area and strong conductivity.

Moreover, graphene's electrical and surface characteristics can be modified via developing heterostructures, heteroatom doping, and defect engineering. Commonly, the electronic states at the graphene's edges are different from those in the basal plane, thus making it effective for electrocatalysis [16]. For example, using density functional theory (DFT) simulations, Deng *et al* discovered that the ORR mechanism is active along graphene's edges [26]. Heteroatom-doped graphene has demonstrated considerable promise for HER and OER applications. The HER studies were reported by Jiapo *et al*, involving a variety of graphenes doped with non-metal heteroatoms, and correlating the theoretically calculated adsorption energies of significant reaction intermediates with the electrochemical reaction rates [27]. In one of the first works to propose an approach for a structure of atomically effective electrocatalysts based on graphene design for more general electrocatalytic processes [28], Sun *et al* looked at the possibility of using N-graphene using a matrix for Pt clusters and monoatomic catalysts [28]. Catalysts based on graphene have also demonstrated an exceptional electrocatalytic response for carbon cycle reactions, which will be described comprehensively below [29]. In addition to the water cycle reaction, N-graphene displayed significant electrochemical CO_2 reduction (ECR) performance toward format production [30]. DFT simulations suggest that the N-atom in a graphene context reduces the free energy obstruction to intermediate adsorption, indicating additional helpful *COOH binding as a possible explanation for this increase in ECR activity.

1.2.2 MXenes

Over the past few years, the need for and attention toward unique 2D materials, for example, graphene, has increased significantly, for use in numerous applications, e.g. memory devices [33, 34], flexible electrical devices [34, 35], and energy storage devices [36, 37]. The 2D field is quite enormous, including metals ($NbSe_2$) [38], semiconductors (MoS_2) [39], and insulators (*h*-BN) [40]. Further, research from 2011 revealed a novel category of materials, called MXenes, with prominent A-group elements, obtained by layer etching (particularly Al, Ge, etc) from the respective bulk counterpart $M_{n+1}AX_n$ (MAX) (see figure 1.2(a)) [41]. The variety of substitutions in different MXene sites has resulted in a great range of $M_{n+1}AX_n$ materials with tunable qualities and new probable applications [42]. MXenes differ from conventional 2D materials in terms of their thermal stability, strong conductivity, and precise composition control.

In a previous study, Mo_2CT_x demonstrated excellent performance for the HER, demonstrating the tremendous potential of MXenes in this domain [43]. The study of materials based on MXene is in its initial stages compared to the vast number of

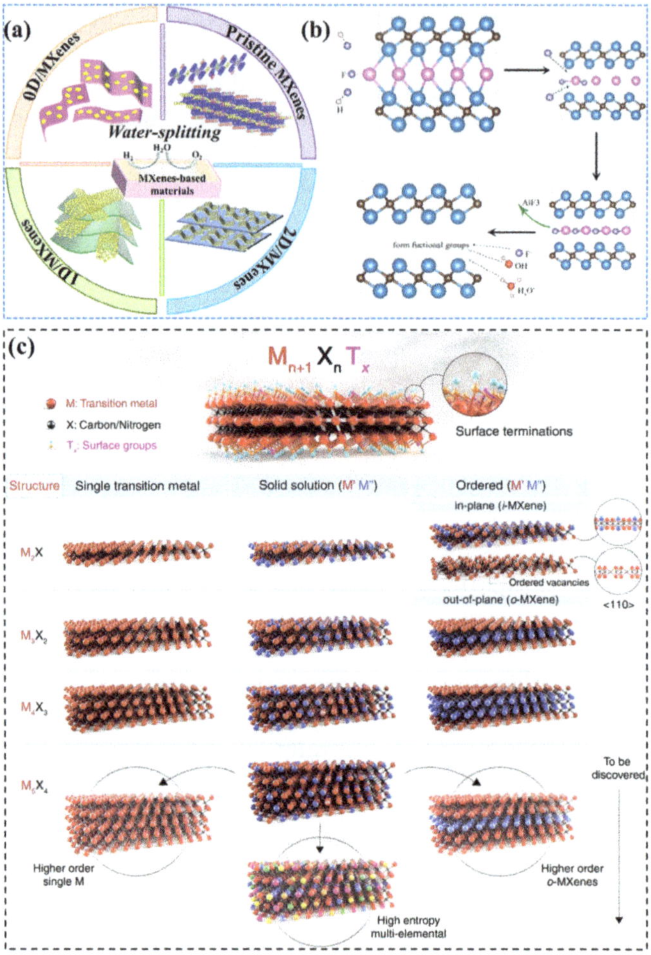

Figure 1.2. (a) Different structures of MXene-based electrocatalysts for water-splitting and (b) illustration of the etching procedure for Cr_2AlC. (Reproduced with permission from [53]. Copyright 2019 American Chemical Society.) (c) Graphic representation of MXene structures. (Reproduced with permission from [54]. Copyright 2021 American Association for the Advancement of Science.)

studies on graphene. Numerous MXenes have been synthesized experimentally to date, including titanium carbide (Ti_3C_2) [44], molybdenum carbide (Mo_2C) [45], vanadium carbide (V_2C) [46], titanium nitride (Ti_3N_2), Ta_4C_3 [47], etc. Additionally, there are reports of experiments on the more complicated monolayer Ti_3CN [48]. Ti_3CN may also be intercalated with ions related to the layered carbides/nitrides [49]. Due to their superior adjustable surface chemistry, resilient metallic conductivity, and obvious active sites, MXenes have received enormous attention in energy storage and conversion applications. Further, the terminal groups over the MXene surface are chemically linked with metal atoms, serving as anchoring sites for external active phases (figure 1.2(b)) [50]. The active phase can modify the electron density of the surface atoms, which is significant in catalysis applications. The environment of the exfoliation solution affects the properties of the terminal groups developed via the exfoliation strategy. MXenes show incredible performance in the domain of electrocatalysis [51]. Due to their unique properties, MXenes can be used as both a substrate and a catalyst to assist the active species. Numerous optimization schemes are employed to stimulate the catalytic response of MXene-based catalysts. MXenes with 2D materials, nanoparticles and single atoms have been introduced; these combinations typically have beneficial effects.

MXenes need to be modified for particular catalytic processes. For example, F-functional groups over the MXene surface may prevent HER activity; however, they are advantageous for the nitrogen reduction reaction (NRR). MXene materials have demonstrated latent capacity in various electrocatalytic processes; however, as mentioned, research in this field is still in its early stages. For MXene-based catalysts, there are many more unexplored domains than for other 2D catalysts, which may be used to boost catalytic behavior through theoretical calculations and experimental techniques. The challenge of MXene production on a larger scale is another issue that needs to be resolved [52]. The definitive goal of investigating reaction pathways is to use fundamental knowledge to create superior catalysts. However, the creation of high-efficiency electrocatalysts is greatly hampered because the discoveries are still in the experimental synthesis phase and are insufficient for considering fundamental catalytic pathways. The design of superior activity electrocatalysts and the investigation of their pathways at the atomic level is now possible through theoretical calculations in conjunction with experiments, which can be attributed to the rapid development of computing technology. Before experimental synthesis, first-principles simulations can evaluate many potential catalyst materials. Theoretical models can help identify the chemical processes occurring at the atomic level, and the combination of theoretical calculations and experiments can reveal the active spots in the materials. Instead of the more time-consuming trial-and-error approach, the design of a catalyst based on theoretical measurements may allow faster and more significant progress.

1.2.3 Transition metal dichalcogenides

Layered TMDs are of interest due to their attractive characteristics and extraordinary potential for various applications [55]. Interestingly, for the HER, the implementation of stacked TMDs as practical and economical electrocatalysts can serve as a replacement for costly Pt-based catalysts [55]. Large energy inputs are a prerequisite to driving the HER, however, TMD electrocatalysts that can reduce the HER overpotential can be selected. Additionally, the Tafel slope and overpotential for the HER are alternative standards to determine its activity. The Tafel slope elucidates that an increment in overpotential is required to produce an order-of-magnitude increase in current density. According to this criterion, smaller Tafel slopes would be given preference in a future HER electrocatalyst. The following steps can prevent the release of H_2 molecules and protons:

1. Adsorption step (Volmer reaction):

$$H_3O^+ + e^- \rightarrow H_{ads} + H_2O, \quad b \approx 120\,\text{mV dec}^{-1}. \tag{1.1}$$

2. Desorption step (Heyrovsky reaction):

$$H_{ads} + H_3O^+ + e^- \rightarrow H_2 + H_2O, \quad b \approx 40 \text{ mV dec}^{-1}. \tag{1.2}$$

3. Desorption step (Tafel reaction):

$$H_{ads} + H_{ads} \rightarrow H_2, \quad b \approx 30 \text{ mV dec}^{-1}. \tag{1.3}$$

Understanding the TMDs' underlying electrochemistry is essential to understanding their properties because the HER electrocatalysis strategy primarily depends upon the surface. A TMDs' redox reaction in favor of a functional redox potential free from analyte is referred to as the TMD inherent electrochemistry. However, factors such as the pH in the experiment could affect the results. The TMD stacked layers of distinct flattened nanosheets allow for electrochemical HER to easily access catalytic sites. Moreover, every TMD layer comprises a single MX_2 unit for a transition metal (M: group IVB–VIIIB) packed between two chalcogens (X: S, Se, and Te). A covalent M–X bond exists between each layer; stacking several TMD layers creates a bulk TMD material (figures 1.3(a) and (b)). Anisotropic qualities manifest at the edges and basal plane of layered TMDs because of their distinct structural characteristic. The van der Waals (vdW) interactions that hold neighboring layers together are weaker. To attain thin sheets, exfoliation techniques are used to overcome these weak bonds between layers [56].

Additionally, TMD electrochemistry is dependent on surface orientation [57]. Other theoretical and experimental findings demonstrate that the MoS_2 basal plane is inert while the edges are catalytically active for the HER (figures 1.3(d) and (e)) [58]. In TMDs, various chalcogen types and transition metal pairings result in various characteristics [59]. The electrical structure of TMDs is determined by occupying d-orbitals in the transition metal. The transition metal of the d-orbital's electron count changes from group IVB to VIIIB TMDs (figure 1.3(c)). The structural coordination of the transitional metal and chalcogen within the TMD can be used evaluate their electrical behavior in addition to the d-electron count [60]. As shown in figure 1.3(b), the TMDs move through the 2H (hexagonal phase), 3R (rhombohedral phase), and 1T (trigonal phase) polymorphs. Crystalline TMDs were found in the 2H- or 1T-phase with trigonal prismatic or octahedral bonds across the transition metal and chalcogen atoms. Using TMDs (2H-phase), the transitional metal non-bonding d-orbitals split into three degenerate states $d_{z^2}(a_1)$, $d_{x^2-y^2, xy}(e)$, and $d_{xz, yz}(e')$. In contrast, 1T-phase TMDs create degenerate $d_{z^2, x^2-y^2}(e_g)$ and $d_{xz, yz, xy}(t_{2g})$ orbitals. Full orbitals have semiconducting conductivity, but metallic properties arise from partially filled orbitals. Normally, TMDs from group IVB are present in 1T (octahedral) phases, whereas group VIB corresponds to 2H (trigonal prismatic) phases. Moreover, the characteristics of TMDs are quite different and dependent on their fundamental properties.

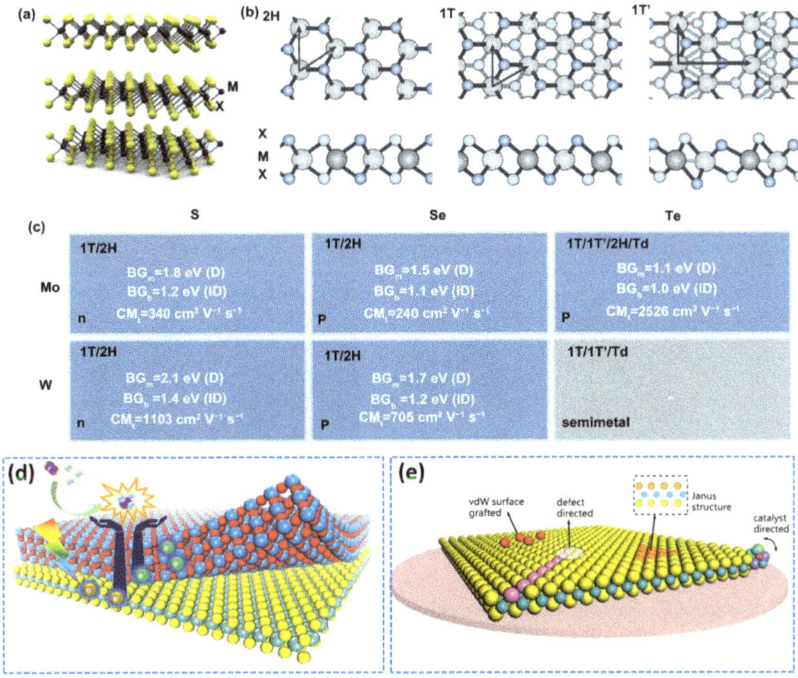

Figure 1.3. (a) Representations of different MX_2 materials. Chalcogen (X) is shown as yellow spheres, while the gray spheres correspond to the transition metal (M). (Reproduced with permission from [61]. Copyright 2011 Springer Nature.) (b) The top and side views of mono-layered TMDs show different phases (i.e. 1T′, 1T, and 2H). (Reproduced with permission from [62]. Copyright 2017 Springer Nature.) (c) TMD electronic parameters: semiconductor, bandgap energy of monolayer (BG_m) and bulk (BG_b), crystal structure, theoretical carrier mobility (CM_t), direct bandgap (D), and indirect bandgap (ID). (Reproduced with permission from [63]. Copyright 2020 Springer Nature, Shanghai Jiao Tong University Press.) (d) 2D vertical heterostructures for electrocatalysis. (Reproduced with permission from [64]. Copyright 2022 American Chemical Society.) (e) Electrochemistry of a 2D nanosheet. (Reproduced with permission from [65]. Copyright 2021 American Chemical Society.)

1.2.4 Hexagonal boron nitride

h-BN is a material that resembles graphite, commonly called ‘white graphite’ [66]. Bulk *h*-BN is described as a stacked structure with a 2.50 Å lattice constant and interplanar distances of 3.33 Å, as shown in figure 1.4(a) [67]. In contrast to graphene, *h*-BN has a wide bandgap energy and behaves like a pure insulator. In optoelectronics devices, *h*-BN nanosheets have been engaged utilizing dielectric substrates for MoS_2 and graphene-based heterostructures [68]. *h*-BN can perform poorly in electrocatalysis due to its weak conductivity and catalytic behavior. Consequently, numerous physical and chemical routes have been carried out to activate *h*-BN. The electrical characteristics of *h*-BN that might be tuned by producing a connection with Au nanoparticles were investigated by Uosaki *et al.* The produced composite demonstrated advanced ORR activity initiated through effective adsorbed O_2 activation; also, it increased the selectivity from O_2–H_2O_2 by

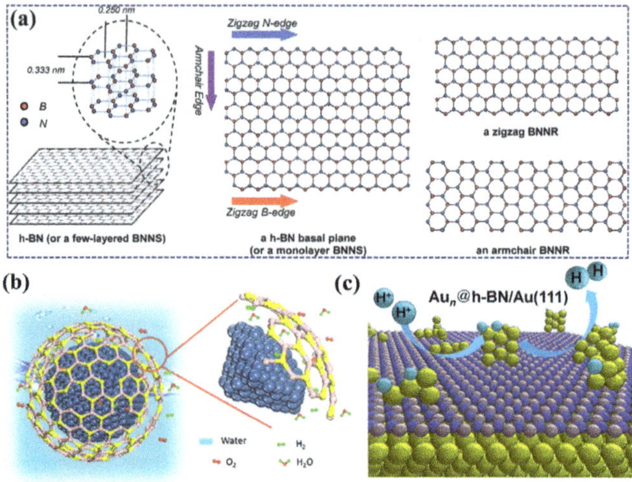

Figure 1.4. (a) Basic structure of 2D *h*-BN nanostructures. (Reproduced with permission from [72]. Copyright 2012 The Royal Society of Chemistry.) (b) Scheme of the proposed electrocatalysis by *h*-BN. (Reproduced with permission from [73]. Copyright 2018 Springer Nature, Tsinghua University Press.) (c) Catalytic response of Au-clusters supported on the *h*-BN/Au(111) surface for the HER. (Reproduced with permission from [74]. Copyright 2021 American Chemical Society.)

means of a $2e^-$ reduction strategy (figures 1.4(b) and (c)). Like other 2D materials, heteroatom doping also proved to be the best method for activating *h*-BN catalytic activity. Li *et al* described the possibility of using heteroatom-doped *h*-BN nanosheets as ORR routes [69]. Their calculations showed that carbon-doped *h*-BN might promote ORR catalysts with better energy transport than Pt-based catalysts (using acidic environments), producing advanced ORR activity. Uosaki *et al* also showed that *h*-BN could change Au catalytic activity for the HER, whereby its catalytic activity was increased by the presence of energetically advantageous sites for H_2 adsorption [70]. Furthermore, carbon-doped *h*-BN nanosheets were developed as stable visible-light-driven photocatalysts for CO_2 reduction [71]. When proper engineering techniques are employed, these nanomaterials have the potential for ECR applications as well (i.e. optimization of electrical conductivity).

1.2.5 Black phosphorus

Black phosphorus (BP) has a century-long history; it was first produced in 1914 [75]. A puckered honeycomb structure is created by the covalent bonding of one phosphorus atom with three others in a single-layer of layered semiconductor BP that contains an orthorhombic crystal structure (figures 1.5(a) and (b)) [76]. Three bonds engage three valence electrons of phosphorus entirely, generating a bandgap energy ($\sim$2 eV). Adjustment of the bandgap energy can be achieved by adjusting the thickness of the bonds [77, 78]. Since each P atom has five electrons in its 3p orbitals, within the phosphorus structure each P atom forms a gathered orthorhombic structure through covalent bonds to three adjacent phosphorus atoms, leaving a

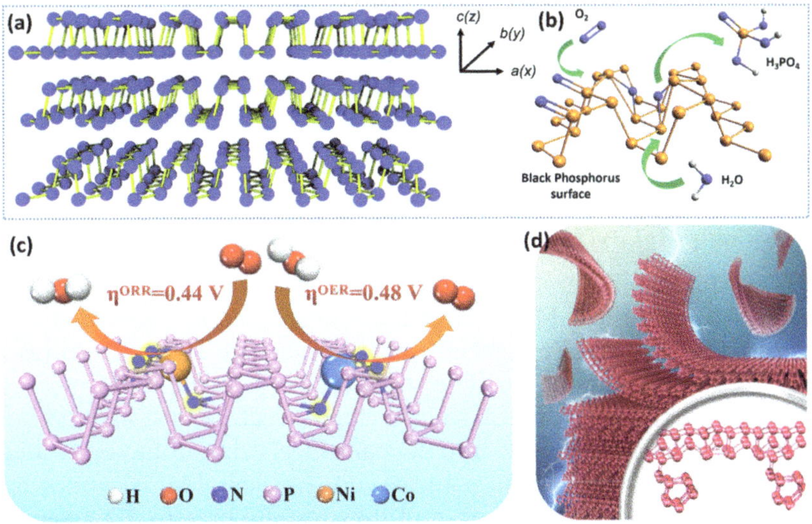

Figure 1.5. (a) Crystal configuration of few-layer BP. (Reproduced with permission from [85]. Copyright 2015 The Royal Society of Chemistry.) (b) Surface chemistry for electrolysis. (Reproduced with permission from [86]. Copyright 2019 American Chemical Society.) (c) Single metal atom supported on N-doped 2D nitride BP. (Reproduced with permission from [87]. Copyright 2021 American Chemical Society.) (d) Violet phosphorus is an additional layered semiconducting phosphorus allotrope. (Reproduced with permission from [110]. Copyright 2020 John Wiley and Sons.)

lone-pair electron [79, 80]. Intense research has been sparked by the intriguing properties of phosphorene, including its layer-dependent direct bandgap, strong carrier mobility, and discrete anisotropic assembly (in-plane). Pioneering research on BP materials has achieved significant growth in a number of sectors [81] and BP-based electrocatalyst research is a vibrant, new, and fast-growing field. BP has so far received a lot of attention for use in electrical and optical equipment. However, only a few works have examined its applicability in electrocatalysis, possibly due to its low electrical conductivity and limited stability under electrocatalytic conditions. Recently, the electrocatalytic activity of several bulk BP forms, including thin films and particles, has been described as analogous to commercial RuO_2-based catalysts for the OER process [82]. However, the low density of active sites in pure BP probably reduces the electrocatalytic performance of this material (figures 1.5(b) and (c)).

Moreover, figure 1.5(d) shows another emerging 2D phosphorus candidate called violet phosphorus, a semiconducting phosphorus allotrope. 2D BP nanosheets created using liquid exfoliation procedures have more visible active edges and surface areas than their bulk form [83]. Due to these inherent advantages, BP development for effective electrocatalysis can be facilitated. As expected, few-layered BP elucidated a better OER performance than the bulk form, obtaining a 1.45 V onset potential versus a reversible hydrogen electrode (RHE) and a 88 mV dec^{-1} Tafel slope using alkaline media [83]. Further, BP also demonstrates synergistic properties for the HER once combined with different composites to form heterostructures. For example, the development of a 0D–2D Ni_2P and BP heterostructure was revealed by Luo *et al*, which resulted in noteworthy HER activity using acidic media compared to commercial Ni_2P [84]. Moreover, excellent electrocatalytic efficiency can be ascribed to the enormous surface area of BP as well as the tuned charge concentration between the BP and Ni_2P interface.

1.2.6 Transition metal oxides

2D TMOs exhibit significant activity during electrocatalysis and their popularity is attributable to their affordable cost, tunable properties and better stability. The basic types of 2D TMOs are stacked and non-stacked TMOs. Various types of TMOs are presented in figure 1.6. The stacked class is made up of metal trioxides (i.e. WO_3, TaO_3, and MoO_3) that have a stacked structure, just like the example of graphene; they can be transformed into 2D nanosheets via exfoliation strategies such as liquid phase exfoliation [89]. In contrast, non-stacked TMOs cannot be synthesized by simple top-down techniques, and they usually present 3D assemblies with strong chemical bonding present between various crystal layers. Consequently, for non-layered 2D nanocomposite fabrication several methods, including 2D precursor-based topotactic conversion approaches and salt-template approaches, have been established [90]. The growth of 2D TMO nanosheets on a large scale (i.e. WO_3, MnO, MoO_2, and MoO_3) by using salt crystals as a template was reported by Xiao *et al.* The growth of TMOs on salt interfaces is directed by the salt crystal geometry and enhanced by lattice matching. Any growth is limited by precursor depletion perpendicular to the salt surface. The results found for TMO nanosheet formation

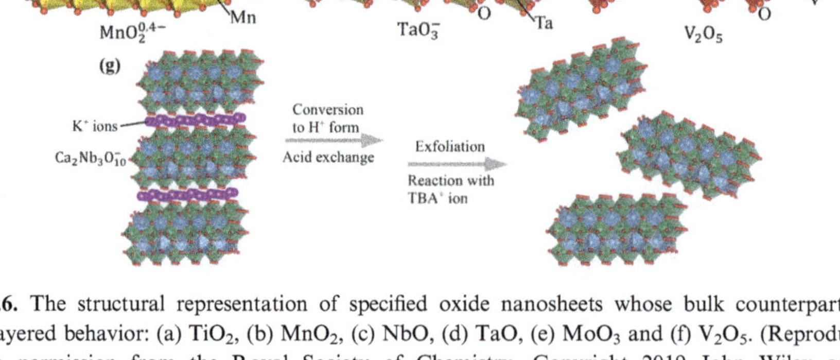

Figure 1.6. The structural representation of specified oxide nanosheets whose bulk counterparts showed natural layered behavior: (a) TiO_2, (b) MnO_2, (c) NbO, (d) TaO, (e) MoO_3 and (f) V_2O_5. (Reproduced from [93] with permission from the Royal Society of Chemistry. Copyright 2019 John Wiley and Sons.) (g) Schematic view of the layered $KCa_2Nb_3O_{10}$ exfoliated into nanosheets using the cation exchange-supported liquid exfoliation procedure. (Adapted with permission from [109]. Copyright 2015 Springer Nature.)

with a thickness in nanometers show that this is even an appropriate method for materials with non-stacked crystal structures [90]. Many reports have show that 2D TMOs are excellent candidates for different electrocatalytic reactions in the carbon and water cycles. Most 2D TMOs can be used directly as electrocatalysts without any modification. Hence, improving their conductivity is the best way to boost their catalytic response. For example, plasma engraved 2D Co_3O4 nanosheets showed effective OER performance because of their sizeable visible surface area, increased conductivity, and vacancies produced in active sites [91]. Moreover, the catalytic activity of 2D TMOs might be enhanced with defect engineering. For example, in contrast to bulk form, 2D CeO_2 nanosheets using surface-confined pits exhibited a rapid response for CO oxidation [92].

1.2.7 Graphitic carbon nitride

Recently, g-C_3N_4 has received significant interest owing to its many benefits, including its distinctive optical qualities, abundance in nature, lack of metals, light sensitivity, and polymeric nature (see figure 1.7 for its crystal structure) [94]. The development of this inorganic polymer dates back to the nineteenth century, when a polymer known as 'melon' was first created in 1834 [94]. Since then, g-C_3N_4 has developed into a desirable choice for many applications [95]. Owing to its porous character, significant surface area (almost 150 m^2 g^{-1}), and high nitrogen content (23.7%), structures based on g-C_3N_4 have proven successful as electrocatalysts in energy conversion reactions (for example, the HER, ORR/OER), as well as electrochemical sensors. Because of their improved compatibility with lithium insertion processes, these structures are valuable as anode materials in electrochemical energy storage systems such as lithium-ion batteries [96]. Various features of the chemical processes in the synthesis and use of different C_3N_4 nanocomposites have been elucidated [97]. Notably, among the various C_3N_4 materials, g-C_3N_4 has the most stable structure [95]. Additionally, g-C_3N_4 materials are considered

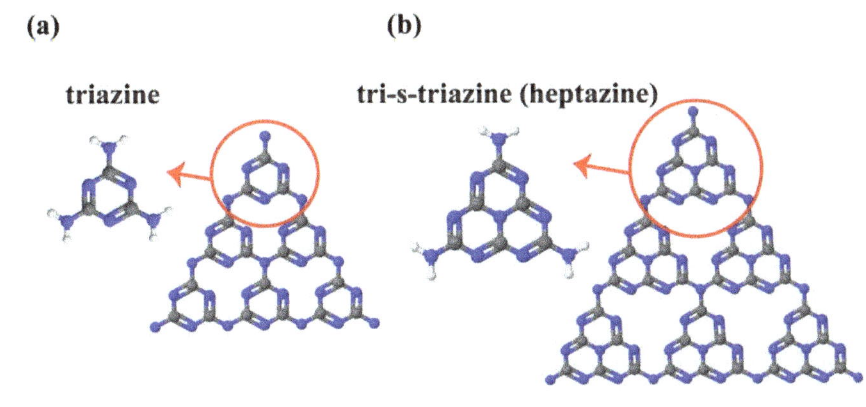

Figure 1.7. (a) Triazine and (b) tri-s-triazine (heptazine) structures of g-C_3N_4. The gray, blue, and white spheres are carbon, nitrogen, and hydrogen, respectively. (Reproduced with permission from [108]. CC BY 4.0.)

excellent prospects for catalysis because of their improved surface properties, electron-rich traits, and H-bonding abilities [98]. Due to its thermal and hydrothermal stability, g-C_3N_4 can function well even at high-temperature rates in both gaseous and liquid phases [99]. Through combination with conducting carbon substrates such as CNTs and graphene through alloying, the electrochemical characteristics of g-C_3N_4 (particularly conductivity) can be improved [100]. Various factors, including a wide surface area, high porosity, and strong electron conductivity, are essential for effective electrocatalytic performance. Interestingly, mesoporous g-C_3N_4 is an excellent option for metal-free catalysis with different semiconductor features because of its favorable physical characteristics [101]. As demonstrated by thermogravimetric analysis under various conditions, g-C_3N_4 differs from other polymeric materials because it has excellent thermal stability and is comparatively inert in an oxygen-rich atmosphere [102]. The effectiveness of an electrocatalyst is also significantly influenced by its chemical stability. g-C_3N_4 is noticeably stable in both water and organic solvents (e.g. toluene, diethyl ether, and dimethylformamide (DMF)).

The semiconducting qualities of g-C_3N_4 make it highly appropriate for photocatalytic reactions due to its bandgap (2.7 eV) [103, 104]. However, the electrical structure and bandgap of g-C_3N_4 are directly influenced by the amount of condensation, which in turn impacts its activity [105, 106]. In this regard, bulk g-C_3N_4 has little activity. Conversely, the many structural flaws and surface terminations crucial to catalytic activation in 2D g-C_3N_4 nanosheets have shown the desired activity. The nitrogen-rich carbon structure of g-C_3N_4 is another crucial component for electrocatalysis in addition to the bandgap. In general, nitrogen doping can increase the reactions between carbon and intermediates in electrocatalytic processes by improving the electron-donor characteristics of graphene and new carbon networks. Two fundamental design ideas are traditionally used to synthesize g-C_3N_4-based electrocatalysts. These are (i) material conductivity optimization and (ii) proper active site tuning. g-C_3N_4 has been composited with conductive carbon compounds to improve conductivity. For example, when g-C_3N_4 was integrated into a mesoporous carbon framework, its ORR activity increased considerably compared to pure g-C_3N_4 [100]. The goal of tuning active sites is to favorably alter the adsorption energy of target intermediates, which can be done in various ways. In one instance, a compound of g-C_3N_4 and N-graphene showed HER efficiency parallel to that of metallic catalysts [107]. The composite with interfacial structure was adjusted for the electronic state and absorption free energy (labeled as ΔG_{H*}), which enabled electron/charge transport and improved catalytic behavior, according to shared atomic imaging, spectroscopic investigation, and DFT measurements [107].

1.3 Conclusion, challenges and perspectives

In recent years, a substantial amount of effort has been put toward improving the electrocatalytic activity of 2D materials for systematic electrochemical processes.

It has become abundantly clear that the 2D interfacial assembly plays a critical role in determining the electrocatalytic activity of these materials. Consequently, numerous breakthroughs have been accomplished, which unequivocally show the benefits of electrocatalysis using 2D materials. Nevertheless, although 2D materials have exciting prospects in electrocatalysis, they are still up against a great deal of competition, and the vast majority of studies produce contradictory findings. One challenge is that the 2D electrocatalytic strategy is not well understood. Another issue is that it cannot effectively track how the reaction happens step by step, making it difficult to figure out the different stages of the process. Finally, the inefficiencies and complexities are also challenging because of the more significant costs linked to the synthesis procedure of 2D materials, which discourages commercialization. To overcome these obstacles, it is necessary to focus on the research areas listed below.

1. In recent years, a substantial amount of work has been done to advance the electrocatalytic performance of 2D materials in various electrochemical processes.
2. The investigation of the electrocatalytic processes of 2D materials via contemporary (*in situ*) characterization approaches is of significant impact. Electrochemical reactions involve several distinct procedures for transferring electrons. Understanding the catalytic pathways requires that these intermediates be identified and characterized first. In this case, the (*in situ*) characterization approaches, including x-ray absorption fine structure spectroscopy (XAFS), provide support in evaluating the valence and structural state of intermediates.
3. The large-scale production of energy-producing and highly efficient 2D electrocatalysts is a prerequisite for developing industrial applications. Currently, most research primarily concentrates on improving the catalytic performance of the components, and they frequently disregard the costs of the different constituents and the difficulty of the synthesis route. As a result, there is an immediate requirement to develop effective solutions for manufacturing high-quality 2D materials at massive scales.
4. The stability and durability of 2D electrocatalysts are essential parameters for achieving effective electrocatalysts. To a large extent, the durability of electrocatalysts is determined by their chemical stability in solution, capacity to aggregate between layers, and correlation with supporting electrodes. The constituents and electrocatalyst structure require stability to conserve their activity throughout the lengthy catalytic process.
5. It is difficult to compare the catalytic activity of 2D materials and other nanomaterials for any particular reaction. This is due to the fact that the electrocatalytic activity of 2D materials is usually based on laboratory data that are frequently erroneous. To satisfy the needs of actual applications, it is necessary to develop a standardized evaluation system to analyse the performance of electrocatalytic materials. This system should take into account factors such as high current density and stability.

Bibliography

[1] Zhou M, Zhang Z, Huang K, Shi Z, Xie R and Yang W 2016 Colloidal preparation and electrocatalytic hydrogen production of MoS_2 and WS_2 nanosheets with controllable lateral sizes and layer numbers *Nanoscale* **8** 15262–72

[2] Wang T, Liu L, Zhu Z, Papakonstantinou P, Hu J, Liu H and Li M 2013 Enhanced electrocatalytic activity for hydrogen evolution reaction from self-assembled monodispersed molybdenum sulfide nanoparticles on an Au electrode *Energy Environ. Sci.* **6** 625–33

[3] Wu S, Hui K S and Hui K N 2018 2D black phosphorus: from preparation to applications for electrochemical energy storage *Adv. Sci.* **5** 1700491

[4] Zheng J, Zhang H, Dong S, Liu Y, Tai Nai C, Suk Shin H, Young Jeong H, Liu B and Ping Loh K 2014 High yield exfoliation of two-dimensional chalcogenides using sodium naphthalenide *Nat. Commun.* **5** 2995

[5] Fuhrer M S and Hone J 2013 Measurement of mobility in dual-gated MoS_2 transistors *Nat. Nanotechnol.* **8** 146–7

[6] Hao Y *et al* 2013 The role of surface oxygen in the growth of large single-crystal graphene on copper *Science* **342** 720–3

[7] Wu Y Q *et al* 2008 Top-gated graphene field-effect-transistors formed by decomposition of SiC *Appl. Phys. Lett.* **92** 092102

[8] Yu Y, Huang S-Y, Li Y, Steinmann S N, Yang W and Cao L 2014 Layer-dependent electrocatalysis of MoS_2 for hydrogen evolution *Nano Lett.* **14** 553–8

[9] He R, Xu D, Cheng B, Yu J and Ho W 2018 Review on nanoscale Bi-based photocatalysts *Nanoscale Horizons* **3** 464–504

[10] Javaid M, Drumm D W, Russo S P and Greentree A D 2017 A study of size-dependent properties of MoS_2 monolayer nanoflakes using density-functional theory *Sci. Rep.* **7** 9775

[11] Li Y *et al* 2018 Water soluble graphitic carbon nitride with tunable fluorescence for boosting broad-response photocatalysis *Appl. Catal.* **B 225** 519–29

[12] Du X *et al* 2019 Modulating electronic structures of inorganic nanomaterials for efficient electrocatalytic water splitting *Angew. Chem. Int. Ed.* **58** 4484–502

[13] Guo Y, Xu K, Wu C, Zhao J and Xie Y 2015 Surface chemical-modification for engineering the intrinsic physical properties of inorganic two-dimensional nanomaterials *Chem. Soc. Rev.* **44** 637–46

[14] Long X, Li G, Wang Z, Zhu H, Zhang T, Xiao S, Guo W and Yang S 2015 Metallic iron–nickel sulfide ultrathin nanosheets as a highly active electrocatalyst for hydrogen evolution reaction in acidic media *J. Am. Chem. Soc.* **137** 11900–3

[15] Yusran Y *et al* 2020 Exfoliated mesoporous 2D covalent organic frameworks for high-rate electrochemical double-layer capacitors *Adv. Mater.* **32** 1907289

[16] Deng D, Novoselov K S, Fu Q, Zheng N, Tian Z and Bao X 2016 Catalysis with two-dimensional materials and their heterostructures *Nat. Nanotechnol.* **11** 218–30

[17] Voiry D *et al* 2013 Enhanced catalytic activity in strained chemically exfoliated WS_2 nanosheets for hydrogen evolution *Nat. Mater.* **12** 850–5

[18] Lee C, Wei X, Kysar J W and Hone J 2008 Measurement of the elastic properties and intrinsic strength of monolayer graphene *Science* **321** 385–8

[19] Walter M G, Warren E L, McKone J R, Boettcher S W, Mi Q, Santori E A and Lewis N S 2010 Solar water splitting cells *Chem. Rev.* **110** 6446–73

[20] Lee Y, Suntivich J, May K J, Perry E E and Shao-Horn Y 2012 Synthesis and activities of rutile IrO_2 and RuO_2 nanoparticles for oxygen evolution in acid and alkaline solutions *J. Phys. Chem. Lett.* **3** 399–404
[21] Allioux F-M, Ghasemian M B, Xie W, O'Mullane A P, Daeneke T, Dickey M D and Kalantar-Zadeh K 2022 Applications of liquid metals in nanotechnology *Nanoscale Horizons* **7** 141–67
[22] Wang D-Y *et al* 2015 Highly active and stable hybrid catalyst of cobalt-doped FeS_2 nanosheets–carbon nanotubes for hydrogen evolution reaction *J. Am. Chem. Soc.* **137** 1587–92
[23] Liu X and Dai L 2016 Carbon-based metal-free catalysts *Nat. Rev. Mater.* **1** 16064
[24] Balandin A A, Ghosh S, Bao W, Calizo I, Teweldebrhan D, Miao F and Lau C N 2008 Superior thermal conductivity of single-layer graphene *Nano Lett.* **8** 902–7
[25] Tan C *et al* 2017 Recent advances in ultrathin two-dimensional nanomaterials *Chem. Rev.* **117** 6225–331
[26] Deng D, Yu L, Pan X, Wang S, Chen X, Hu P, Sun L and Bao X 2011 Size effect of graphene on electrocatalytic activation of oxygen *Chem. Commun.* **47** 10016–8
[27] Jiao Y, Zheng Y, Davey K and Qiao S-Z 2016 Activity origin and catalyst design principles for electrocatalytic hydrogen evolution on heteroatom-doped graphene *Nat. Energy* **1** 16130
[28] Cheng N *et al* 2016 Platinum single-atom and cluster catalysis of the hydrogen evolution reaction *Nat. Commun.* **7** 13638
[29] Duan J, Chen S, Jaroniec M and Qiao S Z 2015 Heteroatom-doped graphene-based materials for energy-relevant electrocatalytic processes *ACS Catal.* **5** 5207–34
[30] Wang H, Chen Y, Hou X, Ma C and Tan T 2016 Nitrogen-doped graphenes as efficient electrocatalysts for the selective reduction of carbon dioxide to formate in aqueous solution *Green Chem.* **18** 3250–6
[31] Zhang X, Gao J, Xiao Y, Wang J, Sun G, Zhao Y and Qu L 2020 Regulation of 2D graphene materials for electrocatalysis *Chem.—Asian J.* **15** 2271–81
[32] Long X, Li D, Wang B, Jiang Z, Xu W, Wang B, Yang D and Xia Y 2019 Heterocyclization strategy for construction of linear conjugated polymers: efficient metal-free electrocatalysts for oxygen reduction *Angew. Chem. Int. Ed.* **58** 11369–73
[33] Sun W-J, Zhao Y-Y, Cheng X-F, He J-H and Lu J-M 2020 Surface functionalization of single-layered $Ti_3C_2T_x$ MXene and its application in multilevel resistive memory *ACS Appl. Mater. Interfaces* **12** 9865–71
[34] Lyu B, Choi Y, Jing H, Qian C, Kang H, Lee S and Cho J H 2020 2D MXene–TiO_2 core–shell nanosheets as a data-storage medium in memory devices *Adv. Mater.* **32** 1907633
[35] Bai S, Guo X, Chen T, Zhang Y, Zhang X, Yang H and Zhao X 2020 Solution processed fabrication of silver nanowire-MXene@PEDOT:PSS flexible transparent electrodes for flexible organic light-emitting diodes *Composites* A **139** 106088
[36] Zheng S, Wang H, Das P, Zhang Y, Cao Y, Ma J, Liu S and Wu Z-S 2021 Multitasking MXene inks enable high-performance printable microelectrochemical energy storage devices for all-flexible self-powered integrated systems *Adv. Mater.* **33** 2005449
[37] Luo J, Matios E, Wang H, Tao X and Li W 2020 Interfacial structure design of MXene-based nanomaterials for electrochemical energy storage and conversion *InfoMat* **2** 1057–76
[38] Al-Amer R M, Barakat F, Aljalham G H, AlQahtani H R, Laref A, Muhammad Alay-e-Abbas S, Mahmood Q, Laref S, Booq Z I and Wu X 2021 Electronic and optical

characteristics of 5D-transition metal doped 2D-$NbSe_2$ monolayer for nanoelectronic device applications: an *ab-initio*-analysis *Mater. Sci. Eng.* B **269** 115155
[39] Permyakov E A, Maximov V V and Kogan V M 2021 Effect of spin polarization and supercell size on specific energy and electronic structure of MoS_2 edge calculated by DFT method in the plane-wave basis *Mendeleev Commun.* **31** 532–5
[40] Yang R and Sun M 2022 Phonon-assisted interfacial charge transfer excitons in graphene/*h*-BN van der Waals heterostructures *Chin. J. Phys.* **76** 110–20
[41] Naguib M, Barsoum M W and Gogotsi Y 2021 Ten years of progress in the synthesis and development of MXenes *Adv. Mater.* **33** 2103393
[42] Wang Y, Nian Y, Biswas A N, Li W, Han Y and Chen J G 2021 Challenges and opportunities in utilizing MXenes of carbides and nitrides as electrocatalysts *Adv. Energy Mater.* **11** 2002967
[43] Lim K R G, Handoko A D, Johnson L R, Meng X, Lin M, Subramanian G S, Anasori B, Gogotsi Y, Vojvodic A and Seh Z W 2020 2H-MoS_2 on Mo_2CT_x MXene nanohybrid for efficient and durable electrocatalytic hydrogen evolution *ACS Nano* **14** 16140–55
[44] Yin H, Yuan C, Lv H, Zhang K, Chen X and Zhang Y 2022 Hierarchical Ti_3C_2 MXene/$Zn_3In_2S_6$ Schottky junction for efficient visible-light-driven Cr(VI) photoreduction *Ceram. Int.* **48** 11320–9
[45] Mei J, Ayoko G A, Hu C, Bell J M and Sun Z 2020 Two-dimensional fluorine-free mesoporous Mo_2C MXene via UV-induced selective etching of Mo_2Ga_2C for energy storage *Sustain. Mater. Technol.* **25** e00156
[46] Lv W, Wu G, Li X, Li J and Li Z 2022 Two-dimensional V_2C@Se (MXene) composite cathode material for high-performance rechargeable aluminum batteries *Energy Storage Mater.* **46** 138–46
[47] Liu M-C, Zhang Y-S, Zhang B-M, Zhang D-T, Tian C-Y, Kong L-B and Hu Y-X 2021 Large interlayer spacing 2D Ta_4C_3 matrix supported 2D MoS_2 nanosheets: a 3D heterostructure composite towards high-performance sodium ions storage *Renew. Energy* **169** 573–81
[48] Li D, Yang C, Rajendran S, Qin J and Zhang X 2022 Nanoflower-like Ti_3CN@TiO_2/CdS heterojunction photocatalyst for efficient photocatalytic water splitting *Int. J. Hydrogen Energy* **47** 19580–9
[49] Huang B, Zhou N, Chen X, Ong W-J and Li N 2018 Insights into the electrocatalytic hydrogen evolution reaction mechanism on two-dimensional transition-metal carbonitrides (MXene) *Chem.-Eur. J.* **24** 18479–86
[50] Dong L, Chu H, Li Y, Ma X, Pan H, Zhao S and Li D 2022 Surface functionalization of Ta_4C_3 MXene for broadband ultrafast photonics in the near-infrared region *Appl. Mater. Today* **26** 101341
[51] Meshkian R *et al* 2018 W-based atomic laminates and their 2D derivative W1.33C MXene with vacancy ordering *Adv. Mater.* **30** 1706409
[52] Shuck C E, Sarycheva A, Anayee M, Levitt A, Zhu Y, Uzun S, Balitskiy V, Zahorodna V, Gogotsi O and Gogotsi Y 2020 Scalable synthesis of $Ti_3C_2T_x$ MXene *Adv. Eng. Mater.* **22** 1901241
[53] Cheng Y, Wang L, Li Y, Song Y and Zhang Y 2019 Etching and exfoliation properties of Cr_2AlC into Cr_2CO_2 and the electrocatalytic performances of 2D Cr_2CO_2 MXene *J. Phys. Chem.* C **123** 15629–36

[54] VahidMohammadi A, Rosen J and Gogotsi Y 2021 The world of two-dimensional carbides and nitrides (MXenes) *Science* **372** eabf1581
[55] Voiry D, Yang J and Chhowalla M 2016 Recent strategies for improving the catalytic activity of 2D TMD nanosheets toward the hydrogen evolution reaction *Adv. Mater.* **28** 6197–206
[56] Lv R, Robinson J A, Schaak R E, Sun D, Sun Y, Mallouk T E and Terrones M 2015 Transition metal dichalcogenides and beyond: synthesis, properties, and applications of single- and few-layer nanosheets *Acc. Chem. Res.* **48** 56–64
[57] Weiss K and Phillips J M 1976 Calculated specific surface energy of molybdenite (MoS_2) *Phys. Rev.* B **14** 5392–5
[58] Jaramillo T F, Jørgensen K P, Bonde J, Nielsen J H, Horch S and Chorkendorff I 2007 Identification of active edge sites for electrochemical H_2 evolution from MoS_2 nanocatalysts *Science* **317** 100–2
[59] Chhowalla M, Shin H S, Eda G, Li L-J, Loh K P and Zhang H 2013 The chemistry of two-dimensional layered transition metal dichalcogenide nanosheets *Nat. Chem.* **5** 263–75
[60] Voiry D, Mohite A and Chhowalla M 2015 Phase engineering of transition metal dichalcogenides *Chem. Soc. Rev.* **44** 2702–12
[61] Radisavljevic B, Radenovic A, Brivio J, Giacometti V and Kis A 2011 Single-layer MoS_2 transistors *Nat. Nanotechnol.* **6** 147–50
[62] Manzeli S, Ovchinnikov D, Pasquier D, Yazyev O V and Kis A 2017 2D transition metal dichalcogenides *Nat. Rev. Mater.* **2** 17033
[63] Chen X, Liu C and Mao S 2020 Environmental analysis with 2D transition-metal dichalcogenide-based field-effect transistors *Nano-Micro Lett.* **12** 95
[64] Guo H, Zhang H, Zhao J, Yuan P, Li Y, Zhang Y, Li L, Wang S and Song R 2022 Two-dimensional WO_3-transition-metal dichalcogenide vertical heterostructures for nitrogen fixation: a photo(electro) catalysis theoretical strategy *J. Phys. Chem.* C **126** 3043–53
[65] Yang T, Li L-J, Zhao J and Ly T H 2021 Precision chemistry in two-dimensional materials: adding, removing, and replacing the atoms at will *Acc. Mater. Res.* **2** 863–8
[66] Li L H and Chen Y 2016 Atomically thin boron nitride: unique properties and applications *Adv. Funct. Mater.* **26** 2594–608
[67] Geick R, Perry C H and Rupprecht G 1966 Normal modes in hexagonal boron nitride *Phys. Rev.* **146** 543–7
[68] Lee G-H *et al* 2013 Flexible and transparent MoS_2 field-effect transistors on hexagonal boron nitride-graphene heterostructures *ACS Nano* **7** 7931–6
[69] Li F, Shu H, Liu X, Shi Z, Liang P and Chen X 2017 Electrocatalytic activity and design principles of heteroatom-doped graphene catalysts for oxygen-reduction reaction *J. Phys. Chem.* C **121** 14434–42
[70] Uosaki K, Elumalai G, Dinh H C, Lyalin A, Taketsugu T and Noguchi H 2016 Highly efficient electrochemical hydrogen evolution reaction at insulating boron nitride nanosheet on inert gold substrate *Sci. Rep.* **6** 32217
[71] Huang C, Chen C, Zhang M, Lin L, Ye X, Lin S, Antonietti M and Wang X 2015 Carbon-doped BN nanosheets for metal-free photoredox catalysis *Nat. Commun.* **6** 7698
[72] Lin Y and Connell J W 2012 Advances in 2D boron nitride nanostructures: nanosheets, nanoribbons, nanomeshes, and hybrids with graphene *Nanoscale* **4** 6908–39
[73] Sun M *et al* 2018 Pt@h-BN core–shell fuel cell electrocatalysts with electrocatalysis confined under outer shells *Nano Res.* **11** 3490–8

[74] Gao M, Nakahara M, Lyalin A and Taketsugu T 2021 Catalytic activity of gold clusters supported on the *h*-BN/Au(111) surface for the hydrogen evolution reaction *J. Phys. Chem.* C **125** 1334–44

[75] Bridgman P W 1914 Two new modifications of phosphorus *J. Am. Chem. Soc.* **36** 1344–63

[76] Prasannachandran R, Vineesh T V, Lithin M B, Nandakishore R and Shaijumon M M 2020 Phosphorene-quantum-dot-interspersed few-layered MoS_2 hybrids as efficient bifunctional electrocatalysts for hydrogen and oxygen evolution *Chem. Commun.* **56** 8623–6

[77] Liu Y and Chen X 2014 Mechanical properties of nanoporous graphene membrane *J. Appl. Phys.* **115** 034303

[78] Ling X, Wang H, Huang S, Xia F and Dresselhaus M S 2015 The renaissance of black phosphorus *Proc. Natl Acad. Sci.* **112** 4523

[79] Zhou Y, Zhang M, Guo Z, Miao L, Han S-T, Wang Z, Zhang X, Zhang H and Peng Z 2017 Recent advances in black phosphorus-based photonics, electronics, sensors and energy devices *Mater. Horizons* **4** 997–1019

[80] Qiu M, Sun Z T, Sang D K, Han X G, Zhang H and Niu C M 2017 Current progress in black phosphorus materials and their applications in electrochemical energy storage *Nanoscale* **9** 13384–403

[81] Lei W, Liu G, Zhang J and Liu M 2017 Black phosphorus nanostructures: recent advances in hybridization, doping and functionalization *Chem. Soc. Rev.* **46** 3492–509

[82] Jiang Q, Xu L, Chen N, Zhang H, Dai L and Wang S 2016 Facile synthesis of black phosphorus: an efficient electrocatalyst for the oxygen evolving reaction *Angew. Chem. Int. Ed.* **55** 13849–53

[83] Ren X *et al* 2017 Few-layer black phosphorus nanosheets as electrocatalysts for highly efficient oxygen evolution reaction *Adv. Energy Mater.* **7** 1700396

[84] Luo Z-Z *et al* 2017 Multifunctional 0D–2D Ni_2P nanocrystals–black phosphorus heterostructure *Adv. Energy Mater.* **7** 1601285

[85] Liu H, Du Y, Deng Y and Ye P D 2015 Semiconducting black phosphorus: synthesis, transport properties and electronic applications *Chem. Soc. Rev.* **44** 2732–43

[86] van Druenen M, Davitt F, Collins T, Glynn C, O'Dwyer C, Holmes J D and Collins G 2019 Evaluating the surface chemistry of black phosphorus during ambient degradation *Langmuir* **35** 2172–8

[87] Zhao X and Pei Y 2021 Single metal atom supported on N-doped 2D nitride black phosphorus: an efficient electrocatalyst for the oxygen evolution and oxygen reduction reactions *J. Phys. Chem.* C **125** 12541–50

[88] Zhang W *et al* 2020 Effects of carbon nanotubes on biodegradation of pollutants: positive or negative? *Ecotoxicol. Environ. Saf.* **189** 109914

[89] Balendhran S, Walia S, Nili H, Ou J Z, Zhuiykov S, Kaner R B, Sriram S, Bhaskaran M and Kalantar-zadeh K 2013 Two-dimensional molybdenum trioxide and dichalcogenides *Adv. Funct. Mater.* **23** 3952–70

[90] Xiao X *et al* 2016 Scalable salt-templated synthesis of two-dimensional transition metal oxides *Nat. Commun.* **7** 11296

[91] Xu L, Jiang Q, Xiao Z, Li X, Huo J, Wang S and Dai L 2016 Plasma-engraved Co_3O_4 nanosheets with oxygen vacancies and high surface area for the oxygen evolution reaction *Angew. Chem. Int. Ed.* **55** 5277–81

[92] Sun Y, Liu Q, Gao S, Cheng H, Lei F, Sun Z, Jiang Y, Su H, Wei S and Xie Y 2013 Pits confined in ultrathin cerium(IV) oxide for studying catalytic centers in carbon monoxide oxidation *Nat. Commun.* **4** 2899

[93] Yang T, Song T T, Callsen M, Zhou J, Chai J W, Feng Y P, Wang S J and Yang M 2019 Atomically thin 2D transition metal oxides: structural reconstruction, interaction with substrates, and potential applications *Adv. Mater. Interfaces* **6** 1801160
[94] Wang X and Abdel-Aty M 2008 Modeling left-turn crash occurrence at signalized intersections by conflicting patterns *Accid. Anal. Prev.* **40** 76–88
[95] Wang Y, Wang X and Antonietti M 2012 Polymeric graphitic carbon nitride as a heterogeneous organocatalyst: from photochemistry to multipurpose catalysis to sustainable chemistry *Angew. Chem. Int. Ed.* **51** 68–89
[96] Kim M, Hwang S and Yu J-S 2007 Novel ordered nanoporous graphitic C_3N_4 as a support for Pt–Ru anode catalyst in direct methanol fuel cell *J. Mater. Chem.* **17** 1656–9
[97] Kroke E and Schwarz M 2004 Novel group 14 nitrides *Coord. Chem. Rev.* **248** 493–532
[98] Su F, Antonietti M and Wang X 2012 mpg-C_3N_4 as a solid base catalyst for Knoevenagel condensations and transesterification reactions *Catal. Sci. Technol.* **2** 1005–9
[99] Bai L, Kuang Y, Luo J, Evans D G and Sun X 2012 Ligand-manipulated selective transformations of Au–Ni bimetallic heteronanostructures in an organic medium *Chem. Commun.* **48** 6963–5
[100] Zheng Y *et al* 2011 Nanoporous graphitic-C_3N_4@carbon metal-free electrocatalysts for highly efficient oxygen reduction *J. Am. Chem. Soc.* **133** 20116–9
[101] Davis M E 2002 Ordered porous materials for emerging applications *Nature* **417** 813–21
[102] Govindaraju V R, Sureshkumar K, Ramakrishnappa T, Muralikrishna S, Samrat D, Krishna Pai R, Kumar V, Vikrant K and Kim K-H 2021 Graphitic carbon nitride composites as electro catalysts: applications in energy conversion/storage and sensing system *J. Clean. Prod.* **320** 128693
[103] Zheng Y, Liu J, Liang J, Jaroniec M and Qiao S Z 2012 Graphitic carbon nitride materials: controllable synthesis and applications in fuel cells and photocatalysis *Energy Environ. Sci.* **5** 6717–31
[104] Su F, Mathew S C, Lipner G, Fu X, Antonietti M, Blechert S and Wang X 2010 mpg-C_3N_4-catalyzed selective oxidation of alcohols using O_2 and visible light *J. Am. Chem. Soc.* **132** 16299–301
[105] Lau V W-h, Mesch M B, Duppel V, Blum V, Senker J and Lotsch B V 2015 Low-molecular-weight carbon nitrides for solar hydrogen evolution *J. Am. Chem. Soc.* **137** 1064–72
[106] Hong Z, Shen B, Chen Y, Lin B and Gao B 2013 Enhancement of photocatalytic H_2 evolution over nitrogen-deficient graphitic carbon nitride *J. Mater. Chem.* A **1** 11754–61
[107] Zheng Y, Jiao Y, Zhu Y, Li L H, Han Y, Chen Y, Du A, Jaroniec M and Qiao S Z 2014 Hydrogen evolution by a metal-free electrocatalyst *Nat. Commun.* **5** 3783
[108] Alaghmandfard A and Ghandi K 2022 A comprehensive review of graphitic carbon nitride (g-C_3N_4)–metal oxide-based nanocomposites: potential for photocatalysis and sensing *Nanomaterials* **12** 294
[109] Osada M and Sasaki T 2015 Nanosheet architectonics: a hierarchically structured assembly for tailored fusion materials *Polym. J.* **47** 89–98
[110] Zhang L *et al* 2020 Frontispiece: structure and properties of violet phosphorus and its phosphorene exfoliation *Angew. Chem. Int. Ed.* **59** 961–1353

IOP Publishing

2D Materials as Electrocatalysts

Muhammad Ikram, Ali Raza, Jahan Zeb Hassan and Salamat Ali

Chapter 2

Development protocols and catalyst design

Parts of this chapter have been reprinted with permission from [9]. Copyright 2023 Elsevier.

The outstanding catalytic performance of 2D electrocatalysts depends fundamentally on their layered configuration and the (layer-controllable) growth of 2D materials, and large surface areas allowing for wide-ranging practical applications. Significant successes have been utilized to develop higher-quality monolayers. This chapter will describe commonly adopted synthesis strategies for developing 2D materials that comprise bottom-up schemes, including the hydrothermal (or solvothermal) route, a top-down scheme through exfoliating bulk counterparts, and vapor-phase deposition. We present various approaches for the growth of catalysts based on accessible facile reaction pathways, using the important relationship between (atomic-level) materials engineering and reaction fundamentals. Predominantly, effort is made to highlight the enhancement of inherent electronic structures to achieve the desired interfaces between reactive intermediates and catalysts. Moreover, direct instructions are provided using several encouraging examples of experimental and theoretical design routes for developing electrocatalysts. Finally, this chapter aims to motivate readers to pay enhanced attention towards the development protocols and design of catalysts. At the end of the chapter, the conclusion, challenges and future perspectives are presented.

2.1 Development protocols

2.1.1 Chemical vapor deposition

Chemical vapor deposition (CVD) is a promising and effective technology that can generate high-quality, low-defect TMD nanostructures or thin films on various substrates [1, 2]. Advances in vapor-phase deposition include adjusting the ratio of precursors so that the final product components may be tuned easily, and nanostructure–substrate interfaces with good contact that is beneficial for charge

doi:10.1088/978-0-7503-5291-8ch2

migration. The investigation of the HER mechanism can also be carried out using high-quality catalysts generated by the CVD approach, which offers a platform that is not only straightforward but also unadulterated. The metal starting materials, including metal films [3], metal oxides [4], metal halides [5, 6], or metal-organics [7], are deposited in a furnace in the middle of the heating element while Te, Se or S powder is sited upstream in terms of the directional flow of the gas carrier. This is the standard CVD synthesis method (usually Ar or N_2 combined with a particular concentration of H_2). The vapor of sulfur, selenium, or tetraethyl will travel downstream when T increases, causing the metal precursor to be converted into the TMDs corresponding to the high T zone (figures 2.1(a)–(c)) [1]. Recent studies have shown that it is possible to create 2D monolayer heterostructures by simply reversing the direction that the carrier gas is flowing [8, 9].

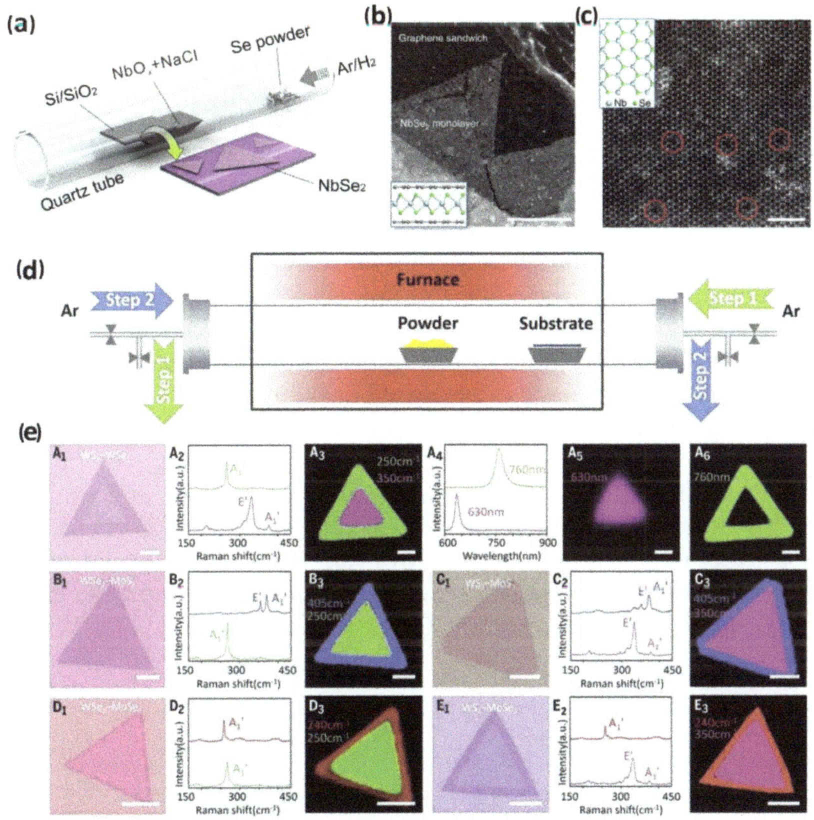

Figure 2.1. (a) CVD arrangement for fabricating $NbSe_2$ nanosheets. (b) Annular dark-field (ADF)-STEM photographs of monolayer $NbSe_2$ nanosheets captured in a graphene sandwich. (c) Hexagonal structure for $NbSe_2$ shown using atomic resolution ADF-STEM. (Reproduced with permission from [1]. Copyright 2017 Springer Nature.) (d) Adjusted CVD arrangements (having reversible Ar movement) for the epitaxial development of lateral TMD heterostructures. (e) Illustration of optical microscopy photographs for the development of numerous 2D nanosheets. (Reproduced with permission from [8]. Copyright 2017 American Association for the Advancement of Science.)

In most cases, the lateral heterostructures that contain two distinct TMDs monolayers are fragile, and it may be difficult for them to endure multistep development. A solution to this issue was offered by Zhang *et al* by providing an innovative step-by-step synthesis technique for building numerous in-plane 2D TMDs micro heterostructures. This strategy involves switching the direction in which the gas is flowing. The monolayer TMD nanosheets needed to be produced on the substrate using the CVD technique before implementing this strategy. For the sequential procedure, the previously formed monolayer was positioned downstream of Ar flow, and Ar moved from the opposite orientation throughout the *T* fluctuation, cooling the current single-layer TMD substances and preventing thermal deterioration. During this period, the argon flow in the opposite direction helped prevent uncontrolled nucleation from occurring before the sequential development stage (figure 2.1(d)). This powerful CVD approach may be utilized to build a range of in-phase 2D adjacent heterostructures (i.e. WSe_2–MoS_2 and WS_2–WSe_2, etc, figure 2.1(d)), as well as various heterojunctions (i.e. WS_2–$MoSe_2$–WSe_2 and WS_2–WSe_2–MoS_2). Illustrations of optical microscopy photographs for developing numerous 2D nanosheets are presented in figure 2.1(e). In conclusion, the fabrication of superior TMDs or heterostructures by the CVD process necessitates the careful control of several crucial factors [9].

In most cases, a higher *T* will result in improved crystallinity, however, the nanostructure may become unstable under extreme conditions. As a result, *T* is an essential characteristic and needs to be selected with attention. Another factor is the distance between the chalcogen source and the metal. Changing the distance would not be possible if the CVD apparatus comprised two independent heat sources. The flow of carrier gas, typically either argon or N, should then be adjusted to a proper value, thus, TMDs nanosheets can be placed over substrates. On the other hand, if only one heating source exists, it is necessary to calibrate the *T* gradients inside the tube to identify where the reactants should be placed. In this particular scenario, carrier gas flow and distance must be controlled meticulously and simultaneously. Different publications merely show a schematic layout of CVD devices when explaining experimental techniques. However, they omit the precise distance between the substrates and reactants, making it difficult to duplicate experimental results owing to minute changes in the CVD set-up.

2.1.2 Exfoliation

The un-exfoliated TMDs, such as MoS_2 [10], $MoSe_2$ [11, 12], and WS_2 [13], are semiconducting materials consisting of the 2H stage. In particular, the remarkable active sites were found in these substances, which significantly restricts their catalytic activity. By exfoliating the bulk TMDs, it is possible to generate layered TMD nanosheets with a greater surface area and a plentiful supply of active sites [14, 15]. Mechanical exfoliation is a direct approach that may be used to create single-layer TMD nanosheets. The synthetic process of mechanical exfoliation is comparable to the creation of 2D graphene [9, 16]. This technique produces single- or multi-layer TMD nanosheets at yield rates that are too low for electrocatalysis. Li insertion has been a practical approach for obtaining layered TMD materials on a large scale, to fulfill the requirements of practical catalytic applications. Li insertion reduces the layer number of bulk TMDs and changes their crystal structure, enhancing their catalytic activity toward the HER. To exfoliate the considerable TMD substance, three different lithiation processes are often utilized. The initial method is the chemical exfoliation procedure, which utilizes organolithium compounds such as butylLi (BuLi) [17, 18], MeLi [19], or $LiBH_4$ [20] (figures 2.2(a)–(c)). The TMD powder is saturated in a solution containing Li bases and an associated organic solvent in this technique. Simultaneously, continuous ultrasonication is applied to

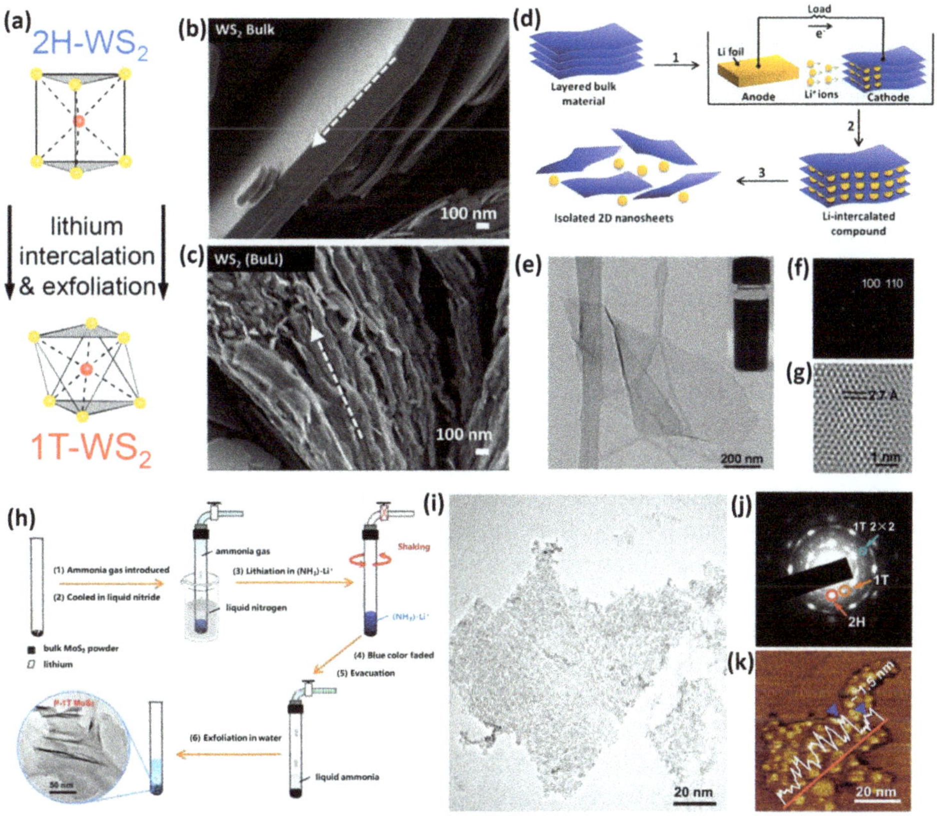

Figure 2.2. Visualization of the frequently utilized Li intercalation and exfoliation approach to acquire layered TMDs nanosheets. (a) Exfoliation of bulk WS_2 (with organolithium compounds) to gain 1T WS_2 nanosheets. (Reproduced with permission from [13]. Copyright 2014 The Royal Society of Chemistry.) SEM micrographs for (b) bulk WS_2 and (c) after the BuLi method with subsequent exfoliation in H_2O. (Reproduced with permission from [19]. Copyright 2014 American Chemical Society.) (d) Production of 2D nanosheet bulk substance with the electrochemical Li intercalation approach. (e) TEM photograph of an exfoliated MoS_2 nanosheet. Inset: digital image. (f) SAED patterns of a monolayer MoS_2 nanosheet. (g) HR-TEM photograph of an exfoliated monolayer MoS_2 nanosheet. (Reproduced with permission from [21]. Copyright 2011 John Wiley and Sons.) (h) Schematic view of LAAL routes. (i) Morphological analysis of mesoporous 1T-MoS_2 nanosheets. (j) Equivalent SAED patterns. (k) AFM photograph of exfoliated MoS_2 nanosheets. (Reproduced with permission from [10]. Copyright 2016 American Chemical Society.)

the mixture for an extended period (usually more than two days), which increases the efficacy of the exfoliation process (figure 2.2(c)).

Therefore, there is still much work to be done to obtain a significant quality yield of single-layer nanosheets and better control over the Li insertion process. A large quality yield electrochemical Li inclusion approach was presented to address the above limitation to create mono-layered 2D material via a regulated lithiation process [9, 21]. The most significant distinction is that the Li incorporation was carried out in a Li battery cell (figure 2.2(d)). During the discharge process, large TMDs at the cathode of a cell were progressively exfoliated to layers of nanosheets when Li was introduced into the cathode from the anode. Numerous nanosheets consisting of monolayers or a few layers were recovered after washing, sonicating, and centrifuging the samples (figures 2.2(e)–(g)). Compared to the yield obtained by chemical exfoliation (typically 10%–20%), the yield obtained through electrochemical Li insertion is significantly greater and can reach over 90% for MoS_2, TaS_2, and TiS_2 [22]. However, this technology still has certain downsides, such as a complicated procedure requiring assembling battery cells. These drawbacks make this method less desirable than others.

Additionally, extra additives that are often utilized throughout the process of fabricating electrodes have the potential to introduce contaminants into the products that are ultimately manufactured [23]. Recently, Jin *et al*'s group devised a novel methodology known as liquid ammonia-assisted lithiation (LAAL) for TMD exfoliation. This method proved to be an effective means of obtaining ultrathin 2D nanosheets [10]. Before beginning the LAAL procedure (shown in figure 2.2(h)), Li metal must first be enclosed in a quartz tube and shielded from the atmosphere with argon (Ar). After that, the tube is emptied of its contents and submerged in a bath of liquid N. Simultaneously, remarkably pristine ammonia gas is injected, slowly transforming into a liquid state during the process. The lithiation process starts as soon as the powder is submerged into the liquid ammonia, and the color of the liquid progressively changes as well (from a blue tint to colorless), which works as a signal for monitoring the reaction's progress as it occurs. Evaporation is used to gently remove the ammonia gas from the tube once the blue hue has been eliminated. In addition, ultrathin 2D nanosheets may be made by including water in the Li intercalated sample (figures 2.2(i)–(k)). The LAAL route revealed three clear advantages in contrast with the other two techniques defined in the earlier paragraph:

(i) The time required to complete this procedure is shorter, typically within an hour. In addition, the noticeable shift in hue made it possible to intuitively estimate the response process without needing any other signal.
(ii) 1T phase TMDs nanosheets with a single layer or a few layers with a large quality yield (~82%) can be achieved.
(iii) The intense lithiation method will result in an abundance of sulfur vacancies (S-vacancies) and a large number of edges, both of which will increase the electrochemical efficacy of the exfoliated TMDs nanosheets.

Because of the strong reaction that occurs when water and metal Li come into contact with one another, as well as the usage of liquid ammonia, each stage of the process must be carried out with extreme caution to guarantee the user's safety. It is possible to create a thinner sheet of TMD material by exfoliating bulk substances. Different methods are effective, particularly the lithiation process. Due to the complicated reaction method in a liquid environment, it is challenging to create TMD nanosheets with desired features, including a regular structure and regulated layer quantity, making electrocatalysis challenging to study.

2.1.3 Hydro/solvothermal method

The hydro/solvothermal method is a cost-effective and appropriate technique to synthesize TMD-based nanomaterials at a high level. Modifying the temperature, reaction duration, metal precursors, surfactants, and other factors may produce different structures and phases, making the hydro/solvothermal method a suitable approach for manufacturing nanostructured materials [24]. Recently, $MoSe_2$ nanosheets produced using a hydrothermal treatment were shown to be an effective HER catalytic material. By modifying the reaction temperature, combined with the ratio of $NaMoO_4{\cdot}2H_2O$ and Se starting materials to the reductant ($NaBH_4$), the products display diverse crystal structures and degrees of disarrangement (figure 2.3(a)) [11]. This shows that an increased concentration of $NaBH_4$ will increase the ratio of 1T $MoSe_2$, which exhibits improved HER performance

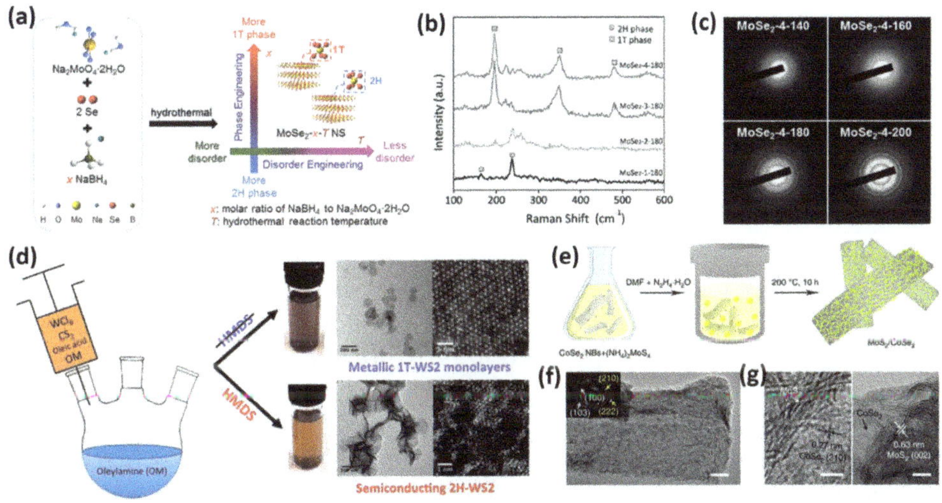

Figure 2.3. (a) The effect of phase and disorder engineering on $MoSe_2$ nanosheets triggered via controlling the preparation T and reductant ratio (x) using the hydrothermal strategy. (b) Raman spectra of various synthesized $MoSe_2$ nanosheet samples (with different ratios of $NaBH_4$). (c) SAED patterns, unlike those for the $MoSe_2$ models. (Reproduced with permission from [11]. Copyright 2017 John Wiley and Sons.) (d) Representation of the solvothermal process and parallel TEM and HAADF-HRSTEM photographs of WS_2 nanosheets with distinct phases. (Reproduced with permission from [28]. Copyright 2014 American Chemical Society.) (e) Synthesis method for a two-component $CoSe_2/MoS_2$ hybrid catalyst. (f) Corresponding TEM and SAED photographs (50 nm scale bar) and (g) HR-TEM images. (Reproduced with permission from [32]. Copyright 2015 Springer Nature.)

(figure 2.3(b), SAED rings in figure 2.3(c)). In contrast, a slower reaction *T* produced high active sites, apart from 1T surface creation. The $MoSe_2$ nanosheets with the best HER catalytic efficacy were created by carefully controlling the degree of disarrangement and the 1T surface ratio. Although the hydrothermal process has frequently been used to create nanostructured materials, it has the drawback that the materials might easily be oxidized throughout the process, either by the environment or the solution, which could affect the purity of the product. Fabrication of TMDs involves the solvothermal process to prevent potential oxidation [9, 25–27]. The selectivity of the final products is a further advantage of solvothermal reactions. For example, 2H WS_2 would be created by carefully adding hexamethyldisilazane. If not, 1T WS_2 would eventually be produced (figure 2.3(d)) [28]. Generally, because of a lower preparation temperature, less crystalline material is formed than under high temperatures, including solid-state reactions. However, this could result in products with lots of active sites. Numerous combined varieties of reactants will give an advantage in the construction of heterostructure nanomaterials, such as MoS_2/CuS [29], MoS_2/CdS [30], MoS_2–graphene [31], and MoS_2/$CoSe_2$ [32], with abundant interfaces and acceptable level structures (figures 2.3(e)–(g)).

2.1.4 Other methods

Other synthesis processes in addition to the ones discussed above have been employed in fabricating TMD materials. For example, thermal annealing is a primary method that may be used to create materials based on TMDs [33]. However, with this method it is challenging to make TMDs with precisely controlled forms and layer numbers. Furthermore, appropriate temperature may frequently result in material aggregation that causes restricted activity toward the HER. Despite this, some research has been dedicated to this approach for generating electrocatalysts with adequate catalytic activity for the HER [9, 34, 35]. In most cases, chemical vapor transport creates high-quality bulk single crystals. In this method, the raw materials are first subjected to a reaction in the hot zone with transference agents, typically pristine halogen elements. After this step, the materials are moved into a cold environment, where the formation of single crystals may be detected. The primary factors that need to be adjusted carefully to produce a single crystal of superior quality TMD are the temperature, reaction time, and the transference agent [36, 37]. Molecular beam epitaxy is an additional efficient way of producing high-quality films [38, 39]. These samples are in particular ideal for analysing the source of the catalytic activity in TMD catalysts because of the extraordinary purity of the films after they have been produced [40, 41].

2.2 Catalyst design

2.2.1 Controlling the electronic structure of electrocatalysts

The electronic structure is the single feature that controls a heterogeneous catalyst's adsorption capabilities in the direction of all sorts of intermediates [42, 43]. Thus, disrupting the electrical catalyst band is an excellent technique to influence its catalytic activity. For an alkaline HER, for example, altering the electronic arrangement of a catalyst, which mainly seeks to advance its hydrogen binding energy (HBE), is comparable to approaches used in acidic conditions. Adsorptive hydrogen bonds are strengthened when the d-band center of a metallic catalytic substance moves closer to the Fermi energy and vice versa when the center of the d-band shifts away from the Fermi level [43, 44]. The most efficient method for producing a d-band vacancy of a metallic catalyst is to alloy one metal with another. As shown in figure 2.4(a), the HBE of the majority of metallic catalysts can be modified via alloying with another metal, commonly due to mass electron shifting between two separate metal sites [45, 46]. This significant electron shifting can also

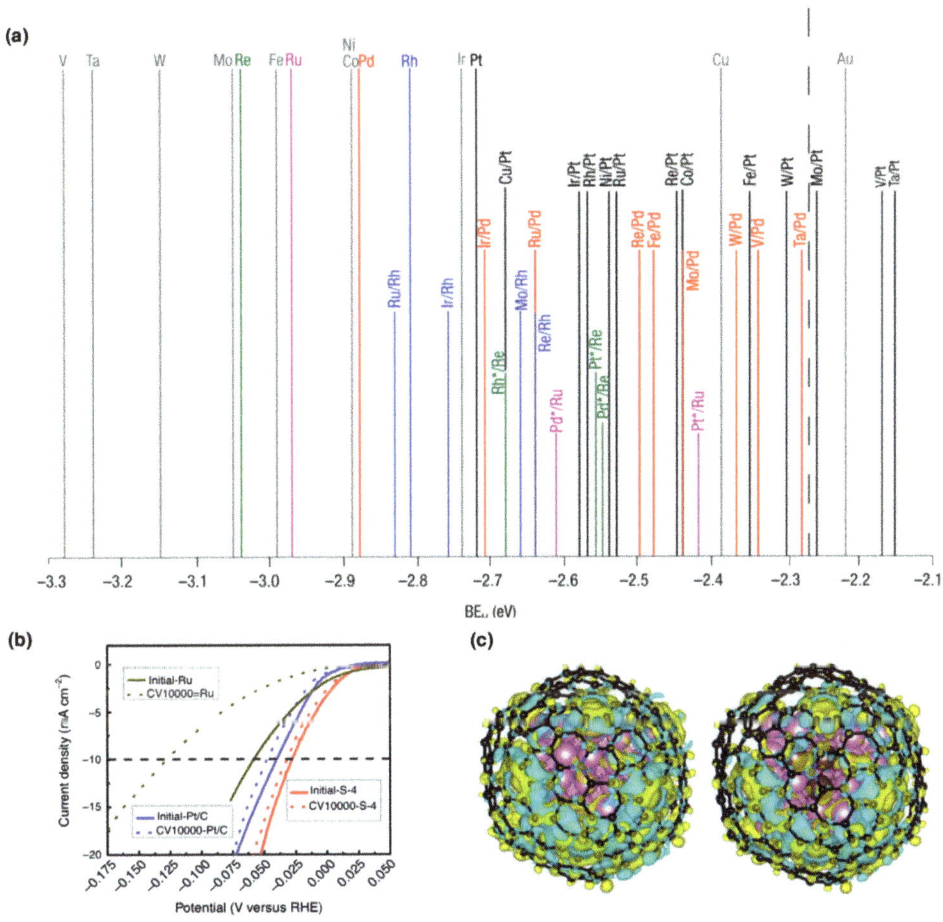

Figure 2.4. (a) A comparison of HBE for specific alloy and metal-based materials typically utilized to design catalysts. (Reproduced with permission from [45]. Copyright 2004 Springer Nature.) (b) Results of an HER experiment for RuCo@NC hybrid compared with other materials before and after 10 000 CV cycles. (c) Charge density variances for Co models on the left and Co_3Ru models on the right. Enhanced and reduced charge distributions are shown by the yellow and cyan regions, respectively. (Reproduced with permission from [47]. Copyright 2017 Springer Nature.)

contribute to the electronic structure of nonmetallic sites, which work synergistically with metal alloys to improve overall catalytic activity.

An example is a succession of RuCo alloys encased in an N-doped graphene sheet [47]. The alkaline HER is catalyzed using active transition metallic Ru powder, resulting in advanced activity and stability of the RuCoC alloy (figure 2.4(b)). DFT studies revealed that high activity is strongly connected to a specific electron arrangement of the carbon shell; due to the alloying of RuCo, the modification of electron transport from atoms to the carbon shell occurred (figure 2.4(c)). Consequently, the C–H bonds present in the carbon assembly increased, decreasing the HBE of the catalyst and enhancing the overall efficiency. Comparable results were obtained for various alloy catalysts. In contrast to Pt/C, a PtNi/NiS nanowire developed by Huang *et al* demonstrated a 5.58 times greater current density at −0.07 V versus RHE [48]. The PtNi alloy (a hexagonally packed nano-multipod) developed by Zheng *et al* exhibited significantly higher HER performance than Pt/C [49]. Notably, alloying can activate non-noble metal substances that typically have poor HER performances by changing the electronic structure [50, 51]. A Cu–Ti bimetallic electrocatalyst that showed exceptional alkaline HER activity was described by Chen *et al* [52]. Altering Ti's content in the catalyst caused an adjustment in the electronic arrangement of CuTi; further, the HBE of the catalytic material was suitably tuned by placing them at the peak point of the volcano plot. Because of this, even though neither Cu nor Ti are promising HER candidates, the Cu–Ti catalyst showed incredible activity [52]. A similar scenario was described [53] for a $MoNi_4$ alloy, that exhibited a significantly improved electronic structure as well as surprising water dissociation capability owing to the rebuilt electronic arrangement of the alloy, in contrast to Mo, Ni, and MoO_2, which display comparatively slow alkaline HER kinetics. These features of $MoNi_4$ confirmed its remarkable HER efficacy in alkaline conditions. Earth-rich and reasonably priced TMDs, phosphides, chalcogenides, and nitrides with various structures have been produced in large quantities during the past few decades thanks to rapidly evolving synthesis techniques [54, 55].

It was observed that several transition metal (TM) based materials exhibit high interaction with H with correct electronic morphological engineering [54, 56]. In particular, molybdenum-based catalysts such as MoS_2 and molybdenum dicarbonyl (Mo_2C) are the most often used because of their exceptional performance in acidic conditions [54]. However, in alkaline conditions, the design of TM-based catalytic activity is made more difficult due to interference with the process of water dissociation. This makes the process more challenging to engineer. It was observed that the active sites of HER in 2D MoS_2 and $MoSe_2$ depended on whether the environment was acidic or alkaline [9, 57, 58]. 2D $MoSe_2$ and MoS_2 were used as examples to demonstrate this. Although it is common knowledge that the Mo edge sites and/or catalytic vacancies of S/Se may be modified to increase their HER in conditions when pH is less than 7, the same method does not work well in primary electrolytes. This is because alkaline electrolytes are more fundamental than acidic due to the significant OH adsorption capacity displayed by the metal-terminated edge sites in an alkaline environment. Most of the H* (hydrogen intermediates, * represents an adsorption site) active sites for basic HER are S(Se)-rich groups on the catalytic materials' terrace [58, 59] because OH intermediates are located on the edge sites. Markovic *et al* [60] established the significance of S sites on transition metal sulfides (TMSs) for basic HER by examining the catalytic procedure of metal sulfides on the equally bulk model catalyst (S island on Au (111)) and TMS_x (TM = Mo, Co and $x = 4$–6) components. Markovic *et al* also demonstrated that S sites are essential for alkaline HER. It was discovered that the permanently absorbed S ($S_{ad}^{\delta-}$) on Au(111) works as a supporter to modify the HER efficacy through $S_{ad}^{\delta-}$–cation^{n+}–H_2O bonding (figures 2.5(a) and (b)). This kind of link can bring in hydrated cations, which can speed up the process of water dissociation and, as a result, increase the alkaline HER process. This process was then utilized in the development of TMS_x catalysts, which exhibit a behavior shown by $S_{ad}^{\delta-}$–TM^{n+} that is analogous to that of $S_{ad}^{\delta-}$–cation^{n+}. In addition, it was shown that the adjustable substrate–adsorbate binding energy and the number of flaws present in the catalytic substance dictate the link between the efficacy and stability of the TMS_x catalysts. Blocks composed of highly active CoS_x material with excessive flaws are less stable than those made of less active MoS_x material. An innovative region may

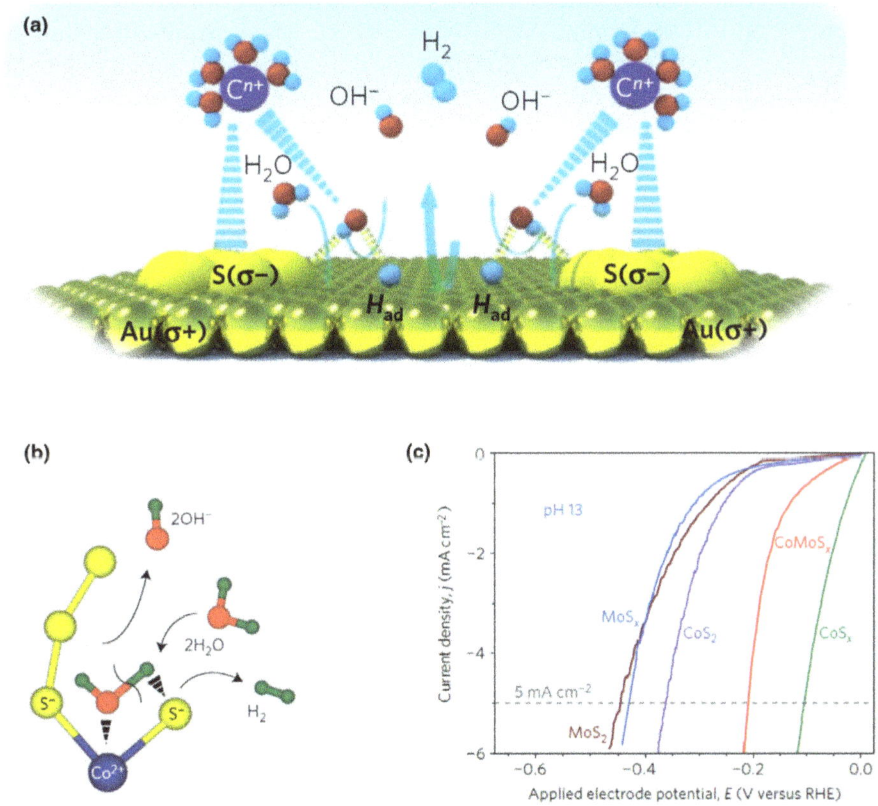

Figure 2.5. (a) Schematic view of HER pathways over a Au surface with $S_{ad}^{\delta-}$. (b) Demonstration of alkaline HER pathways for $CoMoS_x$. (b) Alkaline HER experiment for a chain of Co/Mo sulfides. (Reproduced with permission from [60]. Copyright 2016 Springer Nature.)

be developed for water dissociation by incorporating Co into the initial TMS_x, which will function in concert with H-active S sites. A catalyst with building blocks of increased stability, large efficacy, and higher water dissociation capability was generated by creating a material with the composition $CoMoS_x$ (figure 2.5(c)). It has been demonstrated that utilizing this well-balanced design is an operative approach toward catalyst improvement based on TMs. Defect engineering is a very efficient technique for altering the electrical morphology of 2D TMs [9, 61]. Using 2D MoS_2 as a reference, defects can form spontaneously at the boundaries between various phases; in conjunction with a substantial modification, even the boundary in the inert basal surface may become active for basic HER [62]. Considering this, we can say that flaws can occur on the boundary between various phases. According to additional DFT research, different types of borders have the potential to bring about dramatic variations in the Gibbs's free energy (ΔG) of H*. Additionally, heteroatom doping is a typical technique for altering the electrical arrangement approaching acidic HER technologies (for 2D TMs) and defect engineering [63]. This technique may also be utilized to construct catalysts for alkaline environments.

For instance, carbon-doped MoS_2 (C–MoS_2, which the sulfurization of Mo_2C may generate) has significant alkaline HER efficacy that is quite analogous to that of Pt/C (figure 2.6(a)) [64]. Doping MoS_2 with carbon, which results in a shift in that material's electrical structure, is responsible for the exceptional activity. The carbon in the catalyst does not function as a dual active site; instead, it induces vacant 2p orbitals perpendicular to the MoS_2 basal plane. This creates an environment that is advantageous for the adsorption of water (figure 2.6(b)). Consequently, the catalyst demonstrates better overall activity and an increased rate of water dissociation while operating in alkaline settings. Notable successful techniques for carbon-possessing catalysts include defect engineering and the incorporation of heteroatoms [42, 65, 66]. Significantly, the incorporation of heteroatoms and defect engineering can be directly utilized to improve the valence orbital energy for carbon sites, which ultimately improves the interaction between carbon and H* intermediates. This has been explicitly demonstrated for alkaline HER; however, the efficiency of carbon-based materials is often improved when combined with additional catalysts rather than functioning as the only catalytic center in alkaline HER [67, 68]. Because of this, this evaluation will not provide an in-depth examination of these resources. Altering the charge density of the catalytic surface is yet another frequently used method to enhance the natural capacity of TMs to absorb electrons. Combining catalysts based on TMs with other substances is one of the most frequently utilized to reach this objective. A valuable example of regulating the charge shifting of Mo_2C-based HER catalytic materials was published by Zou *et al* [69]. Coating Mo_2C with N-rich carbon increased the HER activity of the resulting products using all pH ranges (figure 2.6(c)). This scheme improved the H adsorption capability of the catalyst, which was achieved only by covering Mo_2C with N-rich carbon.

In contrast to the case of C doping, as discussed earlier, the significant electron-extracting capacity of N sites produced an electron shift that drew electrons from Mo_2C to N via C sites. This transfer took place because Mo_2C sites could attract electrons. This mechanism activated the C atoms next to the N sites, resulting in the

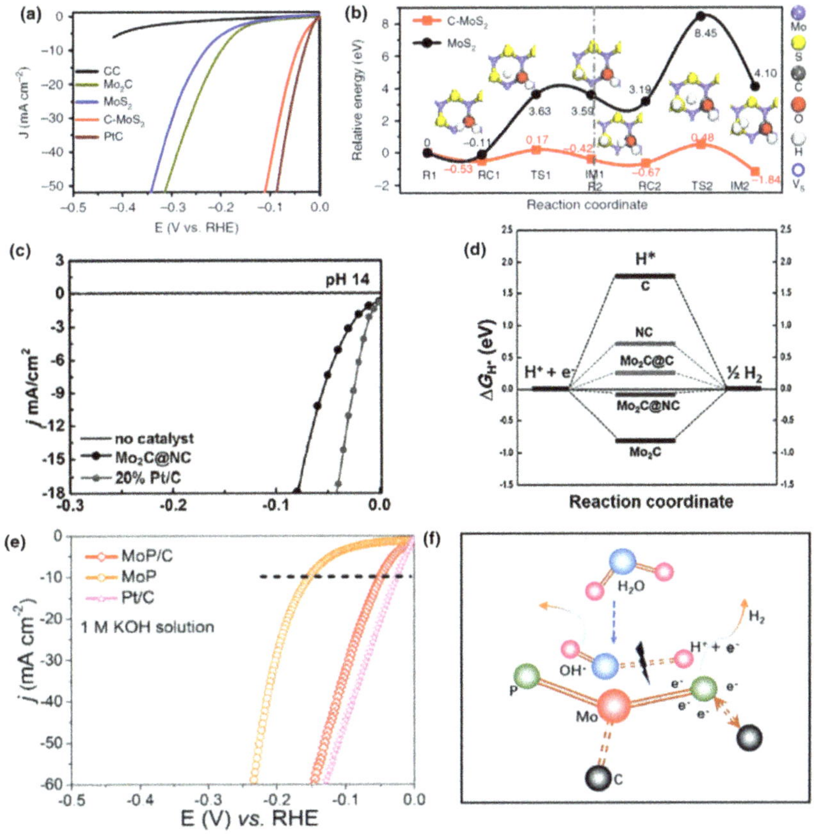

Figure 2.6. (a) HER experiment of C–MoS_2 compared with several related catalysts and the benchmark Pt/C (in 1 M KOH solution). (b) Visualization of the reactive energy with reaction coordinates signifying the water dissociation strategy on the basal surface of MoS_2 and C–MoS_2. (Reproduced with permission from [64]. Copyright 2019 Springer Nature.) (c) HER experiment of Mo_2C@NC compared with Pt/C. (d) Estimated free energy value of H_2 adsorption for different designs of Mo_2C-based catalysts. (Reproduced with permission from [69]. Copyright 2015 John Wiley and Sons.) (e) HER experiment of MoP@C compared with Pt/C and MoP (in 1 M KOH). (f) Representation of the HER pathways for MoP@C. (Reproduced with permission from [71]. Copyright 2018 John Wiley and Sons.)

creation of HER-active C sites capable of working in conjunction with Mo_2C (figure 2.6(d)). As a consequence of this, the total activity of HER might potentially be enhanced in all situations regardless of pH. Similar tactics were used for other basic HER catalytic substances. These strategies are efficient for altering the H adsorption capability of catalysts and enhancing the interaction between OH and the catalyst [70, 71]. The MoP@C catalyst is a good illustration of a typical case because it demonstrated outstanding HER efficacy in an alkaline environment (figure 2.6(e)) [71]. The formation of the Mo–C bond in the catalyst was caused by the presence of carbon on the surface of MoP, which in turn had a significant influence on the electronic arrangement of the element molybdenum. Consequently, the newly upgraded Mo site was now suitable for dissociating water, while the adjacent P sites were responsible for the recombination of hydrogen (figure 2.6(f)). It was demonstrated that such a structure provided a suitable platform for processing an alkaline HER [9]. Its straightforward construction, which may speed up two processes simultaneously, makes it the most effective method for producing high-quality HER catalytic activity.

2.2.2 Creating dual active sites

Even though the precise part that OH^- plays in alkaline HER has not been figured out yet, it is abundantly evident that the interaction between the catalyst and OH^- is one of the most significant aspects in determining how active the catalyst is. However, because of the poorly understood scaling connection between the binding energy of OH* and H* intermediates, it is interesting to realize an equilibrium between the adsorption and desorption abilities toward OH* and H* on a single site. Creating dual active sites that can individually feature unique functionalities is one of the most widely acknowledged ways of gaining control over OH* and H* in a manner that is distinct from one another. Markovic *et al* [72] described a sequence of catalysts by decorating bulk Pt with metal hydroxide sites or supplementary metal; this was done using the method as described above. The overarching goal was to develop a hybrid material with satisfactory HBE for HER, by marrying a material on top of an acid HER volcano plot, for example, Pt with an oxophilic metal that can give adequate OH interaction sites for water dissociation. From the assessment of the oxophilicity of several regularly used materials, a larger oxophilicity predicts a more potent OH binding energy, and vice versa (figure 2.7(a)) [73].

In the same way as with HBE, the OH binding energy of a catalyst must be neither too strong nor excessively weak to produce an appropriate interaction between the catalyst and the water. As a result, there are a few choices that stand out for their adequate oxophilicity; these include Ni, Ru, and Co. These materials show an exceptional level of performance and, as a result, they are utilized extensively in the development of catalysts that have dual active sites [67, 74, 75]. Based on the achievements of the alloys with dual sites, non-noble materials have been employed in place of Pt as H* interaction sites to develop more cost-effective catalysts. In the HER process, it has been demonstrated that MoS_2 and g-C_3N_4 can absorb hydrogen in an ideal manner, similar to Pt. It was reported by Yang *et al* that they employed MoS_2 as the H-active material and LDH as the OH-active sites for superior HER performance (figure 2.7(b)) [76]. According to the results of the DFT calculations, the higher performance of MoS_2/LDH heterostructures may be attributed to the decreased activation energy (figure 2.7(c)), which is the result of the adjusted water dissociation process. Similar to the Pt/$Ni(OH)_2$ system, the MoS_2/LDH system offers separate sites for hydrogen adsorption (on MoS_2) and the dissociation of water (on LDH). These sites contribute complementarily to the overall alkaline HER process by acting as synergists (figure 2.7(d)) [76].

The OH interaction sites may be designed or replaced with several different materials, much like the H adsorption sites. In a different line of research, 2D MoN was employed as a OH* adsorption site paired with C_3N_4 to increase the alkaline

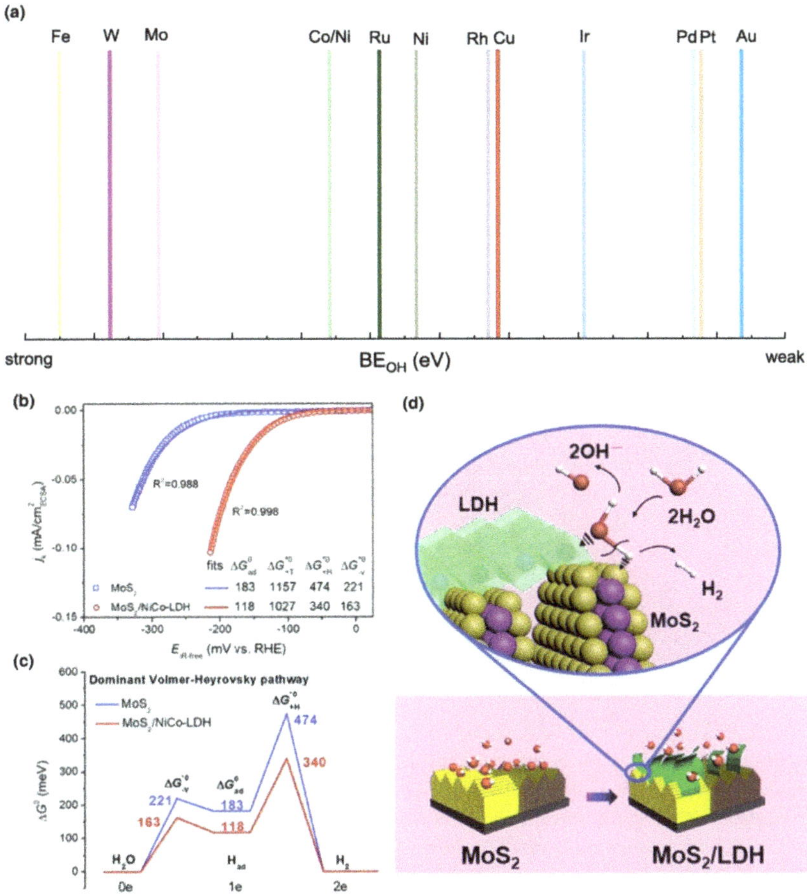

Figure 2.7. (a) Association of the oxophilicities of certain metallic materials recognized as catalysts. (Reproduced with permission from [100]. Copyright 2020 Elsevier.) (b) Evaluation of an HER experiment for MoS_2 and MoS_2/NiCo-LDH (in 1 M KOH solution) using fitted standard activation free energies (meV). (c) Illustration of the free energy value concerning the reaction mechanism of both materials. (d) Display of alkaline HER reaction pathways over MoS_2/LDH catalyst. (Reproduced with permission from [76]. Copyright 2017 Elsevier.)

HER [77]. Using a salt-templated technique accomplished the fabrication of 2D MoN, which was attached to C_3N_4 through electrostatic contact (figure 2.8(a)). Interestingly, such 2D MoN–C_3N_4 exhibits comparable HER activity in electrolytes containing 1 M KOH and 0.5 M H_2SO_4 (figure 2.8(b)), which is a highly unusual occurrence in the context of HER catalysts. As validated by DFT calculations, the better water dissociation capacity of 2D MoN implies that the reaction obstacles created through the water dissociation procedure do not weaken activity for prepared material (2D MoN–C_3N_4), as visualized in figure 2.8(c). The volcano plot demonstrates that adjusting the ratio of g-C_3N_4 to MoN could directly control the activity. This finding provides further evidence of the importance of maintaining a healthy equilibrium between water dissociation (at the MoN site) and the

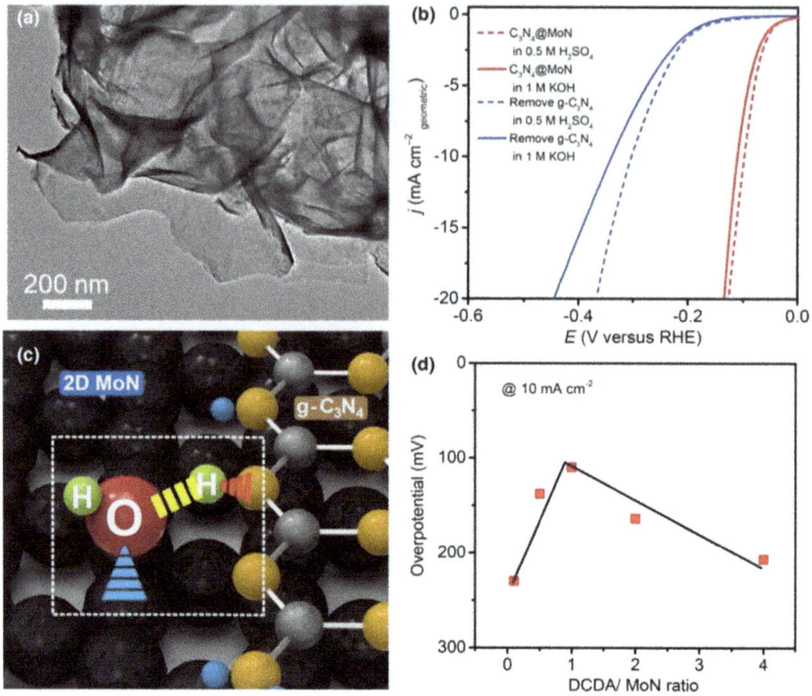

Figure 2.8. (a) A TEM photograph of a C_3N_4@MoN nanosheet. (b) Evaluation of HER experiments (using different electrolytes) for C_3N_4@MoN before/after the elimination of C_3N_4. (c) Schematic illustration of the water dissociation approach for C_3N_4@MoN. (d) The relationship between a variety of DCDA (dicyandiamide as a precursor for C_3N_4)/MoN ratios and the reaction overpotential at a current density of 10 mA cm^{-1}. (Reproduced with permission from [77]. Copyright 2018 Elsevier.)

hydrogen adsorption (at the C_3N_4 site) capabilities of a catalyst (figure 2.8(d)). Similar dual-site catalysts have been described and the outcomes are higher alkaline HER activity, for example, Ni-doped MoS_2 (also Ru, Co, and Fe) and $Ni(OH)_2$/MoS_2 [78, 79]. The effective strategies are grounded in suitably harmonizing hydrogen adsorption processes and water dissociation in every instance.

2.2.3 Adjusting the surface geometry of electrocatalysts

It has been shown that the alkaline HER is very sensitive to catalyst structure [80]. In the early 1990s, HER/hydrogen oxidation reaction (HOR) activity was evaluated by Markovic *et al*, which is exhibited by many Pt single crystals and varies considerably [81]. In alkaline conditions, increasing performance may be seen in the sequence Pt (1 1 1) < Pt (1 0 0) < Pt (1 1 0). Even though DFT studies have provided information on how various Pt crystal planes might alter the catalyst's electronic structure and its hydrogen adsorption ability, the actual surface-sensitive behavior of the alkaline HER remains unclear. To date, it is unknown how the structure of the catalysts affects underpotential deposited hydrogen (H_{upd}) and overpotential deposited hydrogen (H_{opd}) separately [82]. Lattice strain engineering has been shown to be one of the most valuable techniques for altering the surface structure of a catalyst in practical applications. It has been demonstrated that even small variations in the lattice strain can cause significant changes in the interaction between the catalyst and the intermediates. For example, 1% lattice strain on the surface of the Pt atom can affect a ~0.1 eV shift on the d-band center of the Pt atom, which can significantly impact the Pt atom's HBE [43]. Compared to the contact between H and the catalyst, the strain significantly affects the relations between the O-bound intermediates (i.e. OH) and the catalyst. In late TMs such as Ru and Pt, it was shown that the tensile lattice strain could induce improved contact by way of O-based intermediates and so on [75, 83, 84]. One of the complex challenges in material engineering is successfully inducing the necessary strain over the catalyst surface. By altering the lattice, one can often change the surface strain of several materials [85], developing specially angled or defective morphologies [86, 87] and building particular core–shell structures [88]. Particularly for nanostructured catalysts, morphology manipulation techniques such as generating a lattice, crystal shaping, and the de-alloying discrepancy can alter a material's surface into a more elongated shape [89]. These tactics have been used in the process of increasing the performance of ORR catalysts. Furthermore, it has been demonstrated that these strategies are useful for HER catalysts in acidic environments, i.e. WS_2 and MoS_2 [9, 85].

Recently, many attempts have been made to enhance the applicability of this approach in alkaline HER settings. The Pt_3Ni alloy described by Stamenkovic *et al* [90] may create a predominantly active alkaline HER catalyst by adopting an organized procedure, i.e. in sense of tempting approach. Pt_3Ni nanoframes with Pt skin were manufactured by first creating Pt_3Ni polyhedra, then removing any excess Ni, and annealing the result in an environment of Ar (figure 2.9(a)). The fact that through the existence of dual active sites for water dissociation, Ni is encased in Pt, precludes the possibility that exceptional alkaline HER catalytic activity of Pt_3Ni nano-frame was caused. The only explanation for high alkaline HER activity may be present in the refined electronic structure of Pt. Pt_3Ni nanoframes are made of nickel and also provide a range of different Pt crystal structures and crystal contacts. On the Pt surface, compressive strain is caused due to the core–shell structure of Pt_3Ni nanoframes, which can be observed from another perspective. These characteristics work together to mold the electronic band arrangement of the Pt_3Ni nano-frame catalytic material, resulting in an exceptional alkaline HER efficacy (figure 2.9(b)). However, strain engineering by alloying is generally regarded as the most effective method for changing the surface catalytic characteristics of noble metals. Specific queries aimed at enhanced catalytic activity may arise from electronic shifting to strain modifying or between two metals. After all, alloying has historically been considered the most efficient strain engineering form. The analysed electrical and geometric effects for the core–shell organized RuPt catalysts can be seen in figure 2.9(c) [91]. Because of a lattice mismatch, the core–shell bimetallic Ru@Pt displays 3% higher compressive strain (compared to RuPt alloy), as shown in figure 2.9(d). It was observed that the catalyst exhibited a reduced capacity to adsorb hydrogen and an enhanced ability to interact with oxygen when subjected to a compressive strain, which led to a significant increase in activity

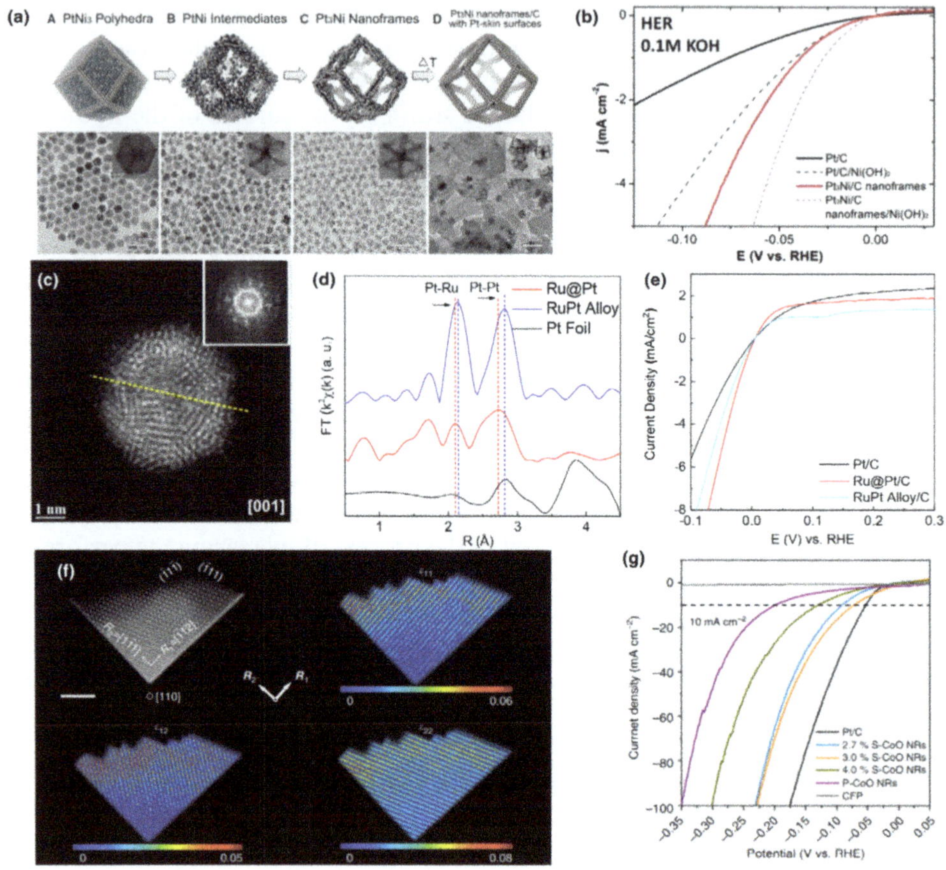

Figure 2.9. (a) Schematic views of altered phases that appeared during the development route of Pt_3Ni nanoframes with corresponding TEM photographs below. (b) Results of HER experiments for Pt_3Ni nanoframes compared with further benchmark catalysts (in 0.1 M KOH solution). (Reproduced with permission from [90]. Copyright 2014 American Association for the Advancement of Science.) (c) Experimental TEM photograph of Ru@Pt nanoparticles. (d) Extended x-ray absorption fine structure (EXAFS) spectra for Ru@Pt compared with other catalysts. (e) Results of HER experiments for Ru@Pt/C compared with RuPt alloy/C and Pt/C (in 0.1 M KOH solution). (Reproduced with permission from [91]. Copyright 2018 American Chemical Society.) (f) Contour plots for strain constituent ε_{xy} compared to mentioned values for nano sawtooth, where R_1 and R_2 are lattice vectors for strain examination. (g) Results of HER experiments for S–CoO nanorods compared with certain additional catalysts (in 1 M KOH solution). (Reproduced with permission from [92]. Copyright 2017 Springer Nature.)

(figure 2.9(e)). This strain-encouraged increase in HER efficacy was seen for metal catalysts and TM oxides, demonstrating that it is not limited to metal catalysts. The S–CoO sawtooth-rich nanorods with 3% surface tensile strain [92] are a case material that serves as an excellent example of this type of material. As can be seen, the strain component *e*, which is related to the contraction and expansion of the corresponding lattice vectors R_1 and R_2, demonstrates a biaxial strain on the surface of the nano-tooth for each of the small sawtooth structures (figure 2.9(f)).

A significant number of O-vacancies are produced. This tensile tension allows the water breakdown obstruction and the HBE of the catalyst to be adjusted. Consequently, tensile strain of S–CoO nanorods was applied to achieve an exceptional HER activity similar to commercial noble metal catalysts (figure 2.9(g)).

2.2.4 Lateral size and thickness regulation

For the following reasons, the thickness and lateral size of 2D materials greatly influences the electrocatalytic capabilities of these materials [93]:

- When the lateral size of 2D materials is smaller, more flaws and edge locations are exposed.
- The width of 2D materials can be changed to alter their electrical structures, which results in adjustable catalytic activity.
- Catalytic activity augmentation may arise from in-plane imperfections caused by structural disarrangement when the thickness is reduced to the atomic level.
- Mono-layered 2D materials have the most extensive hypothetical surface area, which gives them the most potential for catalytic activity.

Kibsgaard *et al*, in particular, revealed how morphological control of MoS_2 at the nanoscale might drastically affect the surface structure at the nano level, improving HER performance [94]. Moreover, the lattice alteration that occurs in many 2D materials provides many vacancies for catalytic activity when thicknesses are reduced to sub-nanometer scales. Another benefit of thickness management is that it enables the use of 2D morphologies, which can expose the maximal area of a catalyst and produce a higher electrochemical surface area (ECSA). The examination of plasmonic and catalytic approaches employed by Huang *et al* allowed the development of freestanding Pd nanosheets [9, 95]. Electrocatalytic oxidation of formic acid causes the Pd nanosheets to offer superior ECSA; thus, higher density was revealed compared to commercial Pd black. As a result, various techniques have been devised to raise ECSA values for metal and metal alloy nanosheets (for example, Pt–Cu, Rh, and Pd–Cu) [96, 97].

2.3 Comparative analysis via experiments and computations

The discovery of electrocatalytic active materials for HER and ORR was described by Jiao *et al* through nonmetallic heteroatom-incorporated graphene products using DFT calculations, precise structural characterizations, and electrochemical experiments [66, 98]. They created five types of graphene, each doped with a specific element (oxygen, phosphorus, boron, sulfur, or nitrogen), to represent a wide range of conditions that can occur during experiments. Characterizations of the shape, surface area, and flaws of the various heteroatom-doped materials were carried out using several different physicochemical tests to demonstrate that these characteristics were similar amongst the samples (figure 2.10(a)). Following that, a sequence of surface characterizations approaches, such as the unique doping structures present in different materials, can be accessed via near edge x-ray absorption fine

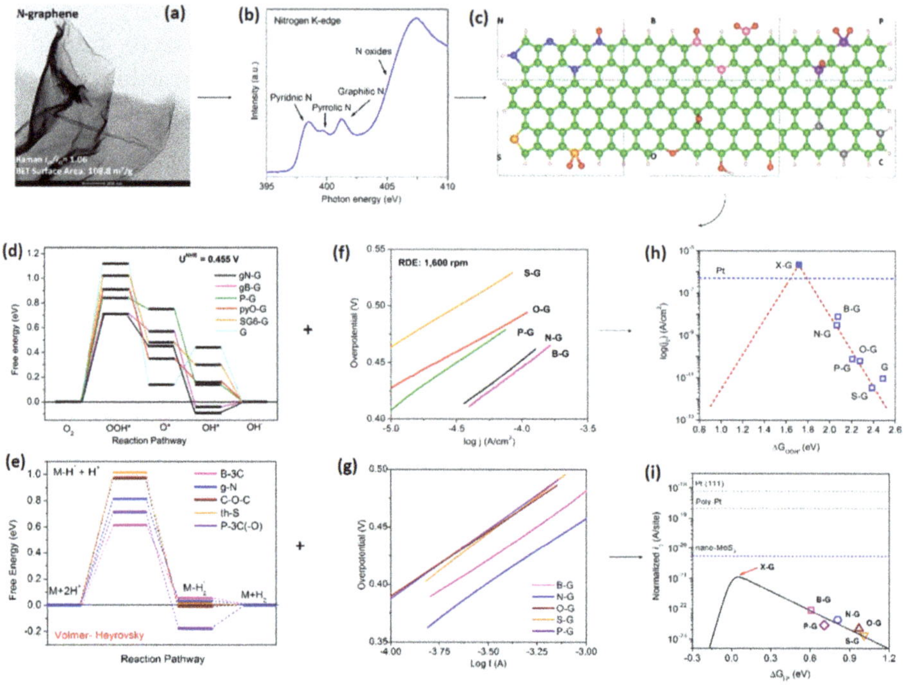

Figure 2.10. (a) TEM photograph of N-graphene as an example (200 nm scale bar); inset: Illustration of I_D/I_G ratios, N-adsorption isotherms, and Raman analysis for specific surface areas. (b) Nitrogen K-edge NEXAFS spectrum for N-graphene as an example. (c) Visualization of doping patterns for heteroatom. (d) ORR free energy values for different materials at equilibrium potential and (e) HER. (f) ORR Tafel plots for different synthesized materials observed using consistent environments and (g) HER. (h) ORR and (i) HER volcano plots comprising experimentally evaluated i_o (y-axis) and theoretically calculated free energy (ΔG) for primary intermediates (x-axis). ((a), (d), (f), and (h) reproduced from [98]. Copyright 2014 American Chemical Society. (b), (c), (e), (g), and (i) reproduced from [66]. Copyright 2016 Springer Nature.)

structure (NEXAFS) (figure 2.10(b)). They meticulously developed a series of characteristic molecular models to compare the computational examination with an experimental investigation (figure 2.10(c)). Free energy schematics were realized from these simulations for HER and ORR, and experimental capacities were approved to determine the apparent electrocatalytic activity of the produced samples (figures 2.10(d)–(g)). A graph in the shape of a volcano was created by linking the appropriate theoretical and experimental characteristics (for example, ΔG and i_o, respectively). These volcano plots illustrate the natural activity pattern of the materials in question and, more crucially, they predict the best performance that an ideal electrocatalyst can achieve (figures 2.10(h) and (i)). The volcano plot is linked to the Sabatier principle [99], which asserts that optimum catalytic activity may be realized over the catalyst surface with sufficient binding energies for reactive intermediates [99]. The Sabatier concept also applies to the volcano plot as optimal catalytic activity could be acquired through the catalyst surface. If the intermediates only attach loosely to one another, surface activation and reaction will be difficult.

On the other hand, if they attach too tightly, they will 'poison' the reactive surface by taking up all available surface sites and preventing any new ones from forming. As can be observed in figures 2.10(h) and (i), all of the materials are located on the right side of the volcano plot, which indicates that the adsorption of the vital chemical intermediates is not as strong as it might be (for both the ORR and HER). Therefore, it is clear that if a catalyst can obtain a ΔG value that is nearer to the middle of the volcano (that is, a lower value and more significant adsorption for graphene), it will have an immense i_o value from the theoretical as well as experimental point of view. After using the molecular orbital idea, they suggested a catalyst design strategy to improve materials containing graphene. This principle was developed using the molecular orbital concept.

2.4 Conclusion, challenges and perspectives

1. The practical use of catalysis will be impossible without organized and macroscopic manufacturing of 2D materials. Liquid phase exfoliation and CVD are possible synthesis strategies that might be used to fabricate layered 2D materials. Previously, (*in situ*) characterization apparatus, which can sense kinetic and thermodynamic reactions throughout the synthesis route of the material, has already been devised. This equipment can be of great use in the emerging broad comprehension of the development of pathways for 2D materials, as it can detect real-time reactions. The restricted synthesis approach could be a significant development technique that allows understanding of the molecular level in purpose to monitor 2D materials synthesis in the direction of catalytic applications. In the future, large-scale development of controlled structures (i.e. non-layered atomically thick nanosheets) with inherent electrocatalytic activity still needs exploration. This aim poses major hurdles, particularly in obtaining accurate monitoring towards developing uniform and higher-quality 2D materials.
2. Anastomotic models have been developed that link theoretical analysis and experimental findings to better comprehend theoretical measurements for catalytic pathways for 2D materials. This has been done to develop a better theoretical comprehension of the mechanisms of catalysis. However, the current electrocatalytic research findings on 2D materials have focused more on the augmentation of electrocatalytic activity than the fundamental science of electrocatalytic pathways. Additionally, a significant portion of the investigations based on catalytic experiments of 2D materials include many inaccuracies and challenges. When compared to 3D bulk materials, using 2D layered nanocatalysts results in a more complicated reaction system. As a result, recent advancements towards a fundamental understanding of 2D electrocatalysts are valuable for discovery and material design strategies. To date, the theoretical electrocatalytic processes of 2D electrocatalyshave been quite difficult to apply to actual complex reaction arrangements. As a result, supplementary basic as well as theoretical investigations on electrocatalytic pathways would be valuable avenues of investigation.

3. The capacity to shape catalysts is of great interest for their applications, in addition to the technical research for mass manufacturing. The employment of 2D materials as loose powders causes them to have the potential to agglomerate, limiting the applications for which they can be used. The possibility of affixing 2D materials as a substrate offers a feasible strategy for making it easier to employ the materials. To improve electrocatalytic activity, 2D materials, for example, might be grown epitaxially over a supplementary material surface, fabricated as foam, or sustained on nickel foam (NF) or carbon fiber paper. All of these methods are possible.
4. Currently, the primary emphasis of modulation techniques for TMD-based electrocatalysts is on improving the characteristics of the catalysts. However, despite the fast development of a wide variety of TMD-based catalysts, there appears to be a bottleneck in the improvement and development of catalytic activity. The application of external driving fields during the testing procedure is one option that might be considered to further enhance the catalytic effect. According to certain current research, electric as well as magnetic fields might considerably increase the water-splitting effectiveness via adjusting the spin polarization or conductance of catalysts. It has been hypothesized that the HER performance of innovative catalysts might be further enhanced by gaining assistance from an external driving field.

Bibliography

[1] Wang H *et al* 2017 High-quality monolayer superconductor $NbSe_2$ grown by chemical vapour deposition *Nat. Commun.* **8** 394

[2] Jiang J, Li N, Zou J, Zhou X, Eda G, Zhang Q, Zhang H, Li L-J, Zhai T and Wee A T S 2019 Synergistic additive-mediated CVD growth and chemical modification of 2D materials *Chem. Soc. Rev.* **48** 4639–54

[3] Kong D, Wang H, Cha J J, Pasta M, Koski K J, Yao J and Cui Y 2013 Synthesis of MoS_2 and $MoSe_2$ films with vertically aligned layers *Nano Lett.* **13** 1341–7

[4] Shi J *et al* 2014 Controllable growth and transfer of monolayer MoS_2 on au foils and its potential application in hydrogen evolution reaction *ACS Nano* **8** 10196–204

[5] Zhao S, Hotta T, Koretsune T, Watanabe K, Taniguchi T, Sugawara K, Takahashi T, Shinohara H and Kitaura R 2016 Two-dimensional metallic NbS_2: growth, optical identification and transport properties *2D Mater.* **3** 025027

[6] Yuan J *et al* 2015 Facile synthesis of single crystal vanadium disulfide nanosheets by chemical vapor deposition for efficient hydrogen evolution reaction *Adv. Mater.* **27** 5605–9

[7] Kang K, Xie S, Huang L, Han Y, Huang P Y, Mak K F, Kim C-J, Muller D and Park J 2015 High-mobility three-atom-thick semiconducting films with wafer-scale homogeneity *Nature* **520** 656–60

[8] Zhang Z, Chen P, Duan X, Zang K, Luo J and Duan X 2017 Robust epitaxial growth of two-dimensional heterostructures, multiheterostructures, and superlattices *Science* **357** 788–92

[9] Raza A, Rafi A A, Hassan J Z, Rafiq A and Li G 2023 Rational design of 2D heterostructured photo- and electro-catalysts for hydrogen evolution reaction: a review *Appl. Surf. Sci. Adv.* **15** 100402

[10] Yin Y *et al* 2016 Contributions of phase, sulfur vacancies, and edges to the hydrogen evolution reaction catalytic activity of porous molybdenum disulfide nanosheets *J. Am. Chem. Soc.* **138** 7965–72
[11] Yin Y *et al* 2017 Synergistic phase and disorder engineering in 1T-$MoSe_2$ nanosheets for enhanced hydrogen-evolution reaction *Adv. Mater.* **29** 1700311
[12] Najafi L, Bellani S, Oropesa-Nuñez R, Ansaldo A, Prato M, Del Rio Castillo A E and Bonaccorso F 2018 Engineered $MoSe_2$-based heterostructures for efficient electrochemical hydrogen evolution reaction *Adv. Energy Mater.* **8** 1703212
[13] Lukowski M A, Daniel A S, English C R, Meng F, Forticaux A, Hamers R J and Jin S 2014 Highly active hydrogen evolution catalysis from metallic WS_2 nanosheets *Energy Environ. Sci.* **7** 2608–13
[14] Chhowalla M, Shin H S, Eda G, Li L-J, Loh K P and Zhang H 2013 The chemistry of two-dimensional layered transition metal dichalcogenide nanosheets *Nat. Chem.* **5** 263–75
[15] Coleman J N *et al* 2011 Two-dimensional nanosheets produced by liquid exfoliation of layered materials *Science* **331** 568–71
[16] Yin Z, Li H, Li H, Jiang L, Shi Y, Sun Y, Lu G, Zhang Q, Chen X and Zhang H 2012 Single-layer MoS_2 phototransistors *ACS Nano* **6** 74–80
[17] Ambrosi A, Sofer Z and Pumera M 2015 2H $\rightarrow$ 1T phase transition and hydrogen evolution activity of MoS_2, $MoSe_2$, WS_2 and WSe_2 strongly depends on the MX_2 composition *Chem. Commun.* **51** 8450–3
[18] Voiry D *et al* 2013 Enhanced catalytic activity in strained chemically exfoliated WS_2 nanosheets for hydrogen evolution *Nat. Mater.* **12** 850–5
[19] Eng A Y S, Ambrosi A, Sofer Z, Šimek P and Pumera M 2014 Electrochemistry of transition metal dichalcogenides: strong dependence on the metal-to-chalcogen composition and exfoliation method *ACS Nano* **8** 12185–98
[20] Voiry D, Salehi M, Silva R, Fujita T, Chen M, Asefa T, Shenoy V B, Eda G and Chhowalla M 2013 Conducting MoS_2 nanosheets as catalysts for hydrogen evolution reaction *Nano Lett.* **13** 6222–7
[21] Zeng Z, Yin Z, Huang X, Li H, He Q, Lu G, Boey F and Zhang H 2011 Single-layer semiconducting nanosheets: high-yield preparation and device fabrication *Angew. Chem. Int. Edn* **50** 11093–7
[22] Zeng Z, Tan C, Huang X, Bao S and Zhang H 2014 Growth of noble metal nanoparticles on single-layer TiS_2 and TaS_2 nanosheets for hydrogen evolution reaction *Energy Environ. Sci.* **7** 797–803
[23] Tan C *et al* 2017 Recent advances in ultrathin two-dimensional nanomaterials *Chem. Rev.* **117** 6225–331
[24] Zhen C, Zhang B, Zhou Y, Du Y and Xu P 2018 Hydrothermal synthesis of ternary $MoS_{2x}Se_{2(1-x)}$ nanosheets for electrocatalytic hydrogen evolution *Inorg. Chem. Front.* **5** 1386–90
[25] Gao M-R, Chan M K Y and Sun Y 2015 Edge-terminated molybdenum disulfide with a 9.4-Å interlayer spacing for electrochemical hydrogen production *Nat. Commun.* **6** 7493
[26] Ikram M, Raza A, Imran M, Ul-Hamid A, Shahbaz A and Ali S 2020 Hydrothermal synthesis of silver decorated reduced graphene oxide (rGO) nanoflakes with effective photocatalytic activity for wastewater treatment *Nanoscale Res. Lett.* **15** 95
[27] Raza A, Ikram M, Aqeel M, Imran M, Ul-Hamid A, Riaz K N and Ali S 2020 Enhanced industrial dye degradation using Co doped in chemically exfoliated MoS_2 nanosheets *Appl. Nanosci.* **10** 1535–44

[28] Mahler B, Hoepfner V, Liao K and Ozin G A 2014 Colloidal synthesis of 1T-WS_2 and 2H-WS_2 nanosheets: applications for photocatalytic hydrogen evolution *J. Am. Chem. Soc.* **136** 14121–7

[29] Zhang L, Guo Y, Iqbal A, Li B, Gong D, Liu W, Iqbal K, Liu W and Qin W 2018 One-step synthesis of the 3D flower-like heterostructure MoS_2/CuS nanohybrid for electrocatalytic hydrogen evolution *Int. J. Hydrogen Energy* **43** 1251–60

[30] Zhang J, Zhu Z and Feng X 2014 Construction of two-dimensional MoS_2/CdS p–n nanohybrids for highly efficient photocatalytic hydrogen evolution *Chem. Eur. J.* **20** 10632–5

[31] Maitra U, Gupta U, De M, Datta R, Govindaraj A and Rao C N R 2013 Highly effective visible-light-induced H_2 generation by single-layer 1T-MoS_2 and a nanocomposite of few-layer 2H-MoS_2 with heavily nitrogenated graphene *Angew. Chem. Int. Ed.* **52** 13057–61

[32] Gao M-R, Liang J-X, Zheng Y-R, Xu Y-F, Jiang J, Gao Q, Li J and Yu S-H 2015 An efficient molybdenum disulfide/cobalt diselenide hybrid catalyst for electrochemical hydrogen generation *Nat. Commun.* **6** 5982

[33] Zong X, Han J, Ma G, Yan H, Wu G and Li C 2011 Photocatalytic H_2 evolution on CdS loaded with WS_2 as cocatalyst under visible light irradiation *J. Phys. Chem.* C **115** 12202–8

[34] Tian Y, Ge L, Wang K and Chai Y 2014 Synthesis of novel MoS_2/g-C_3N_4 heterojunction photocatalysts with enhanced hydrogen evolution activity *Mater. Charact.* **87** 70–3

[35] Wei L, Chen Y, Lin Y, Wu H, Yuan R and Li Z 2014 MoS_2 as non-noble-metal co-catalyst for photocatalytic hydrogen evolution over hexagonal $ZnIn_2S_4$ under visible light irradiations *Appl. Catalysis* B **144** 521–7

[36] Wang J, Zheng H, Xu G, Sun L, Hu D, Lu Z, Liu L, Zheng J, Tao C and Jiao L 2016 Controlled synthesis of two-dimensional 1T-$TiSe_2$ with charge density wave transition by chemical vapor transport *J. Am. Chem. Soc.* **138** 16216–9

[37] Jiang J *et al* 2017 Signature of type-II Weyl semimetal phase in $MoTe_2$ *Nat. Commun.* **8** 13973

[38] Zeng L, Zhe F, Wang Y, Zhang Q, Zhao X, Hu X, Wu Y and He Y 2019 Preparation of interstitial carbon doped BiOI for enhanced performance in photocatalytic nitrogen fixation and methyl orange degradation *J. Colloid Interface Sci.* **539** 563–74

[39] Zhang Z, Yang P, Hong M, Jiang S, Zhao G, Shi J, Xie Q and Zhang Y 2019 Recent progress in the controlled synthesis of 2D metallic transition metal dichalcogenides *Nanotechnology* **30** 182002

[40] Sugawara K, Nakata Y, Shimizu R, Han P, Hitosugi T, Sato T and Takahashi T 2016 Unconventional charge-density-wave transition in monolayer 1T-$TiSe_2$ *ACS Nano* **10** 1341–5

[41] Yan M *et al* 2017 High quality atomically thin $PtSe_2$ films grown by molecular beam epitaxy *2D Mater.* **4** 045015

[42] Wang X, Vasileff A, Jiao Y, Zheng Y and Qiao S-Z 2019 Electronic and structural engineering of carbon-based metal-free electrocatalysts for water splitting *Adv. Mater.* **31** 1803625

[43] Hammer B and Nørskov J K 2000 Theoretical surface science and catalysis—calculations and concepts *Advances in Catalysis* (New York: Academic) pp 71–129

[44] Nørskov J K, Bligaard T, Logadottir A, Kitchin J R, Chen J G, Pandelov S and Stimming U 2005 Trends in the exchange current for hydrogen evolution *J. Electrochem. Soc.* **152** J23

[45] Greeley J and Mavrikakis M 2004 Alloy catalysts designed from first principles *Nat. Mater.* **3** 810–5
[46] Stamenkovic V R, Mun B S, Arenz M, Mayrhofer K J J, Lucas C A, Wang G, Ross P N and Markovic N M 2007 Trends in electrocatalysis on extended and nanoscale Pt-bimetallic alloy surfaces *Nat. Mater.* **6** 241–7
[47] Su J, Yang Y, Xia G, Chen J, Jiang P and Chen Q 2017 Ruthenium-cobalt nanoalloys encapsulated in nitrogen-doped graphene as active electrocatalysts for producing hydrogen in alkaline media *Nat. Commun.* **8** 14969
[48] Wang P, Zhang X, Zhang J, Wan S, Guo S, Lu G, Yao J and Huang X 2017 Precise tuning in platinum-nickel/nickel sulfide interface nanowires for synergistic hydrogen evolution catalysis *Nat. Commun.* **8** 14580
[49] Cao Z, Chen Q, Zhang J, Li H, Jiang Y, Shen S, Fu G, Lu B-a, Xie Z and Zheng L 2017 Platinum-nickel alloy excavated nano-multipods with hexagonal close-packed structure and superior activity towards hydrogen evolution reaction *Nat. Commun.* **8** 15131
[50] Li Z, Yu C, Wen Y, Gao Y, Xing X, Wei Z, Sun H, Zhang Y-W and Song W 2019 Mesoporous hollow Cu–Ni alloy nanocage from core–shell Cu@Ni nanocube for efficient hydrogen evolution reaction *ACS Catal.* **9** 5084–95
[51] Zhang Q, Li P, Zhou D, Chang Z, Kuang Y and Sun X 2017 Superaerophobic ultrathin Ni–Mo alloy nanosheet array from *in situ* topotactic reduction for hydrogen evolution reaction *Small* **13** 1701648
[52] Lu Q *et al* 2015 Highly porous non-precious bimetallic electrocatalysts for efficient hydrogen evolution *Nat. Commun.* **6** 6567
[53] Zhang J, Wang T, Liu P, Liao Z, Liu S, Zhuang X, Chen M, Zschech E and Feng X 2017 Efficient hydrogen production on $MoNi_4$ electrocatalysts with fast water dissociation kinetics *Nat. Commun.* **8** 15437
[54] Gao Q, Zhang W, Shi Z, Yang L and Tang Y 2019 Structural design and electronic modulation of transition-metal-carbide electrocatalysts toward efficient hydrogen evolution *Adv. Mater.* **31** 1802880
[55] Yuan C, Wu H B, Xie Y and Lou X W 2014 Mixed transition-metal oxides: design, synthesis, and energy-related applications *Angew. Chem. Int. Ed.* **53** 1488–504
[56] Chen W-F, Muckerman J T and Fujita E 2013 Recent developments in transition metal carbides and nitrides as hydrogen evolution electrocatalysts *Chem. Commun.* **49** 8896–909
[57] Yan Y, Xia B, Xu Z and Wang X 2014 Recent development of molybdenum sulfides as advanced electrocatalysts for hydrogen evolution reaction *ACS Catal.* **4** 1693–705
[58] Wiensch J D, John J, Velazquez J M, Torelli D A, Pieterick A P, McDowell M T, Sun K, Zhao X, Brunschwig B S and Lewis N S 2017 Comparative study in acidic and alkaline media of the effects of pH and crystallinity on the hydrogen-evolution reaction on MoS_2 and $MoSe_2$ *ACS Energy Lett.* **2** 2234–8
[59] Li X S, Xin Q, Guo X X, Grange P and Delmon B 1992 Reversible hydrogen adsorption on MoS_2 studied by temperature-programmed desorption and temperature-programmed reduction *J. Catal.* **137** 385–93
[60] Staszak-Jirkovský J *et al* 2016 Design of active and stable Co–Mo–S_x chalcogels as pH-universal catalysts for the hydrogen evolution reaction *Nat. Mater.* **15** 197–203
[61] Ye G, Gong Y, Lin J, Li B, He Y, Pantelides S T, Zhou W, Vajtai R and Ajayan P M 2016 Defects engineered monolayer MoS_2 for improved hydrogen evolution reaction *Nano Lett.* **16** 1097–103

[62] Zhu J *et al* 2019 Boundary activated hydrogen evolution reaction on monolayer MoS_2 *Nat. Commun.* **10** 1348

[63] Deng J, Li H, Xiao J, Tu Y, Deng D, Yang H, Tian H, Li J, Ren P and Bao X 2015 Triggering the electrocatalytic hydrogen evolution activity of the inert two-dimensional MoS_2 surface via single-atom metal doping *Energy Environ. Sci.* **8** 1594–601

[64] Zang Y *et al* 2019 Tuning orbital orientation endows molybdenum disulfide with exceptional alkaline hydrogen evolution capability *Nat. Commun.* **10** 1217

[65] Zhou W, Jia J, Lu J, Yang L, Hou D, Li G and Chen S 2016 Recent developments of carbon-based electrocatalysts for hydrogen evolution reaction *Nano Energy* **28** 29–43

[66] Jiao Y, Zheng Y, Davey K and Qiao S-Z 2016 Activity origin and catalyst design principles for electrocatalytic hydrogen evolution on heteroatom-doped graphene *Nat. Energy* **1** 16130

[67] Mahmood J, Li F, Jung S-M, Okyay M S, Ahmad I, Kim S-J, Park N, Jeong H Y and Baek J-B 2017 An efficient and pH-universal ruthenium-based catalyst for the hydrogen evolution reaction *Nat. Nanotechnol.* **12** 441–6

[68] Zheng Y, Jiao Y, Zhu Y, Li L H, Han Y, Chen Y, Jaroniec M and Qiao S-Z 2016 High electrocatalytic hydrogen evolution activity of an anomalous ruthenium catalyst *J. Am. Chem. Soc.* **138** 16174–81

[69] Liu Y, Yu G, Li G-D, Sun Y, Asefa T, Chen W and Zou X 2015 Coupling Mo_2C with nitrogen-rich nanocarbon leads to efficient hydrogen-evolution electrocatalytic sites *Angew. Chem. Int. Edn* **54** 10752–7

[70] Chen Y-Y, Zhang Y, Jiang W-J, Zhang X, Dai Z, Wan L-J and Hu J-S 2016 Pomegranate-like N,P-doped Mo_2C@C nanospheres as highly active electrocatalysts for alkaline hydrogen evolution *ACS Nano* **10** 8851–60

[71] Li G *et al* 2018 Carbon-tailored semimetal MoP as an efficient hydrogen evolution electrocatalyst in both alkaline and acid media *Adv. Energy Mater.* **8** 1801258

[72] Strmcnik D, Uchimura M, Wang C, Subbaraman R, Danilovic N, van der Vliet D, Paulikas A P, Stamenkovic V R and Markovic N M 2013 Improving the hydrogen oxidation reaction rate by promotion of hydroxyl adsorption *Nat. Chem.* **5** 300–6

[73] Kepp K P 2016 A quantitative scale of oxophilicity and thiophilicity *Inorg. Chem.* **55** 9461–70

[74] Yoon D, Lee J, Seo B, Kim B, Baik H, Joo S H and Lee K 2017 Cactus-like hollow $Cu_{2-x}S$@Ru nanoplates as excellent and robust electrocatalysts for the alkaline hydrogen evolution reaction *Small* **13** 1700052

[75] Qumar U, Hassan J Z, Bhatti R A, Raza A, Nazir G, Nabgan W and Ikram M 2022 Photocatalysis vs adsorption by metal oxide nanoparticles *J. Mater. Sci. Technol.* **131** 122–66

[76] Hu J, Zhang C, Jiang L, Lin H, An Y, Zhou D, Leung M K H and Yang S 2017 Nanohybridization of MoS_2 with layered double hydroxides efficiently synergizes the hydrogen evolution in alkaline media *Joule* **1** 383–93

[77] Jin H, Liu X, Jiao Y, Vasileff A, Zheng Y and Qiao S-Z 2018 Constructing tunable dual active sites on two-dimensional C_3N_4@MoN hybrid for electrocatalytic hydrogen evolution *Nano Energy* **53** 690–7

[78] Liu J, Zheng Y, Zhu D, Vasileff A, Ling T and Qiao S-Z 2017 Identification of pH-dependent synergy on Ru/MoS_2 interface: a comparison of alkaline and acidic hydrogen evolution *Nanoscale* **9** 16616–21

[79] Zhang J, Wang T, Liu P, Liu S, Dong R, Zhuang X, Chen M and Feng X 2016 Engineering water dissociation sites in MoS_2 nanosheets for accelerated electrocatalytic hydrogen production *Energy Environ. Sci.* **9** 2789–93
[80] Kitchin J R, Nørskov J K, Barteau M A and Chen J G 2004 Role of strain and ligand effects in the modification of the electronic and chemical properties of bimetallic surfaces *Phys. Rev. Lett.* **93** 156801
[81] Markovića N M, Sarraf S T, Gasteiger H A and Ross P N 1996 Hydrogen electrochemistry on platinum low-index single-crystal surfaces in alkaline solution *J. Chem. Soc., Faraday Trans.* **92** 3719–25
[82] Schmidt T J, Ross P N and Markovic N M 2002 Temperature dependent surface electrochemistry on Pt single crystals in alkaline electrolytes: part 2. The hydrogen evolution/oxidation reaction *J. Electroanal. Chem.* **524–5** 252–60
[83] Mavrikakis M, Hammer B and Nørskov J K 1998 Effect of strain on the reactivity of metal surfaces *Phys. Rev. Lett.* **81** 2819–22
[84] Hammer B and Nørskov J K 1995 Electronic factors determining the reactivity of metal surfaces *Surf. Sci.* **343** 211–20
[85] Wang H *et al* 2016 Direct and continuous strain control of catalysts with tunable battery electrode materials *Science* **354** 1031–6
[86] Luo M *et al* 2018 Stable high-index faceted Pt skin on zigzag-like PTFE nanowires enhances oxygen reduction catalysis *Adv. Mater.* **30** 1705515
[87] Huang H, Jia H, Liu Z, Gao P, Zhao J, Luo Z, Yang J and Zeng J 2017 Understanding of strain effects in the electrochemical reduction of Co_2: using Pd nanostructures as an ideal platform *Angew. Chem. Int. Ed.* **56** 3594–8
[88] Strasser P *et al* 2010 Lattice-strain control of the activity in dealloyed core–shell fuel cell catalysts *Nat. Chem.* **2** 454–60
[89] You B, Tang M T, Tsai C, Abild-Pedersen F, Zheng X and Li H 2019 Enhancing electrocatalytic water splitting by strain engineering *Adv. Mater.* **31** 1807001
[90] Chen C *et al* 2014 Highly crystalline multimetallic nanoframes with three-dimensional electrocatalytic surfaces *Science* **343** 1339–43
[91] Wang X, Zhu Y, Vasileff A, Jiao Y, Chen S, Song L, Zheng B, Zheng Y and Qiao S-Z 2018 Strain effect in bimetallic electrocatalysts in the hydrogen evolution reaction *ACS Energy Lett.* **3** 1198–204
[92] Ling T *et al* 2017 Activating cobalt(II) oxide nanorods for efficient electrocatalysis by strain engineering *Nat. Commun.* **8** 1509
[93] Tao L, Wang Q, Dou S, Ma Z, Huo J, Wang S and Dai L 2016 Edge-rich and dopant-free graphene as a highly efficient metal-free electrocatalyst for the oxygen reduction reaction *Chem. Commun.* **52** 2764–7
[94] Kibsgaard J, Chen Z, Reinecke B N and Jaramillo T F 2012 Engineering the surface structure of MoS_2 to preferentially expose active edge sites for electrocatalysis *Nat. Mater.* **11** 963–9
[95] Huang X, Tang S, Mu X, Dai Y, Chen G, Zhou Z, Ruan F, Yang Z and Zheng N 2011 Freestanding palladium nanosheets with plasmonic and catalytic properties *Nat. Nanotechnol.* **6** 28–32
[96] Duan H *et al* 2014 Ultrathin rhodium nanosheets *Nat. Commun.* **5** 3093
[97] Saleem F, Xu B, Ni B, Liu H, Nosheen F, Li H and Wang X 2015 Atomically thick Pt–Cu nanosheets: self-assembled sandwich and nanoring-like structures *Adv. Mater.* **27** 2013–8

[98] Jiao Y, Zheng Y, Jaroniec M and Qiao S Z 2014 Origin of the electrocatalytic oxygen reduction activity of graphene-based catalysts: a roadmap to achieve the best performance *J. Am. Chem. Soc.* **136** 4394–403

[99] Sabatier P 1911 Hydrogénations et déshydrogénations par catalyse *Ber. Dtsch. Chem. Ges.* **44** 1984–2001

[100] Wang X, Zheng Y, Sheng W, Xu Z J, Jaroniec M and Qiao S-Z 2020 Strategies for design of electrocatalysts for hydrogen evolution under alkaline conditions *Mater. Today* **36** 125–38

IOP Publishing

2D Materials as Electrocatalysts

Muhammad Ikram, Ali Raza, Jahan Zeb Hassan and Salamat Ali

Chapter 3

Characteristics and performance of 2D electrocatalysts

The distinguishing electronic and anisotropic characteristics of 2D materials have inspired great attention toward a wide range of applications in electrochemistry. Investigations of prototype 2D materials have provided an extensive collection of additional ultrathin layered structures. In this chapter, we examine the similarities of these 2D materials and point out the differences in their electrocatalytic and electrochemical characteristics. A combined investigation will provide surface characteristics and dimensionality; these two important aspects will reflect the design and development of compounds to obtain specific characteristics for different electrocatalytic applications. The mechanisms of reactions and a number of strategies utilized to boost the activity, efficiency, and selectivity of low-dimensional catalysts are also discussed.

3.1 Electrochemical reactions at catalyst surfaces

The transport of electrons, often over an electrode's surface, is the fundamental mechanism behind electrochemical processes. These reactions are connected to the development of energy, denoted by the symbol ΔG°, and the activation energy, denoted by ΔG_{act}. Following the equation $\Delta G^\circ = -nFE^\circ$, the establishment of energy is converted into the standard potential, E°, during the process of electrocatalysis. In the case of transport of a single electron ($n = 1$), the reformation energy of the solvent determines whether or not an electron is transported, and the activation energy, ΔG_{act}, is determined by the reformation energy of the neighboring solvent, λ_{out}, as well as the ligating energy coordinates sphere, λ_{in} [1]. This is per the Marcus–Hush theory:

$$E_a = (\lambda_{out} + \lambda_{in} + \eta)^2/4(\lambda_{out} + \lambda_{in}). \tag{3.1}$$

doi:10.1088/978-0-7503-5291-8ch3

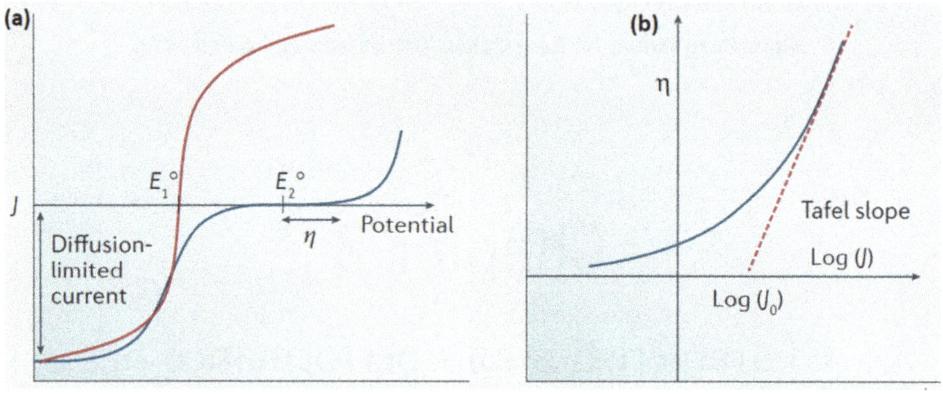

Figure 3.1. Representations of the relationship between (a) current density and overpotential (applied to the electrode) and (b) Tafel slope. (Reproduced with permission from [1]. Copyright 2018 Springer Nature.)

In this expression η denotes the overpotential. The Marcus–Hush hypothesis fails to account for the significantly increased activation energy of electrochemical processes, including forming new chemical bonds or breaking existing ones. In these circumstances, the application of a catalyst and the impact of the electronic structure of a catalytic substance develop a position of critical importance. At a certain potential E, the development energy ΔG may be stated as $\Delta G = -nFE$, and the overpotential, η is taken as $\eta = E - E^{\circ}$. These expressions are based on the supposition of constant potential [1]. In thermodynamics, the overpotential of the reaction can be well-defined as the potential difference between the reaction standard form and the reaction potential that has the fewest favorable conditions; this difference is written as $\eta_{\mathrm{T}} = E_{\mathrm{Limiting}} - E^{\circ}$. The minimum value of η_{T} is considered as 0 V for a reaction when n is taken as two and one intermediate; however, for reactions with higher than one intermediate, the value of η_{T} is greater than 0 V because of the varied binding energies of consecutive reaction intermediates. The overpotential given to the electrode affects the current density measured at the electrode (see figure 3.1(a)). For example, when a reduction occurs at the cathode, the current density is represented as $J_{\mathrm{c}} = -J_{\mathrm{o}}e^{-\eta/b}$. In this expression, J_{o} represents the exchange current density that exists when the potential difference is equal to 0 V, and figure 3.1(b) represents the slope of the Tafel diagram. Another key factor contributing to the electrocatalysis process is Faradaic efficacy (ε), which can be defined as the ratio of charge utilized during the synthesis of a particular material to the entire charge traveling via an electrode. The formula for calculating this ratio is $\varepsilon = n\alpha F/Q$; in this formula, α represents the number of electrons taking part in the reaction, n corresponds to the number of moles, and Q represents the entire amount of charge [1].

3.2 Electrochemical stability of 2D materials as electrodes

Despite significant research progress, our understanding of the electrochemical applications of 2D materials remains limited. This understanding is necessary because the materials may undergo chemical or structural changes when used in applications. Their effectiveness in electrochemical devices may suffer as a result of such events. The stability of electrodes may be understood in terms of their innate electrochemistry and their propensity toward catalytic activity depending on the electrolyte and the applied potential window. When it comes to electrocatalysis, the term 'intrinsic electrochemistry' refers to the innate redox behavior of the electrode material in response to the application of an electrochemical potential [2]. The electrochemical potentials discussed in this section are provided in relation to an Ag/AgCl electrode. Chemically improved graphene with significant carbon–O_2 bonds exhibits a wide variety of electrochemical reductions that originate from their electroactive O_2 functionalities. In contrast, for graphene in its natural state, reduction of O_2 functionalities, including the peroxyl and aldehyde groups, occurred under moderate circumstances, but the reduction of carboxyl groups occurred at low potentials ~2.0 V [3]. The chemical reaction of O_2 reduction functional groups in graphene oxide (GO) cannot be reversed. At −0.7 V, a powerful cathodic wave peaks between −0.9 and −1.5 V during the starting sweep [4]. The disappearance of the cathodic peaks in successive scans indicates that all electro reducible groups have reduced completely by the end of initial scan. The intrinsic electrochemistry of TMDs, which is complex compared to the electrochemistry of graphene products, is governed by the production of inevitable surface oxides and binary elemental elements, which are the TM and the chalcogen. Bonde *et al* were the first to report the intrinsic anodic signals produced by MoS_2 and WS_2 when subjected to acidic conditions [5]. Studies using XPS show that molybdenum disulfide (MoS_2) may be oxidized to produce species such as molybdenum oxide (MoO_3), sulfur disulfide S_2^{2-}, and sulfur dioxide SO_4^{2-}. The bulk and exfoliated TMD materials from group VIB display oxidative wave ranges from 1.0 to 1.2 V under neutral conditions confirmed by a sequence of electrochemical investigations. Metal center oxidation originated these waves in the range of +4 to +6, which was validated by comparisons to the intrinsic electrochemistry of the related TM oxides [6–8]. In contrast to the diselenides and disulfides, the anodic wave having a potential of 0.5 V that is shared by bulk and exfoliated $MoTe_2$ and WTe_2 may be traced back to the tellurium electrochemistry that ultimately results in the formation of TeO_2 [9].

TMDs' tendency toward oxidation contributes to electrochemistry that is already present. On the other hand, the oxidation potential of TMDs from group VIB reveals an upward sequence $WSe_2 > MoSe_2 > WS_2 > MoS_2$ [6, 7]. These results are supported by earlier research, which found that diselenides are more stable than disulfides and the oxidation of tungsten dichalcogenides occurred more quickly than for molybdenum dichalcogenides [10]. Additionally, it is well known that ditellurides are the most susceptible to oxidation [10, 11]. Coleman and colleagues observed that in liquid phase exfoliated TMDs, oxide impurities did not exist except for $MoTe_2$ and WTe_2 [12]. Raman spectra revealed that there was TeO_2 present in the ditellurides, and WTe_2 had a greater quantity of oxides than the other two. Vanadium and Pt-based dichalcogenides also show the chalcogen dependency of the intrinsic electro-oxidative waves. In these two types of dichalcogenides, the basic oxidation potentials drop as one moves farther, slowing the chalcogen group [13, 14].

Because the beginning of the hydrogen evolution process takes place in the negative potential zone, the propensity of TMDs to undergo oxidation is not a primary factor that must be considered when the HER is conducted under acidic circumstances. The inherent electroactivity of g-C_3N_4 is almost nonexistent, which stands in sharp contrast to the intrinsic electroactivity of other 2D materials [15, 16]. The cyclic voltammograms of g-C_3N_4 made using a variety of processes are featureless [16], although the compound contains atoms of N. As a result, there are no electroactive functional groups present on the g-C_3N_4 that have been synthesized. The g-C_3N_4 exhibits a large operating window, similar to that of graphene in its purest form, which reduces the interference caused by intrinsic noise in electrochemical applications. Recent research has looked into the electrochemistry in the prototype MXene $Ti_3C_2T_x$. The $Ti_3C_2T_x$ displays a significant irreversible oxidation signal at 430 mV in an electrolyte with a pH of 7.0; however, this signal is missing in subsequent sweeps [17]. The electrochemical oxidation of nicotinamide adenine dinucleotide is helped along by this signal, which is favorable [17]. When it comes to electrochemical applications, a stable window of operation is presented by electrochemically oxidized $Ti_3C_2T_x$ that cannot be reduced.

3.3 Electron transfer at 2D materials in electrochemical reactions

After elucidating the intrinsic electrochemistry of the 2D materials, in which the electrode products undergo inherent oxidation or reduction, we will turn our focus to the kinetics of the electrodes themselves. In its most fundamental sense, electrode kinetics refers to the transport of a heterogeneous electron that occurs between solid-state electrodes and electroactive molecular probe products. Further, electrical, surface, and anisotropic characteristics are recognized as significant ramifications for the kinetics of electron transport. The heterogeneous electron transfer (HET) rates reported on basal and edge planes are two instances of how anisotropy in graphene materials can be discovered. Fast HET rates are observed for $[Fe(CN)_6]^{3-/4-}$ redox pair related edge planes of a pyrolytic graphite electrode (graphite-based electrode) [18], while the HET rates are noticeably reduced when seen from the basal planes [18]. A similar pattern is observed when examining a single layer of graphene: the HET rates in the presence of molecular probes, including ascorbic acid, $[Fe(CN)_6]^{3-/4-}$ and nicotinamide adenine dinucleotide are faster for edge planes compared to the basal plane [19].

O_2-containing functional groups on graphene considerably affect HET rates in the direction of specific electroactive probes. Moreover, improved HET rates are acquired via flaws in graphene materials that can be ascribed to a superior electronic density of states (DOS). In a disordered pristine sp^2 structure, the energy level of defective states is located in the gap between the conduction band and the valence band. This completes the DOS within a small distance of the Fermi level [20]. Instead, since there is only a slight overlap between the valence band and conduction band of pure graphene, a considerably lower DOS was attained at the Fermi level. When equated to the basal plane HET rates, they are in the order of higher magnitude at defective sites of CVD-deposited graphene, which holds chemical and mechanically induced faults. A study that used GO and reduced GO found that a boosted HET rate was achieved as the carbon to O_2 ratio was enhanced [21]. The results of the investigation supported this finding. Reduced GO, in which the O_2 groups are removed, demonstrates a minor charge transfer resistance and higher HET rate in an $[Fe(CN)_6]^{3-/4-}$ redox probe than GO [22]. Additionally, electrostatic repulsion between O_2-comprising groups in GO and the negatively charged $[Fe(CN)_6]^{3-/4-}$ redox probe impedes electron transport across the electrode–electrolyte edge, slowing down the HET rates of GO. Further, surface-insensitive redox probes, including $[Ru(NH_3)_6]^{3+/2+}$ of the O_2-comprising moieties on HET rates, are unable to affect the surface-insensitive redox probes such as $[Ru(NH_3)_6]^{3+/2+}$. These effects are exclusive to surface-sensitive redox probes such as $[Fe(CN)_6]^{3-/4-}$ [23].

Because of their anisotropic nature, TMDs are analogous to graphene because their edge and basal planes give rise to noticeable electron transport features. By utilizing $[Fe(CN)_6]^{3-/4-}$, $Fe^{3+/2+}$, and $Cu^{2+/+}$ as redox probes, Gerischer *et al* studied the process of electron transport at these twice orthogonal planes of MoS_2 and demonstrated a connection for an electronic arrangement of MoS_2 [24]. Moreover, the HET rate for the MoS_2 basal surface is lower than the edge plane, which can also be observed in the case of graphene. This is something that one could expect. Considerable correspondence occurs between the d_{xy} and $d_{x^2-y^2}$ orbitals of redox probes and the Mo conduction band is thought to be the cause of this occurrence. These results provide the edge surface of macroscopic MoS_2 (shown in figure 3.2) that exhibits a rapid HET rate by computing 1.1×10^{-3} cm s^{-1} for the $[Ru(NH_3)_6]^{3+/2+}$ probe and $k^o_{obs} = 4.96 \times 10^{-5}$ cm s^{-1} for the $[Fe(CN)_6]^{3-/4-}$ redox probe. In contrast, the uncontaminated basal surface exhibited low HET rates, almost zero for redox probes. Lastly, the exfoliation route for bulk TMD might increase or exacerbate the HET rates toward $[Fe(CN)_6]^{3-/4-}$. This is possible depending on the direction of the exfoliation. Exfoliated $MoSe_2$ and WS_2 showed variable responses, being faster in some situations and lower in others, depending on the intercalant [7, 8].

The exfoliated $MoSe_2$ and WS_2 had faster HET rates across standard organolithium reagents and aromatic intercalants. This was in contrast to their bulk counterparts, which showed slower HET rates. Exfoliation can occasionally result in the introduction of oxides, inhibiting the transmission of electrons between an electrode and an electrolyte. The material can undergo structural and electrical changes due to electrochemical treatment, which can also be used to customize the HET rates of TMDs. Zhang and his team produced exfoliated MoS_2 nanosheets from molybdenite crystal using electrochemical Li-intercalation. The increased conductivity of the material brought about by the electroreduction of the exfoliated MoS_2 nanosheets led to quicker HET rates for redox probes compared to prior to treatment [26]. After undergoing a preliminary reductive treatment [27], the results of this research for bulk and exfoliated MoS_2 demonstrated greater HET rates for $[Fe(CN)_6]^{3-/4-}$. Preliminary oxidation, on the other hand, reduced the HET rates of both bulk and exfoliated MoS_2, which was disappointing. Calculations based on the DFT show that the incorporation of electrons throughout the electroreduction is the key contributor to stabilizing the 1T stage. This results in improved electron transfer

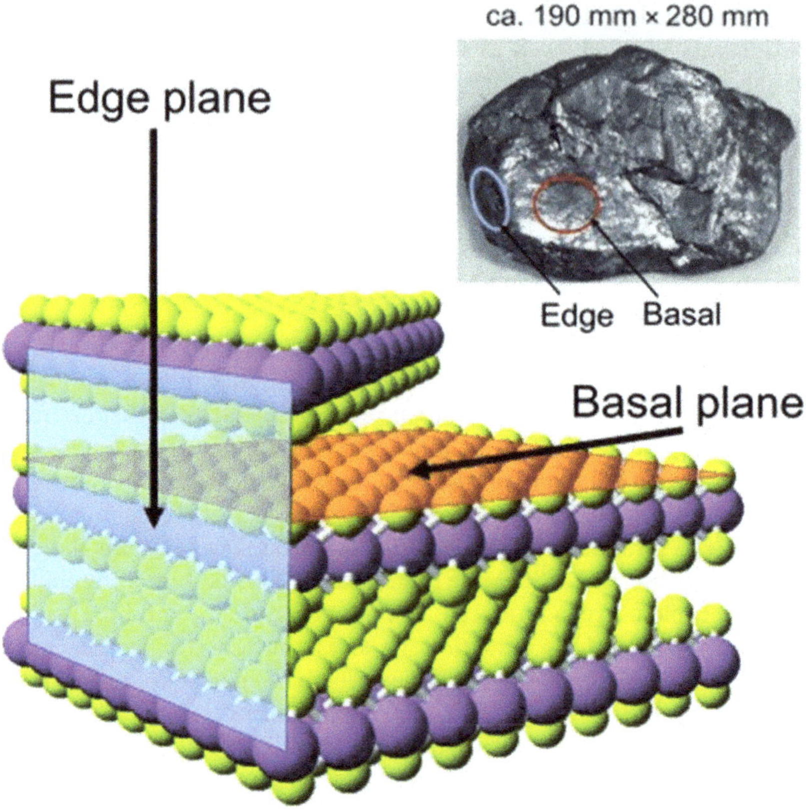

Figure 3.2. MoS_2 molecular model and a representation of the basal and edge planes. Inset: Optical photograph of macroscopic molybdenite crystal, from the Sörumsaasen mine in Norway. Basal planes are indicated by the red circle and the edge planes by the blue circle. (Reproduced with permission from [25]. Copyright 2015 John Wiley and Sons.)

along with catalytic activity capabilities for the HER. This investigation was expanded to study the effect that electro-treatment has on the HET capabilities of additional groups of stacked TMDs, specifically [6, 14, 28].

Electro-oxidation can lower HET rates toward $[Fe(CN)_6]^{3-/4-}$, in contrast to electroreduction, which can speed up those rates, but electro-oxidation can also slow down the electron transfer. This is a characteristic shared by all TMDs (figure 3.3). It is valuable to discuss that the dopants and impurities modify the electron transport characteristics of materials. Electron-donating N dopants in N-doped graphene were implanted into the material through graphite oxide in an ammonia-saturated condition by thermal exfoliation. As a consequence, the N-doped graphene exhibited faster HET rates in the direction of $[Fe(CN)_6]^{3-/4-}$ redox in comparison to pristine graphene [29]. In another scenario, N-doped graphene that was produced by subjecting graphene to an N plasma treatment exhibited more catalytic activity in the reduction of H_2O_2 than graphene that had not been subjected to this treatment [30]. Numerous investigations also suggest that metal-based impurities in graphene

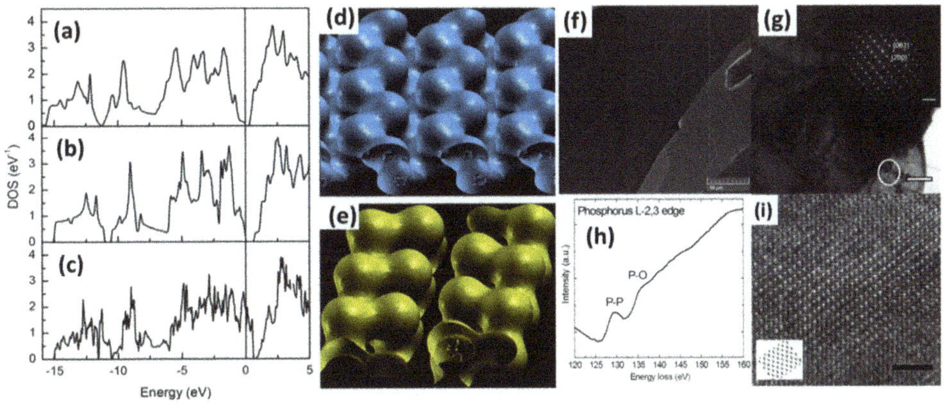

Figure 3.3. Graph of the DOS for BP with a variety of structures: (a) bulk, (b) monolayer, and (c) As slab, which has its surface perpendicular to the layers. (d, e) Valence electron density of a single layer of black phosphorus, shown in two different orientations: (d) perpendicular to the (010) direction and (e) perpendicular to the (100) direction. (f) SEM photograph for surface examination of BP. (g) Parallel TEM photograph (500 nm scale bar) with corresponding SAED configuration (5 nm^{-1} scale bar). (h) EEL spectrum. (i) HR-TEM photograph. (Reproduced with permission from [34]. Copyright 2016 John Wiley and Sons.)

materials, such as Co, Fe, and Ni (parts per billion), are responsible for increased electrocatalytic efficacy for NaHS and hydrazine, known as analytes [31, 32]. Similarly, TM dopants in TMDs affect the HET behavior of the device when exposed to a $[Fe(CN)_6]^{3-/4-}$ redox probe. Deprived of undoped MoS_2, the HET rates for Nb- and Ta-doped MoS_2 were somewhat slower, but when compared to undoped MoS_2, the HET of WS_2 doped with Nb or Ta was much quicker [33]. A further factor contributing to the disparity in electron shifting rates between the basal and edge planes is the anisotropic nature of BP. A faster electron transfer rate was identified for the redox probes $[Ru(NH_3)_6]^{3+/2+}$ and $[Fe(CN)_6]^{3-/4-}$ for the edge plane; on the other hand, slower electron shifting rates on the basal plane manifest as weakly defined redox signals for the basal plane (figures 3.3(a)–(e)) [34]. This was the case for both of these probes. Similarly, BP with respect to the edge plane is extra sensitive to ascorbic acid analyte oxidation. This is seen from the higher current that flows through the edge plane, in contrast to the mild current that flows through the basal surface. High-quality BP crystals were grown specifically to manufacture electrodes using vapor transport growth.

The layered BP structure can be seen in SEM images, as shown in figures 3.3(f)–(i). The use of EDS does not reveal the presence of any other contaminants or elements, such as oxygen. In addition to TEM, SAED and electron energy loss spectroscopy were also carried out [34]. Shear exfoliation appears beneficial for other elements of the pnictogen family in improving their electron shifting rates. Boosted HET rates were obtained following shear exfoliation of bulk pnictogens, while bismuthene demonstrated the most marked increase in contrast to its bulk condition [35]. Every shear-exfoliated pnictogen has improved catalytic activity compared to ascorbic acid oxidation. As an example, antimonene demonstrated a very significant reduction in the oxidation of ascorbic acid by 0.1 V in comparison with bulk antimony [36]. The surface features of *h*-BN influence its ability to transport electrons since distinct current signals are produced when *h*-BN is immobilized on different carbon-based substrates. A $[Ru(NH_3)_2]^{3+/2+}$ redox probe reduces the HET rate with the amount of *h*-BN mass deposited, which indicates that the electrode kinetics on the *h*-BN surfaces are slower than the carbon-based substrates below them. The strength of cathodic current signal effectively changes due to the modification of *h*-BN on smooth substrates (i.e. glassy carbon electrode).

Furthermore, cathodic current was considerably amplified due to the introduction of *h*-BN modified on a screen-printed electrode that has a ridged and rough surface; nevertheless, this increase was not significant. In addition, MXene-based materials have also been investigated for HET rates recently. Ti_3C_2, the prototypical MXene completed with fluorine and O_2, slowed down the electron shifting kinetics of $[Fe(CN)_6]^{3-/4-}$. $[Fe(CN)_6]^{3-/4-}$ is an iron cyanide complex [37]; observation of a high charge transfer resistance value for Ti_3C_2 employing electrochemical impedance spectroscopy lends credence to this assertion. Thus, electron transfer is sluggish on a halogen-terminated diamond electrode because of faint interaction between the electronegative fluorine and hydrophobic surface as well as anionic $[Fe(CN)_6]^{3-/4-}$ [38]. As a result, electron transport is delayed. Thus, electronegative fluorine functional groups were exchanged with hydroxyl groups, which have low electronegativity, during the alkalization of Ti_3C_2. As a result, the alk-Ti_3C_2 demonstrates a quicker electron transfer rate than it did in the past, followed by a decrease in charge transfer resistance [37].

3.4 Parameters for evaluating the performance of catalytic activity

In particular, a few fundamental factors are often utilized to evaluate a particular material's catalytic activity. A comprehensive understanding of these criteria is of the utmost importance to developing an effective catalyst. In the following sections, specific well-established criteria will be discussed.

3.4.1 Gibbs free energy and overpotential

The fact that the value of ΔG_{H*} is either more than or less than zero in practice demonstrates the requirement for a thermodynamic η in order to drive the HER method. There are three probable causes for the rise in η: (i) the activation potential affected by an inherent feature of the catalyst; (ii) the concentration potential affected by the consistency coefficient of the electrolyte; and lastly, (iii) the production of resistance potential via an electrochemical edge. A voltage that must be provided to obtain a particular current density level might be an approximate indicator of how well the catalyst will function for HER. In this particular scenario, the onset potential and the gamma value are crucial factors that must be considered. The former represents a potential value when the linear sweep voltammetry (LSV) curve displays noteworthy twisting [39–41]. The latter is recognized as a standard to evaluate the efficacy of electrochemical water-splitting systems in economically similar systems, as well as a significant criterion for the rating of electrocatalysts aimed at water-splitting reactions (current up to 10 mA cm^{-2}) [42–44]. It should be no surprise that a suitable electrocatalyst can produce H_2 gas at very low η. Because the IR loss can be traced back to its origins η, a few parameters should be standardized to reduce IR loss. These include selecting an effective catalyst, compensating for IR loss by ohmic drop compensation, and stirring the solution while testing it [45].

3.4.2 Tafel slope and exchange current density

The commonly employed parameter to determine whether there exists any association between the current density or not, is called the Tafel slope, and is information that can be acquired using the Butler–Volmer kinetic model [46, 47]:

$$|\eta| = \frac{2.3\,RT}{\alpha nF}\log\frac{j}{j_0}. \tag{3.2}$$

In this equation, F and R correspond to the Faraday constant (96 485 C mol^{-1}) and ideal gas constant (8.314 J mol^{-1} C^{-1}), respectively. Further, n, α, and T signify transferred electrons, the electrochemical transfer coefficient, and Kelvin temperature, respectively. Finally, J_o and j correspond to the exchange current and catalytic densities.

One such expression for the Tafel slope b (mV dec^{-1}) is

$$b = \frac{2.3\,RT}{\alpha nF}. \tag{3.3}$$

The value of b can be calculated via linear fitting of the Tafel plot. This measured value can be additionally employed for proposing the catalytic mechanism that is taking place at an electrode [48]. It was determined that the Tafel slops for the Volmer, Heyrovsky, and Tafel stages were about 120, 40, and 30 mV dec^{-1}, respectively, when measured at room temperature [49]. Above 120 mV dec^{-1} the value of b corresponds to the rate-determining step (RDS), the Volmer reaction. Thus, hydrogen atoms travel at slower rates to be dissolved on the catalytic surface. If the value of b is 40–120 mV dec^{-1}, the Heyrovsky reaction will change into the RDS, and the hydrogen atom is readily adsorbed on the catalytic surface, but it will be desorbed more gradually. On the other hand, if b is determined to be anywhere around 30 mV dec^{-1}, the Tafel process will be considered the RDS. In most cases, one will notice this circumstance when observing a Pt/C catalyst. Therefore, the major HER route for a given catalyst may be approximately deduced by reference to the Tafel slope. J_o is another descriptor that can be used to estimate the capability of charge shifting from the electrode toward the electrolyte. This ability is an essential metric that must be considered when determining the HER activity of electrocatalysts. If the value of J_o is larger, it implies that the electrode material possesses a greater intrinsic activity under equilibrium circumstances. In general, J_o may be determined by projecting the Tafel plots along the plot's linear component until the point where η becomes zero [50].

3.4.3 Turnover frequency

According to equation (3.4), the turnover frequency (TOF) is understood as the number of product molecules that are created at active sites in a given amount of time [51, 52]. To determine the total quantity of H_2 molecules, it is necessary to accumulate some H_2 according to the definition. If we assume that Faraday's laws of electrolysis apply, we can use equation (3.5) to compute the theoretical number of H_2 depending on the charge flowing through the circuit. Here, n is the material concentration (mol), z corresponds to the number of electrons transported per molecule, I is the current (A), and lastly, F is the Faraday constant (96485 C mol^{-1}). Then, a TOF against an overpotential curve (equation (3.6)) may be obtained by using equation (3.4) in conjunction with equation (3.5). As a result, measuring the TOF is dependent primarily on knowing the number of active sites, which may be determined using a variety of methods, such as the copper underpotential deposition approach [53], which computes the number of molecules on the uncovered plane [54–56], or quantification from cyclic voltammetry (CV) experiments [57–59]. Defining and evaluating active sites is essential for determining an accurate TOF or TOF–overpotential curve. When reporting TOF values, the overpotential value should always be provided since the TOF increases as overpotentials increase and vice versa. This is a crucial point that should be taking into account.

$$\mathrm{TOF} = \frac{\text{molecules number of product}}{\text{number of active sites}} \times \frac{1}{\text{unit time}} \tag{3.4}$$

$$n = \frac{It}{zF} \tag{3.5}$$

$$\mathrm{TOF} = \frac{I \times N_A}{zF \times \text{number of active sites}}. \tag{3.6}$$

3.4.4 Stability

In terms of real-world applications, the significant characteristics of a catalyst are its durability and stability. Stability refers to a catalyst's capacity to keep its initial activity for a prolonged period, while durability refers to its ability to withstand wear and tear. Stability can be measured by tracking how the overpotential changes at a particular current density or how the cathodic current density changes when overpotential is applied over time. Because LSV curves are frequently obtained both before and after the stability test, if the catalyst loses its activity in a short time, the overpotential will rise noticeably. Continuous CV cycling is yet another method that might be utilized to assess stability. It is essential to point out that the current methods concentrate on determining stability in work settings, and it is impossible to find useful methods for determining stability under ideal situations.

3.5 Conclusion, challenges and perspectives

1. Despite the progress made in TMD-based catalysts for HER, catalytic pathways for the majority of these catalysts require a thorough understanding. A comprehensive analysis of catalytic pathways offers sensible advice toward developing novel TMD-based HER catalysts. In order to provide appropriate advice for developing future HER catalysts, it is therefore essential to conduct in-depth research into the underlying mechanism. Emerging *in situ* and operando characterization techniques, for instance, operando XRD, operando Raman, and operando XANES, can assist us in gaining a profound comprehension of the actual catalytic procedure and the composition–performance association of TMD-based electrocatalysts.
2. TMD-based materials have shown almost endless potential for use in the HER. This is due to these materials' myriad of intriguing chemical and physical characteristics. However, only a select number of TMD-based multilayer catalysts (i.e. $MoSe_2$, MoS_2 (a significantly investigated TMD-based electrocatalyst), and WS_2) have been examined thoroughly. In point of fact, layer-structured TMDs make up a sizable family, of which many members have not yet been investigated. It would be beneficial to explore not just the 2D TMDs, but also other materials closely linked to them. For example, TaS_2, NbS_2 and VS_2 all indicate their potential for electrochemical hydrogen production.

   For this reason, for future commercial use in water-splitting, it is vital to seek new materials based on TMDs with advantageous qualities including higher proficiency, excellent durability, cost efficiency, and pH independence. In addition to the quest for single-phase TMDs electrocatalysts, the construction of appropriate heterostructures is an additional effective technique for optimizing the electrochemical characteristics of TMD associated systems. Conversely, the integration of two distinct materials often depends on the formation of chemical bonds and close contact. Because of this, a comparable lattice constant is required, restricting the options of raw materials that may be used. In recent times, innovative vdW heterostructures have garnered much interest in various domains. Such heterostructures are physically constructed with various 2D materials via vdW forces that do not rely on chemical bonds to hold the components together. The straightforward combination of many 2D materials will allow for the inheritance of their individual qualities and functionalities, which will significantly broaden the scope of use for layered materials. Some recent research has shown the addition of supplementary layered materials with 2D TMDs to considerably boost the HER activity, particularly in alkaline conditions. However, the present tactics that are utilized to produce vdW heterostructures often entail the use of liquid or CVD procedures. Because of this, the product's physical and chemical characteristics (hierarchical porosity, homogeneity and layer number) cannot be controlled. In order to manufacture vdW heterostructures in a manner that can be controlled, further research has to be done on methods that are both flexible and easy to use.

3. It is quite difficult to compare the HER activities of electrocatalysts produced by different studies directly because of the varying testing settings. Even for the same substance, several actions were observed to be performed. For example, the majority of current densities that are given are normalized by the geometrical area of the electrodes. This ignores the actual operative catalytic area and the loading quantity of electrocatalysts, which clearly affects how the catalytic activity is evaluated. Therefore, the application of a systematic testing procedure is recommended to accurately assess HER activity and profit from comparisons of other TMD-based catalysts. It is necessary to carry out several necessary tests, such as those pertaining to the catalyst loading, the substrate that was applied, and the current densities, which should be standardized relative to the ECSA and the geometrical area. Some aspects remain outside of our control, such as the number of layers, the hierarchical porosity, and the homogeneity. Thus, more research should be done on flexible and easy-to-use methods to build vdW heterostructures that can be controlled.

Bibliography

[1] Voiry D, Shin H S, Loh K P and Chhowalla M 2018 Low-dimensional catalysts for hydrogen evolution and CO_2 reduction *Nat. Rev. Chem.* **2** 0105

[2] Chia X, Eng A Y S, Ambrosi A, Tan S M and Pumera M 2015 Electrochemistry of nanostructured layered transition-metal dichalcogenides *Chem. Rev.* **115** 11941–66

[3] Chua C K, Sofer Z and Pumera M 2012 Graphite oxides: effects of permanganate and chlorate oxidants on the oxygen composition *Chem.—Eur. J.* **18** 13453–9

[4] Zhou M, Wang Y, Zhai Y, Zhai J, Ren W, Wang F and Dong S 2009 Controlled synthesis of large-area and patterned electrochemically reduced graphene oxide films *Chem. Eur. J.* **15** 6116–20

[5] Bonde J, Moses P G, Jaramillo T F, Nørskov J K and Chorkendorff I 2009 Hydrogen evolution on nano-particulate transition metal sulfides *Faraday Discuss.* **140** 219–31

[6] Chia X, Ambrosi A, Sofer Z, Luxa J and Pumera M 2015 Catalytic and charge transfer properties of transition metal dichalcogenides arising from electrochemical pretreatment *ACS Nano* **9** 5164–79

[7] Eng A Y S, Ambrosi A, Sofer Z, Šimek P and Pumera M 2014 Electrochemistry of transition metal dichalcogenides: strong dependence on the metal-to-chalcogen composition and exfoliation method *ACS Nano* **8** 12185–98

[8] Tan S M, Sofer Z, Luxa J and Pumera M 2016 Aromatic-exfoliated transition metal dichalcogenides: implications for inherent electrochemistry and hydrogen evolution *ACS Catal.* **6** 4594–607

[9] Luxa J, Vosecký P, Mazánek V, Sedmidubský D, Pumera M, Lazar P and Sofer Z 2017 Layered transition-metal ditellurides in electrocatalytic applications—contrasting properties *ACS Catal.* **7** 5706–16

[10] Jaegermann W and Schmeisser D 1986 Reactivity of layer type transition metal chalcogenides towards oxidation *Surf. Sci.* **165** 143–60

[11] Kautek W and Gerischer H 1982 Anisotropic photocorrosion of n-type MoS_2 $MoSe_2$, and WSe_2 single crystal surfaces: the role of cleavage steps, line and screw dislocations *Surf. Sci.* **119** 46–60

[12] Gholamvand Z *et al* 2016 Comparison of liquid exfoliated transition metal dichalcogenides reveals $MoSe_2$ to be the most effective hydrogen evolution catalyst *Nanoscale* **8** 5737–49
[13] Chia X, Adriano A, Lazar P, Sofer Z, Luxa J and Pumera M 2016 Layered platinum dichalcogenides (PtS_2, $PtSe_2$, and $PtTe_2$) electrocatalysis: monotonic dependence on the chalcogen size *Adv. Funct. Mater.* **26** 4306–18
[14] Chia X, Ambrosi A, Lazar P, Sofer Z and Pumera M 2016 Electrocatalysis of layered group 5 metallic transition metal dichalcogenides (MX_2, M = V, Nb, and Ta; X = S, Se, and Te) *J. Mater. Chem.* A **4** 14241–53
[15] Fu Y, Zhu J, Hu C, Wu X and Wang X 2014 Covalently coupled hybrid of graphitic carbon nitride with reduced graphene oxide as a superior performance lithium-ion battery anode *Nanoscale* **6** 12555–64
[16] Yew Y T, Lim C S, Eng A Y S, Oh J, Park S and Pumera M 2016 Electrochemistry of layered graphitic carbon nitride synthesised from various precursors: searching for catalytic effects *ChemPhysChem* **17** 481–8
[17] Lorencova L *et al* 2017 Electrochemical performance of $Ti_3C_2T_x$ MXene in aqueous media: towards ultrasensitive H_2O_2 sensing *Electrochim. Acta* **235** 471–9
[18] Davies T J, Hyde M E and Compton R G 2005 Nanotrench arrays reveal insight into graphite electrochemistry *Angew. Chem. Int. Ed.* **44** 5121–6
[19] Yuan W, Zhou Y, Li Y, Li C, Peng H, Zhang J, Liu Z, Dai L and Shi G 2013 The edge- and basal-plane-specific electrochemistry of a single-layer graphene sheet *Sci. Rep.* **3** 2248
[20] McCreery R L 2008 Advanced carbon electrode materials for molecular electrochemistry *Chem. Rev.* **108** 2646–87
[21] Ambrosi A, Chua C K, Bonanni A and Pumera M 2014 Electrochemistry of graphene and related materials *Chem. Rev.* **114** 7150–88
[22] Chua C K, Ambrosi A and Pumera M 2012 Graphene oxide reduction by standard industrial reducing agent: thiourea dioxide *J. Mater. Chem.* **22** 11054–61
[23] Ji X, Banks C E, Crossley A and Compton R G 2006 Oxygenated edge plane sites slow the electron transfer of the ferro-/ferricyanide redox couple at graphite electrodes *ChemPhysChem* **7** 1337–44
[24] Ahmed S M and Gerischer H 1979 Influence of crystal surface orientation on redox reactions at semiconducting MoS_2 *Electrochim. Acta* **24** 705–11
[25] Tan S M, Ambrosi A, Sofer Z, Huber Š, Sedmidubský D and Pumera M 2015 Pristine basal- and edge-plane oriented molybdenite MoS_2 exhibiting highly anisotropic properties *Chem.—Eur. J.* **21** 7170–8
[26] Wu S, Zeng Z, He Q, Wang Z, Wang S J, Du Y, Yin Z, Sun X, Chen W and Zhang H 2012 Electrochemically reduced single-layer MoS_2 nanosheets: characterization, properties, and sensing applications *Small* **8** 2264–70
[27] Chia X, Ambrosi A, Sedmidubský D, Sofer Z and Pumera M 2014 Precise tuning of the charge transfer kinetics and catalytic properties of MoS_2 materials via electrochemical methods *Chem.—Eur. J.* **20** 17426–32
[28] Chia X, Sofer Z, Luxa J and Pumera M 2017 Layered noble metal dichalcogenides: tailoring electrochemical and catalytic properties *ACS Appl. Mater. Interfaces* **9** 25587–99
[29] Poh H L, Šimek P, Sofer Z, Tomandl I and Pumera M 2013 Boron and nitrogen doping of graphene via thermal exfoliation of graphite oxide in a BF_3 or NH_3 atmosphere: contrasting properties *J. Mater. Chem.* A **1** 13146–53

[30] Wang Y, Shao Y, Matson D W, Li J and Lin Y 2010 Nitrogen-doped graphene and its application in electrochemical biosensing *ACS Nano* **4** 1790–8
[31] Wong C H A, Sofer Z, Kubešová M, Kučera J, Matějková S and Pumera M 2014 Synthetic routes contaminate graphene materials with a whole spectrum of unanticipated metallic elements *Proc. Natl Acad. Sci.* **111** 13774–9
[32] Chee S Y and Pumera M 2012 Metal-based impurities in graphenes: application for electroanalysis *Analyst* **137** 2039–41
[33] Chua X J, Luxa J, Eng A Y S, Tan S M, Sofer Z and Pumera M 2016 Negative electrocatalytic effects of p-doping niobium and tantalum on MoS_2 and WS_2 for the hydrogen evolution reaction and oxygen reduction reaction *ACS Catalysis* **6** 5724–34
[34] Sofer Z, Sedmidubský D, Huber Š, Luxa J, Bouša D, Boothroyd C and Pumera M 2016 Layered black phosphorus: strongly anisotropic magnetic, electronic, and electron-transfer properties *Angew. Chem. Int. Ed.* **55** 3382–6
[35] Gusmão R, Sofer Z, Bouša D and Pumera M 2017 Pnictogen (As, Sb, Bi) nanosheets for electrochemical applications are produced by shear exfoliation using kitchen blenders *Angew. Chem. Int. Ed.* **56** 14417–22
[36] Gusmão R, Sofer Z, Bouša D and Pumera M 2017 Pnictogen (As, Sb, Bi) nanosheets for electrochemical applications are produced by shear exfoliation using kitchen blenders *Angew. Chem.* **56** 14417–22
[37] Zhu X, Liu B, Hou H, Huang Z, Zeinu K M, Huang L, Yuan X, Guo D, Hu J and Yang J 2017 Alkaline intercalation of Ti_3C_2 MXene for simultaneous electrochemical detection of Cd(II), Pb(II), Cu(II) and Hg(II) *Electrochim. Acta* **248** 46–57
[38] Kondo T, Ito H, Kusakabe K, Ohkawa K, Einaga Y, Fujishima A and Kawai T 2007 Plasma etching treatment for surface modification of boron-doped diamond electrodes *Electrochim. Acta* **52** 3841–8
[39] Zou X and Zhang Y 2015 Noble metal-free hydrogen evolution catalysts for water splitting *Chem. Soc. Rev.* **44** 5148–80
[40] Wang J, Xu F, Jin H, Chen Y and Wang Y 2017 Non-noble metal-based carbon composites in hydrogen evolution reaction: fundamentals to applications *Adv. Mater.* **29** 1605838
[41] Wei X, Akbar M U, Raza A and Li G 2021 A review on bismuth oxyhalide based materials for photocatalysis *Nanoscale Adv.* **3** 3353–72
[42] Yu P, Wang F, Shifa T A, Zhan X, Lou X, Xia F and He J 2019 Earth abundant materials beyond transition metal dichalcogenides: a focus on electrocatalyzing hydrogen evolution reaction *Nano Energy* **58** 244–76
[43] Zhang H and Lv R 2018 Defect engineering of two-dimensional materials for efficient electrocatalysis *J. Materiomics* **4** 95–107
[44] Chen F, Zhu Q, Wang Y, Cui W, Su X and Li Y 2016 Efficient photoelectrochemical hydrogen evolution on silicon photocathodes interfaced with nanostructured NiP_2 cocatalyst films *ACS Appl. Mater. Interfaces* **8** 31025–31
[45] Anantharaj S, Ede S, Sakthikumar K, Karthick K, Mishra S and Kundu S 2016 Recent trends and perspectives in electrochemical water splitting with an emphasis on sulfide, selenide, and phosphide catalysts of Fe, Co, and Ni: a review *ACS Catal.* **6** 8069
Yan Y, Xia B Y, Zhao B and Wang X 2016 A review on noble-metal-free bifunctional heterogeneous catalysts for overall electrochemical water splitting *J. Mater. Chem.* A **4** 17587

[46] Zhang X, Jiang Z H, Yao Z P, Song Y and Wu Z D 2009 Effects of scan rate on the potentiodynamic polarization curve obtained to determine the Tafel slopes and corrosion current density *Corros. Sci.* **51** 581–7
[47] Mann R F, Amphlett J C, Peppley B A and Thurgood C P 2006 Application of Butler–Volmer equations in the modelling of activation polarization for PEM fuel cells *J. Power Sources* **161** 775–81
[48] Conway B E and Tilak B V 2002 Interfacial processes involving electrocatalytic evolution and oxidation of H_2, and the role of chemisorbed H *Electrochim. Acta* **47** 3571–94
[49] Tian X, Zhao P and Sheng W 2019 Hydrogen evolution and oxidation: mechanistic studies and material advances *Adv. Mater.* **31** 1808066
[50] Zhao G, Rui K, Dou S X and Sun W 2018 Heterostructures for electrochemical hydrogen evolution reaction: a review *Adv. Funct. Mater.* **28** 1803291
[51] Thomas J M and Thomas W J 2014 *Principles and Practice of Heterogeneous Catalysis* (New York: Wiley)
[52] Costentin C, Drouet S, Robert M and Savéant J-M 2012 Turnover numbers, turnover frequencies, and overpotential in molecular catalysis of electrochemical reactions. cyclic voltammetry and preparative-scale electrolysis *J. Am. Chem. Soc.* **134** 11235–42
[53] Voiry D, Salehi M, Silva R, Fujita T, Chen M, Asefa T, Shenoy V B, Eda G and Chhowalla M 2013 Conducting MoS_2 nanosheets as catalysts for hydrogen evolution reaction *Nano Lett.* **13** 6222–7
[54] Li H *et al* 2016 Activating and optimizing MoS_2 basal planes for hydrogen evolution through the formation of strained sulphur vacancies *Nat. Mater.* **15** 48–53
[55] Shin S, Jin Z, Kwon D H, Bose R and Min Y-S 2015 High turnover frequency of hydrogen evolution reaction on amorphous MoS_2 thin film directly grown by atomic layer deposition *Langmuir* **31** 1196–202
[56] Kibsgaard J, Jaramillo T F and Besenbacher F 2014 Building an appropriate active-site motif into a hydrogen-evolution catalyst with thiomolybdate $[Mo_3S_{13}]^{2-}$ clusters *Nat. Chem.* **6** 248–53
[57] Yan Y, Ge X, Liu Z, Wang J-Y, Lee J-M and Wang X 2013 Facile synthesis of low crystalline MoS_2 nanosheet-coated CNTs for enhanced hydrogen evolution reaction *Nanoscale* **5** 7768–71
[58] Mukherjee D, Austeria P M and Sampath S 2016 Two-dimensional, few-layer phosphochalcogenide, $FePS_3$: a new catalyst for electrochemical hydrogen evolution over wide pH range *ACS Energy Lett.* **1** 367–72
[59] Zhang J, Zhao L, Liu A, Li X, Wu H and Lu C 2015 Three-dimensional MoS_2/rGO hydrogel with extremely high double-layer capacitance as active catalyst for hydrogen evolution reaction *Electrochim. Acta* **182** 652–8

IOP Publishing

2D Materials as Electrocatalysts

Muhammad Ikram, Ali Raza, Jahan Zeb Hassan and Salamat Ali

Chapter 4

The electrochemical hydrogen evolution reaction

Parts of this chapter have been reprinted with permission from [16]. Copyright 2023 Elsevier.

The HER has recently gained importance as a critical focus in materials chemistry and surface electrocatalysis. Water electrolysis is considered as the oldest electrochemical approach for developing highly purified hydrogen fuel. Therefore, developing HER electrocatalysts with excellent activity is in high demand to yield energy at significant levels. Thus, 2D materials with attractive electronic and structural characteristics can provide innovative prospects for determining the reaction pathways, particularly for the alkaline HER. Numerous innovative research works have discovered several 2D structured electrocatalysts for the HER in both types of media (acidic and alkaline). This chapter reviews the recent progress in different 2D materials for the HER (i.e. graphene-based electrocatalysts, TMDs, 2D MXenes, etc). Further, changes in enthalpy and band structure are decisive parameters for the HER, which are addressed in various sections throughout this chapter. For reasons of brevity, the main focus of this chapter is the metallic TMDs and characteristic 2D metal-free graphene-based materials that provide a combination of experimental and theoretical measurements toward illuminating the nature of the HER. At the end of the chapter, along with the conclusion, the comprehensive challenges of 2D electrocatalysts for the HER are described.

4.1 Principles and mechanisms of the electrochemical HER

Redox reactions at the electrode/electrolyte edge are at the basis of the electrocatalytic HER, and are fundamental to electrochemical strategies. Hydrogen can be produced using H_2O or proton (H^+) reduction, that comprise a sequence of straightforward phases subject to the pH level of the electrolyte.

doi:10.1088/978-0-7503-5291-8ch4

4.1.1 Acidic media

A widely recognized theory is that the HER at the surface of different catalysts takes place throughout two distinct processes that take place in acidic conditions [1]. The initial step of the HER includes the adsorption of H^+ ions on the catalyst surface to produce adsorbed hydrogen, denoted by the symbol H*, where * denotes an active site on the catalytic material. The first step is then subjected to the discharge phase (equation (4.1)) or Volmer step. The next stage (electrochemical desorption phase or Heyrovsky stage) involves the combination of H* with a H^+ and an electron (e^-) to produce an H_2 molecule. This process is given in equation (4.2). Alternatively, H_2 might be generated by the Tafel step, also called chemical desorption. This step involves an arrangement of two H* on the catalyst, and the equation (4.3) for this process is shown below. Equation (4.3), which represents the entire reaction of the HER, is the standard electrode potential (E°), which serves as a locus for determining the standard electrode potential of electrochemical processes [2].

$$H^+ + e^- \rightarrow H^* \tag{4.1}$$

$$H^* + H^+ + e^- \rightarrow H_2 \tag{4.2}$$

$$2H^* \rightarrow H_2. \tag{4.3}$$

The HER kinetics depends greatly on the interaction between the H* and catalyst, as governed by the Sabatier principle [3]. Therefore, Trasatti developed the first volcano curve for the HER by comparing the HER reaction rate for various catalyst surfaces to the energy required to create hydride [4]. Because of the lack of experimental or hypothetical data for the hydrogen adsorption energy at the time, the hydride formation energy was utilized to characterize the adsorption behavior instead of the hydrogen adsorption energy. The first modern volcano plots were created in 2004 by Nørskov's team, who collected exchange current density data through experiments on the HER with different metals. They used DFT to calculate the proper adsorption energies [5] that reveal the more significant HER progress for Pt; also, they explained the dependence of the HER on the hydrogen binding energy. Furthermore, the edge locations of MoS_2 were projected to have stronger HER progress based on a close-to-neutral ΔG_{H*} of MoS_2 boundaries that were subsequently validated by further studies [6]. This prediction served as a successful paradigm [7]. Ever since, work to uncover more MoS_2 surfaces has dramatically advanced the creation of sophisticated HER catalysts [8], which shows the status of merging theories. It is now possible to obtain ΔG_{H*} in an uncomplicated manner; validation using DFT measurements is considered an essential component of contemporary catalysis science. These calculations have successfully interpreted the electrocatalytic activity inherent to HER catalysts [9].

4.1.2 Alkaline media

Concerning the HER in alkaline environments, the expression ΔG_{H*} may still be used to describe the adsorption performance of hydrogen on the catalytic surface. Interestingly, unlike in an acidic solution, the volcano plot in an alkaline medium only shifts upward and downward rather than to the left and right. This is in contrast to the behavior of the plot in an acidic solution [10]. In alkaline solutions, HER current exchange densities are two to three orders of magnitude lower than in acidic electrolytes [11]. This lower catalytic activity level can be attributed, at least partly, to the fact that the alkaline HER follows a different route to the one followed in acidic solutions. Because of an absence of H^+, alkaline HER commences by breaking H_2O molecules to produce protons, which take part in the Volmer step (equation (4.4)) and the Heyrovsky step (equation (4.5)), while the Tafel step stays the same. The complete course of the reaction is outlined in equation (4.6)), and the E° of this reaction concerning the standard hydrogen electrode (SHE) is determined to be −0.826 V:

$$H_2O + e^- \rightarrow H^* + OH^- \tag{4.4}$$

$$H^* + H_2O + e^- \rightarrow OH^- + H_2 \tag{4.5}$$

$$2H_2O + 2e^- \rightarrow 2OH^- + H_2. \tag{4.6}$$

Since extra energy is needed to produce protons in an alkaline medium, the HER kinetics on most catalysts is slower in alkaline electrolytes because these electrolytes are more basic. It has been observed that the activity of alkaline HER is managed by maintaining a careful equilibrium between ΔG_{H^*} and amount of energy necessary to dissociate H_2O [12]. However, using a number of HER catalysts, such as MoS_2 and CO_2P, is not recommended for the water dissociation process [13]. Therefore, encouraging the process of water-splitting while maintaining a good hydrogen adsorption energy is an operative technique for efficient catalysts for the basic HER.

4.2 Graphene-based electrocatalysts

This section introduces mechanistic studies that employ theoretical and experimental measurements to examine the HER in relation to graphene-containing materials. The following is a summary of recent developments in the development of HER catalysts consisting of graphene, with an emphasis on the management of chemical composition to enhance efficacy.

4.2.1 Single-atom catalyst

Graphene is widely regarded as an appropriate support material for single-atom catalysts because of its huge specific surface area (extraordinary catalyst loading), superior stability (tolerance of harsh working conditions), and outstanding electrical conductivity (simplified electron transport). A method that involves spreading metal catalytic surfaces on graphene atomically, particularly earth-abundant metals, might be a potential strategy for generating alternative HER catalysts instead of using the expensive metal Pt. In this regard, a valuable synthesis approach for producing insulated single clusters and Pt atoms on N-graphene using atomic layer deposition (ALD) was described by Cheng *et al* [14]. N dopants may improve the interaction between Pt and the carbon matrix, resulting in less agglomeration and atomic diffusion. Further, the as-prepared catalyst reveals a dramatically increased mass performance, approximately 37.4 times higher than for most Pt/C catalysts. According to DFT calculations, the outstanding kinetics of the prepared catalyst was attributable to its relatively low activation barriers. A straightforward thermal treatment of GO with trace quantities of cobalt salts in gaseous NH_3 produced precious metals and cobalt on N-incorporated graphene (Co-NG) [15]. The overpotential was found to be ~147 mV to provide a cathodic current density of 10 mA cm^{-2}. This value is significantly lower compared to cobalt-assisted graphene and N-graphene alone. Therefore, resulted active sites of Co-NG are connected to coordinate N and cobalt.

Further, a greater N-doping level enhanced the HER, which lends credence to the theory that N plays an essential role in forming catalytically active sites [16]. In this regard, the development of single-atom Ni catalysts that were attached to nanoporous graphene via the CVD strategy, which was then followed by chemical exfoliation, was described by Qiu *et al* [17]. Figure 4.1(a) shows an SEM image of nanoporous Ni (np-Ni) with a thick graphene layer formed in 2.0 m HCl. The graphene inherited the structure of np-Ni after full np-Ni breakdown (figure 4.1(b)). Ni elements are not apparent in the SEM photograph (figure 4.1(b)), showing a lack of Ni-rich particles. The bright region in the TEM micrograph (figure 4.1(c)) demonstrates the complex 3D structure of np-graphene, which includes convex and concave curves and nanopores. The high-magnification atomic-resolution image (figure 4.1(d), inset) reveals that graphene lattices of Ni type are present in the carbon locations. These solitary-atom Ni catalysts demonstrate an exceptional HER in an acidic solution, with an acceleration voltage of less than 50 mV. Because of the sp–d orbital charge transport between adjacent carbon atoms and Ni dopants, a low ΔG_{H*} (0.1 eV) was acquired by DFT measurements. The contrast between the overpotentials and Tafel slopes of Ni-incorporated graphene results in enormously effective HER catalysts (figure 4.1(e)). Evidently, substitutional Ni addition transforms graphene's chemical action from being inert to one of the greatest HER catalysts. At a 150 mV continuous overpotential, Ni-doped graphene demonstrated exceptional HER activity, with ~90% retention attained after 120 h of operation (figure 4.1(f)). The $|\Delta G_{H*}|$ standards of pristine graphene and Ni_{ab}-incorporated graphene (figures 4.1(g) and (h)) are positive, suggesting that H* cannot be powerfully adsorbed onto the catalysts, limiting their catalytic activity.

4.2.2 Heteroatom-doped graphene

With a reasonably significant positive ΔG_{H*}, the basal phase of bare graphene is regarded as being inert for the HER (1.85 eV) [18]. This indicates the difficulty of hydrogen adsorption that causes the suppression of the HER, in accordance with research. Adding single or compound elements (having variable electronegativity) into the carbon matrix of graphene results in the redistribution of charge and spin in graphene that boosts the graphene's HER activity [18–20]. B-graphene has the most

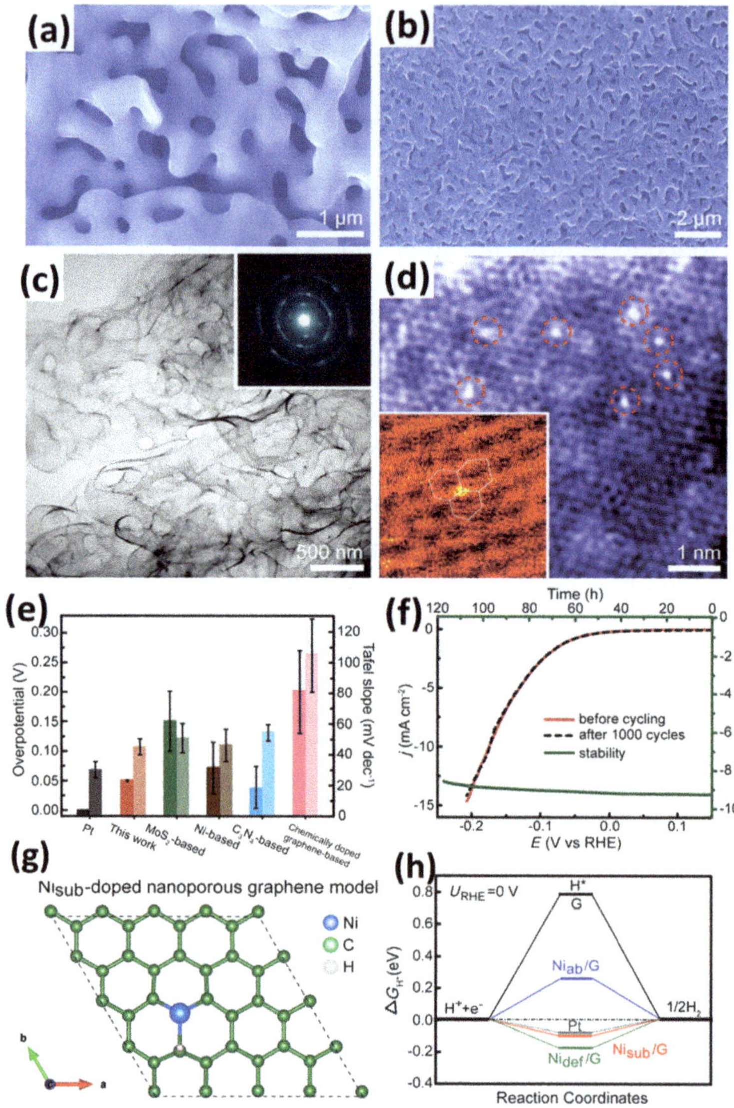

Figure 4.1. SEM micrograph of (a) the synthesized np-Ni/graphene composite and (b) Ni-doped np-G for which the dissolution of Ni was performed for six hours. (c) TEM photograph of Ni-doped np-G and the corresponding SAED pattern (inset). (d) HAADF-STEM photograph of Ni-incorporated graphene. Inset: Enlarged HAADF-STEM micrograph (red circles), that illustrates a substitutional Ni atom (bright orange spot), and these atoms are occupied by carbon sites in the graphene lattice (white lines). (e) Analysis of HER catalysts in an acidic solution using the Tafel slope (right) and overpotential (left). (f) Durability study of Ni-added graphene at a regular overpotential (150 mV) applied for 120 h (green) and reusability assessments (red): polarization curve of Ni-doped graphene for 6 h dissolution; the dashed black line corresponds to after 1000 cycles. (g) Configuration of the Ni_{sub}/G (having $\Delta G_{H*} = -0.10$ eV) and hydrogen adsorption sites. (h) Evaluated G diagram (right) of the HER for Pt catalyst and Ni-doped graphene (Ni_{ab}/G, Ni_{sub}/G, and Ni_{def}/G) samples at equilibrium potential. (Reproduced with permission from [17]. Copyright 2015 John Wiley and Sons.)

notable level of specific HER action compared to other single heteroatom-incorporated graphene materials. However, it is still regarded as low-grade activity compared to Pt-based catalysts and metallic MoS_2 [21]. In contrast, the Tafel slope for every incorporated graphene sample is approximately 120 mV dec^{-1}. This suggests that the Volmer step is the RDS because of their poor hydrogen adsorption capabilities; this conclusion is supported by the findings of DFT calculations (figure 4.2(a)). Although theoretical studies reveal that heteroatom doping might significantly boost graphene's hydrogen adsorption strength, all ΔG_{H*} values are still positive, meaning that adsorption is thermodynamically unfavorable (figure 4.2(b)). As a result, samples are located around the bottom of the branch on the right side of the volcanic graphic. Jiao *et al* employed molecular orbital theory to elucidate an underlying notion, allowing them to climb to the volcano plot's top by further optimizing ΔG_{H*} [21]. They looked at the DOS of the active centers in each model and observed that the maximum DOS peak position for active carbon had the best linear relationship with ΔG_{H*} (figure 4.2(c)). Furthermore, the peak observed closer to the Fermi level of active carbon concerning DOS is vital to accomplish more H* adsorption and, thus, a small ΔG_{H*} value. This is due to the lower hydrogen adsorption strength for the inspected graphene models.

Following this theoretical prediction, a dual doping technique was devised to elevate ΔG_{H*} for graphene materials. By annealing GO with various precursors, one may chemically produce N,P-, N,S-, and N,B-graphene. These graphenes have unique properties [21–24]. Dual incorporated materials were found to have greater

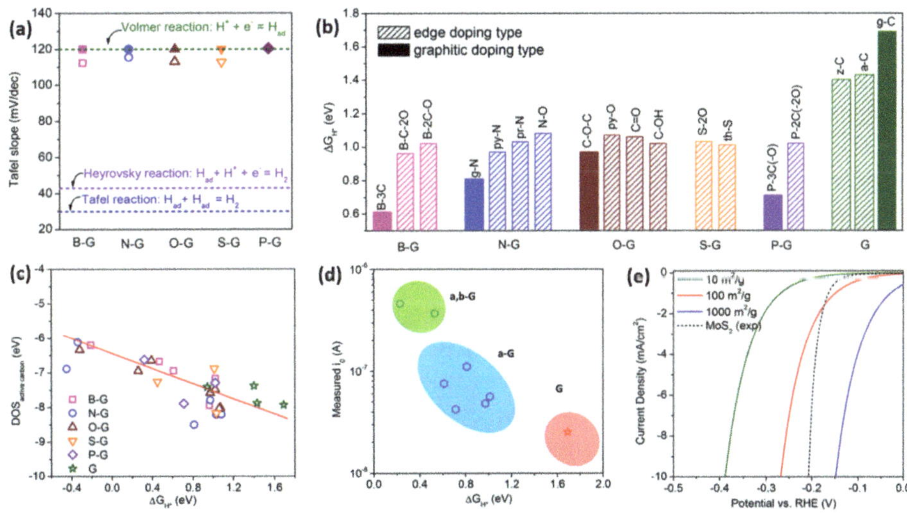

Figure 4.2. (a) Tafel slopes acquired from experimental assessment (solid symbols) as well as theoretical computation (hollow symbols) for several graphene materials. (b) Lowest calculated values for ΔG_{H*} using different samples. (c) Association between ΔG_{H*} and maximum peak position for DOS through active carbon; the straight red line in the graph is a guide. (d) The volcano trend includes doubly incorporated graphene (a,b-G), single-added graphene (a-G), and pristine graphene (G) samples. (e) Evaluated HER polarization curves for G-based materials having changed surface areas ($i_o = 1.10 \times 10^{-21}$ A/site, 40 μg mass loading, and 5% doping level). (Reproduced with permission from [21]. Copyright 2016 Springer Nature.)

HER activity than single-doped samples (except for N,B-graphene) that matched with DFT estimated ΔG_{H^*} values well. This was in line with what was expected (figure 4.2(d)). The perceived activity of a solid catalyst is influenced not only by the inherent activity of each active site but also by the extrinsic physicochemical qualities associated with introducing these active sites. This is because these factors contribute to the catalyst's overall activity. When it comes to doped materials, in particular, the amount of active sites is proportional to doping. Still, the quantity of active sites that are accessible is governed via the specific surface area. In theory, the HER characteristics of graphene-based materials may be enhanced by changing these two factors [21]. Assuming a sample has 5% doping if its exact surface area could be enhanced to 1000 $m^2\ g^{-1}$, the total activity of an optimum graphene material ($\Delta G_{H^*} = 0$ eV) would be superior to MoS_2 from an overpotential perspective (figure 4.2(e)). The validity of this design idea was established by additional research. Ito *et al*, for example, claimed that a N,S co-doped nanoporous graphene with an extraordinarily high specific surface area in the order of 1320 $m^2\ g^{-1}$ achieved a η_{10} of ~280 mV that is analogous to the activity of 2D MoS_2 material [25].

4.3 Transition metal dichalcogenides

Since its initial publication in 1789 [26], electrocatalytic water-splitting to create hydrogen has drawn significant research attention, in particular in recent decades [27]. It is well known that an overpotential, also known as the variance between applied potential and thermodynamic potential, is always essential for an electrochemical process to take place. The overpotential may be significantly reduced by an effective electrocatalyst, which improves the efficiency with which energy is utilized [28]. The Pt group is the greatest effective catalyst for electrocatalytic HER and exhibits minimal overpotential. This has been demonstrated via extensive research [29]. Because of their prohibitively high cost, metals of the Pt group have only a limited number of applications in the field of the electrocatalytic HER. As a result, it is of the utmost importance to create earth-abundant element electrocatalysts for a large efficacy HER. MoS_2 and WS_2 have both been utilized in the hydro-desulfurization (HDS) manufacturing process as catalysts. The HDS process and the HER process are analogous in that the intermediate products are the adsorbed hydrogen atoms; hence, it would be preferable to consider earth-abundant HER electrocatalysts from the pool of HDS candidates [30, 31]. Although MoS_2 was utilized as an electrocatalyst for the HER experiment in the 1970s [32], its bulk counterpart exhibited low catalytic activity because of its restricted active sites and low conducting nature. Recently, 2D TMD nanosheets have received great interest in research due to their remarkable physical and chemical features compared to their bulk counterpart [33]. Since Hinnemann *et al* revealed in 2005 that the MoS_2 under-coordinated edge locations were electrocatalytic HER active locations [6], researchers have looked at MoS_2 and WS_2 as potentially useful catalysts for the electrocatalytic HER [34, 35]. Significant leaps have been made in improving the electrocatalytic performance of TMDs, particularly MoS_2, by phase- and assembly control, surface alterations, and modifications in the arrangement of the material [36, 37]. Hybrid materials can realize exceptional performance in various applications since they combine the benefits of each component [38, 39]. In recent times, much work has been put into hybridizing TMD nanosheets with various nanomaterials to boost the electrocatalytic HER of TMD nanosheets. This section will address the fundamentals of electrocatalytic HER in TMD composites. Finally, a discussion of the strategy for HER activity using various kinds of TMD composites will be presented, along with the highlights of attractive nanosheets and mechanistic research.

4.3.1 TMD–metal electrocatalysts

It is common knowledge that electrocatalysts derived from the Pt group of metals are effective for the HER. Nevertheless, their scope of use is restricted because of their high price and rarity. Even while the electrocatalytic performance of TMDs for the HER has been increasing significantly through structural engineering, it is not yet comparable with the electrocatalytic activity of catalysts made from noble metals, particularly in terms of the onset overpotential and current density. Alternatively, using the abundant TMDs as supports or co-catalysts for Pt group metals would be a workable strategy for lowering the costs while maintaining a high catalytic activity level. Huang's research team has discovered that a variety of nanostructures made of noble metals may be epitaxially formed on the surface of one-layer MoS_2 nanosheets by means of a straightforward chemical reaction [16, 40]. The reduction of K_2PdCl_4 with ascorbic acid results in the epitaxial growth of Pd nanoparticles on one-layer nanosheets; the size of the Pd nanoparticles was around 5 nm.

Moreover, Pt nanoparticles with a size ranging from 1 to 3 nm could grow epitaxially on MoS_2 nanosheets by photochemically reducing K_2PtCl_4 with the help of sodium citrate (figures 4.3(a) and (b)). Additionally, Ag nanoplates in the shape of triangles were produced and epitaxially aligned on the surface of MoS_2. With a Tafel slope of 40 mV dec^{-1}, equivalent to Pt–C catalysts, the as-synthesized Pt–MoS_2 hybrid displayed exceptional electrocatalytic efficacy in the HER direction. This can be seen in figures 4.3(c) and (d). Notably, Pt–MoS_2 displayed a greater current density than Pt–C under an identical potential. This may be due to the epitaxial development's exposure of Pt's highly active {110} and {311} facets and the substantial specific surface area of the MoS_2 support [41]. The Li-intercalation method was recently optimized by Zeng *et al* [42] to develop TiS_2 and TaS_2 nanosheets in a large quality yield. The surfaces of the TiS_2 and TaS_2 nanosheets were then loaded with Pt nanoparticles, which demonstrated good electrocatalytic performance for the HER (figures 4.3(e) and (f)) [43]. Significantly, the deep stimulation of Au plasmon modes was shown to cause single-layer MoS_2 to display a reversible conversion between the 2H and 1T phases. This made it possible for a simple method to be developed for the surface engineering of ultrathin TMD materials [44]. Not only may noble metals be utilized as active components for hybridization with TMD nanosheets, but they can also be utilized as functional supports for hybrid composites [45, 46]. Recent research conducted by Kim and colleagues has shown that the direct reduction of $HAuCl_4$ in water can result in the

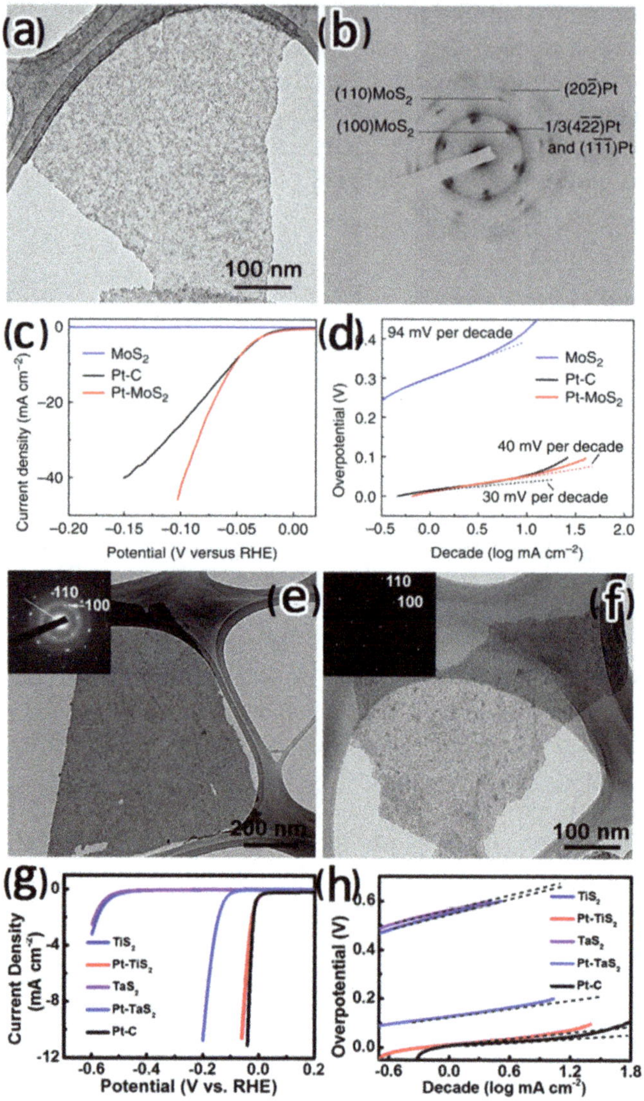

Figure 4.3. (a) TEM micrograph of Pt nanoparticles prepared over a MoS_2 nanosheet. (b) Corresponding SAED pattern (the electron beam is adjusted perpendicular to the MoS_2 nanosheet's basal plane). (c) Polarization curves were acquired from rotating disk glassy carbon electrodes for MoS_2 (0.027 mg cm^{-2}), Pt–C (0.27 mg cm^{-2}), and Pt–MoS_2 (0.075 mg cm^{-2}) loading. (d) Corresponding Tafel plots. (Reproduced with permission from [40]. Copyright 2013 Springer Nature.) (e) TEM photograph of Pt nanoparticles synthesized over the TiS_2 nanosheet; the inset shows a diffraction pattern. (f) TEM photograph of Pt nanoparticles synthesized on a TaS_2 nanosheet; the inset reveals the diffraction pattern. (Reproduced with permission from [43]. Copyright 2014 The Royal Society of Chemistry.)

spontaneous formation of gold nanoparticles on Li-exfoliated MoS_2 and WS_2 nanosheets [47]. It was discovered that a low filling of Au nanoparticles (3.3 mol %) could result in an electron transportation development of approximately 25% by utilizing the acquired WS_2–Au hybrid nanomaterial as an electrocatalyst. With a 230 mV overpotential and a 57 mV dec^{-1} Tafel slope, the improved material containing 21.1 mol% gold demonstrated the highest level of electrocatalytic activity with respect to the HER. The decrease in HER efficiency in WS_2 may be due to the active sites being blocked, which happens when the gold content in the WS_2–Au hybrid exceeds 30 mol%. Therefore, an enhancement in charge transfer between neighboring nanosheets can result in better electrocatalytic efficacy relative to the HER.

Tan *et al* produced single-layer MoS_2 sheets on an absorbent Au substrate using CVD to elucidate the TMD nanosheet catalysis process for the electrocatalytic HER [45]. Figures 4.4(a) and (b), respectively, show schematic depictions of the HER catalyzed using the single-layer-MoS_2–porous-Au hybrid substance and the synthesis technique for this material. During the electrocatalysis process, nanoporous gold with a pore size of around one hundred nanometers functioned as a conductive substance to shift the electrons to the single-layer of molybdenum disulfide (figures 4.4(c) and (d)). The appearance of the single-layer MoS_2 sheets on the Au material is shown clearly in the high-resolution STEM (figure 4.4(e)). The bonding angle (α) between the S and Mo exhibited noticeable fluctuation from one side to the other (figures 4.4(e3) and (e4)), particularly in the bent area, which demonstrated the presence of a lattice alteration in the MoS_2. Thus, it is essential to note that the MoS_2 sheet was entirely monolithic and did not have an adequate 'active edge'.

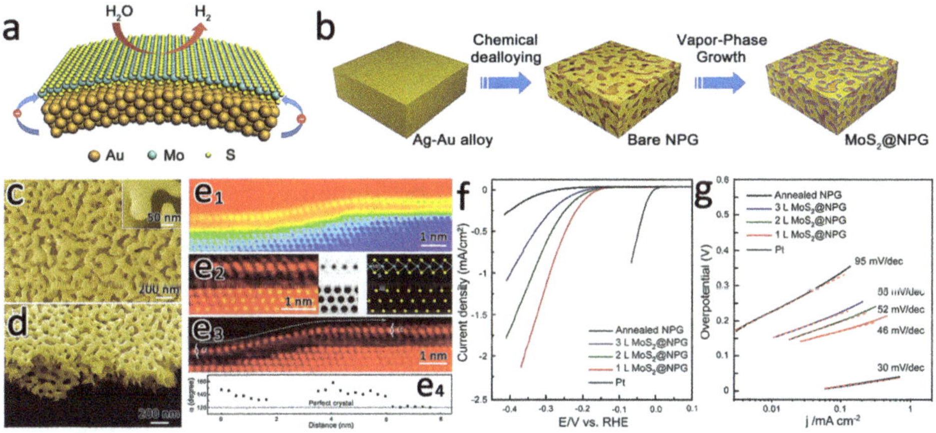

Figure 4.4. (a) Visualization of the HER reaction of single-layer-MoS_2–absorbent-Au hybrid material. (b) The growth procedure for the prepared substance. NPG = nanoporous gold. (c) SEM photograph of the prepared material; a high-magnification photograph is displayed in the inset. (d) SEM photograph (side view) of the prepared material ligament (3D porous structure). (e) Atomic structure of the prepared material: (e1) cross-section HAADF-STEM photograph of the prepared hybrid material; (e2) HR-STEM photograph of the prepared material from a flat section of Au ligament, the standard structure sample for 2H-MoS_2 is shown in the right-hand panel as a simulated STEM photograph; (e3) annular bright-field STEM micrograph of the prepared material taken from a bent area; (e4) distinction of S–Mo–S bonding angles (α) taken from the curved (e3); (f) polarization curves and (g) corresponding Tafel plots. (Reproduced with permission from [45]. Copyright 2014 John Wiley and Sons.)

Moreover, the MoS_2 film only sometimes had less-coordinated stairs, terraces, bends, and/or angle atoms. Exploring the role of lattice strains without disturbing other active sites in the crystal structure was possible. It is important to remember that 118 mV was the preliminary overpotential of the MoS_2 film, which is less than the value reported for numerous MoS_2 nanostructures with significant densities of edge locations and lattice flaws (figure 4.4(f)) [47, 48]. The Tafel slope of the monolayer-layer MoS_2 sheet was similar to that of a good MoS_2-based catalyst, 46 mV dec^{-1} (figure 4.4(g)) [49]. It is possible to conclude that incorporating out-of-surface strains can stimulate the atoms in-phase in MoS_2 nanosheets to function as active catalytic sites. Wang *et al* observed that self-assembled MoS_2 nanoparticles on an Au electrode exhibited a low initial overpotential for hydrogen development (only 90 mV), which is lower than for drop-cast MoS_2 nanoparticle samples (150 mV) [50]. This experimental result suggested that, during the assembly of MoS_2 onto the surface of Au electrodes, the structural rearrangement of the S-rich edge of MoS_2 accounted for an enhancement in catalytic activity. The findings of the experiment supported this hypothesis. Later, Nrskov and colleagues did some theoretical research on the influence of various supports on MoS_2 catalysts for electrocatalytic HERs [46]. Calculations were made utilizing DFT to determine the hydrogen adsorption-free energy at the Mo-edge and S-edge sites of MoS_2 on the various catalytic supports. According to their results, it is feasible to control the catalytic activity of the Mo-edge of MoS_2 by selecting additional support for MoS_2 electrocatalysts. This is something that may be done to obtain the desired results [46].

4.3.2 TMD–carbon electrocatalysts

When TMDs are loaded onto the surface of supports, not only may this greatly enhance the number of edge sites that the TMD possesses, but it can also inhibit the aggregation of TMDs throughout the mechanism of the catalytic process. Conductivity affects HER electrocatalyst performance, as represented by the Tafel slope. Because MoS_2 is a semiconductor with a small conductivity [51], it is essential to couple MoS_2 with a high conductivity support to boost the catalytic activity of MoS_2. Because of their high conductivity and stability, carbon substances, including absorbent carbon [52], carbon nanofibers [53], CNTs [54, 55], and graphene [56], have exhibited excellent potential in applications as electrocatalyst supports [57]. Graphene [56, 58] has also shown great potential. As a result, the hybridization of TMD nanosheets with carbon substances could be a workable and potentially fruitful strategy for enhancing their efficacy. In this section, TMD-nanosheet–carbon hybrid electrocatalysts are described. TMD–carbon hybrid materials may be prepared using a wide variety of different processes. The most straightforward approach is to combine TMD nanosheets with highly conductive carbon materials employing mechanical mixing [51]. When MoS_2 nanosheets and single-walled CNTs are combined, the catalytic activity is enhanced. It is common knowledge that carbon cloth, CNTs, microporous carbon, reduced graphene oxide (rGO), rGO paper, graphene, and 3D graphene networks are all capable of performing the function of highly conductive carbon supports, which are required to attain high dispersal of edge-rich TMD nanosheets. For instance, Dai and colleagues devised a solvothermal approach to manufacture highly disseminated MoS_2 nanoparticles on the surface of rGO nanosheets (figure 4.5(a)) [56]. Because of the interaction between the Mo starting material and the functional groups on GO sheets, MoS_2 nanoparticles with plenty of active edges will coat the rGO nanosheets (figure 4.5(b)). Compared to pure MoS_2 aggregates, the hybrid MoS_2/rGO composite demonstrated much higher electrocatalytic efficacy toward the HER (figure 4.5(c)). The MoS_2/rGO composite showed a slight overpotential of around 100 mV and a Tafel slope of 41 mV dec^{-1} (figure 4.5(d)). This slope was lower than that of the majority of MoS_2 catalysts, which indicates that the standard Volmer–Heyrovsky process was at work. Two main factors improved the activity: first, the abundant edge sites of the highly dispersed MoS_2 nanosheets became exposed and easily accessible in the solution. Second, the enhanced conductivity of the rGO material further boosted the catalytic activity. Zhang *et al* created an electrode that is free-standing, flexible, and long-lasting. This electrode comprises homogeneous MoS_2 nanoflowers grown on rGO paper [59]. Because of the improved conductivity, the electrode overpotential dropped to 190 mV, which is the lowest value achieved. In addition, Cui and colleagues successfully loaded MoS_2 nanoparticles onto 3D carbon fiber paper (CFP) substrates. These nanoparticles have preferentially exposed edge locations [60]. Pyrolysis was the first step in the synthesis process, followed by sulfurization. Notably, it was demonstrated that by utilizing Li electrochemical intercalation, the phase structure of MoS_2 could be changed from 2H to 1T. This demonstrates a simple method for the structure engineering of MoS_2 nanosheets and is a significant discovery. The new 1T-MoS_2–CFP catalyst had high HER activity, as evidenced by the fact that a cathodic current of

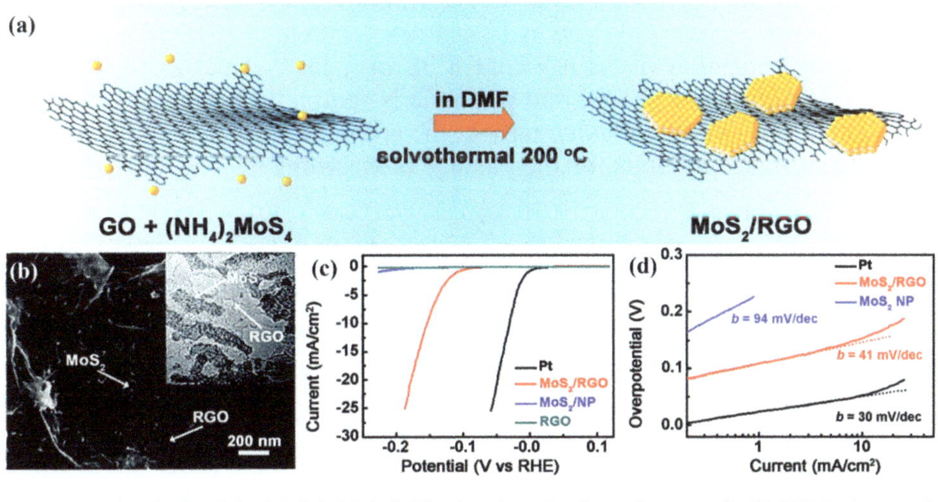

Figure 4.5. (a) Synthesis of the MoS_2/rGO hybrid using the solvothermal approach. (b) SEM micrograph of the prepared hybrid (inset: TEM photograph). (c) Polarization curves and (d) associated Tafel plots (recorded via 0.28 mg cm^{-2} catalyst loading on glassy carbon electrodes). (Reproduced with permission from [61]. Copyright 2011 American Chemical Society.)

200 mA cm^{-2} was achieved with just a 200 mV overpotential and that a Tafel slope of 62 mV dec^{-1} was measured. Moreover, the considerable efficacy of 1T-MoS_2–CFP exhibited a slight degradation after the initial 1000 cycles and continued steadily until 7000 cycles, confirming its good electrochemical constancy. This was seen until the point when 7000 cycles were completed.

The vigorous catalytic activity was caused by many factors, including the numerous edge locations on the MoS_2 nanoparticles, the large conductivity of CFP, and the one-of-a-kind 1T phase structure. The size of the TMD nanostructures formed *in situ* on carbon substrates is often rather large and difficult to control, which is one of the most significant issues arising during this process. To our good fortune, space-confined growth has been shown to be an efficient method for producing TMD-based composites with a morphology that can be controlled and that maintain close interaction with their supports. For example, Zheng *et al* utilized the region between the two graphite oxide sheets next to one another as the reaction zone [56]. The solvent's straightforward evaporation was required to intercalate the molybdenum salt precursor into the interlayer gap. Following the solvothermal treatment, nanoscale MoS_2 nanosheets were produced and, at the same time, GO was decreased and exfoliated. Remarkably, the composite generated using the space-confined technique demonstrated superior electrocatalytic performance (an overpotential of about 140 mV and a Tafel slope of 41 mV dec^{-1}) for the HER in comparison to the typical MoS_2 nanosheets maintained on pre-exfoliated rGO. Moreover, Zhu *et al* made molybdenum salt precursor containing polymer nanofibers by employing the electrospinning technique [53]. The sulfurization and graphitization processes produced one-layer MoS_2 nanosheets encapsulated in N-incorporated carbon fibers. It was also revealed that the one-layer nanosheets had excessive bridging S_2^{2-} and/or apical S^{2-} and terminal S_2^{2-} and/or S^{2-}, which hints at the presence of an S-rich molecular configuration. The resulting hybrid demonstrated exceptional electrocatalytic hydrogen development performance, with a very low initial overpotential of 30 mV and a 38 mV dec^{-1} Tafel slope, due to the synergistic effect of S-rich MoS_2 and pyridinic N-/graphitic N-incorporated carbon fibers. In addition, the high conductivity of the graphitic carbon fibers allowed them to function as quick conduits for the passage of electrons.

4.3.3 TMD–metal-oxide electrocatalysts

In addition to metals and carbon products, metal oxides hybridized with TMD nanosheets have recently been developed for the electrocatalytic HER. Huang *et al* found that MoS_2 nanosheets coated SnO_2 nanotubes in a 3D configuration through electrospinning coupled with a one-step solvothermal method [62]. The homogeneous coating of the internal and external surfaces of the SnO_2 nanotubes with MoS_2 nanosheets enables good electrocatalytic performance for the HER (an overpotential of c. 150 mV and a 59 mV dec^{-1} Tafel slope). Chen *et al* [63] constructed a MoO_3@MoS_2 nanowire array using a two-stage procedure in another fascinating study. Initially, sub-stoichiometric MoO_3 nanorods were deposited on the substrate with a diameter between 20–50 nm utilizing CVD. The obtained MoO_3 nanorods were heated further in a combination of H_2S and H_2 to form $MoO_3@MoS_2$ nanowires (figure 4.6(a)). Throughout the heating procedure, the outer phase of the MoO_3 converted into a 2–5 nm thick MoS_2 shell (figure 4.6(b)). The exceptional catalytic activity of the obtained $MoO_3@MoS_2$ nanowire can be attributed to the conductive center of reduced MoO_3 and the HER catalytic product of the MoS_2 shell with 150–200 mV overpotential and a 50–60 mV dec^{-1} Tafel slope (figures 4.6 (c) and (d)). Moreover, the MoS_2 shell might function as a shielding layer to boost the composite's acid electrolyte stability.

Recently, Hu *et al* carried out the edge modulation of MoS_2/metal oxide heterostructures for proficient hydrogen development electrocatalysis [64]. They reap the benefits of the earth-abundant MoS_2, which is beneficial for the adsorption and recombination of hydrogen intermediates (H_{ad}) and metal oxides, that is, Ni_2O_3H, Co_3O_4, and Fe_2O_3, which are efficacious for cleaving the HO–H bond [65], to create a class of MoS_2/metal oxide heterostructures for the objective of promoting several steps of the entire multistep HER method (the morphology can be seen in figures 4.7(a)–(e)). Because of a careful selection of the transition oxide, it is possible to achieve flexible modulation of HER performance of composite interfaces. This is made possible by the fact that the 3d-band of Ni in MoS_2/Ni_2O_3H has been substantially activated as the 'electron pump', allowing for remarkable HER performance in an alkaline environment. The optimized MoS_2/Ni_2O_3H composite

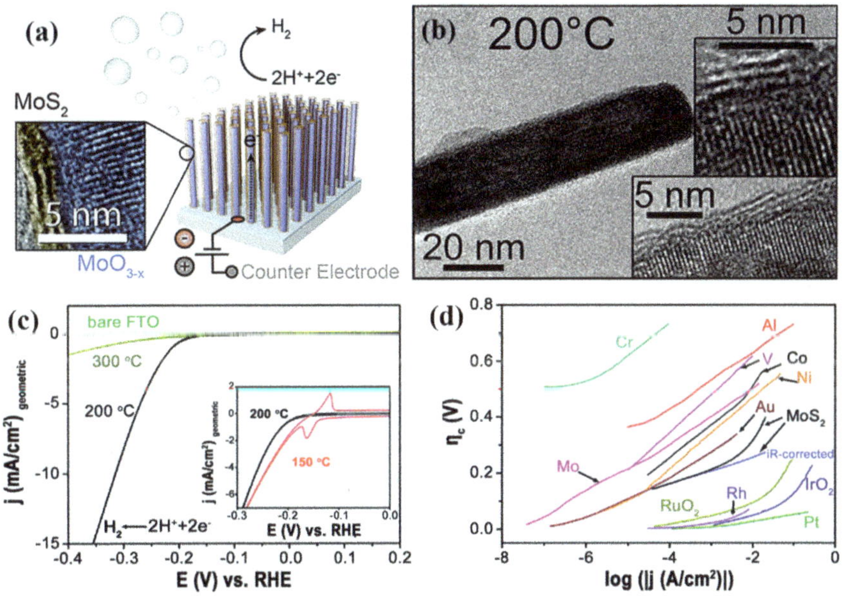

Figure 4.6. Visualization of the HER reaction for a core–shell $MoO_3@MoS$ nanowire array. (b) TEM photograph of the prepared material's sulfurization at 200 °C. (c) CV measurements of nanowire sulfurization at 200 °C and 300 °C; the reference material adopted for assessment was pristine fluorine-doped tin oxide without nanowires after sulfurization at 200 °C (the association between nanowire sulfurization at 200 °C and 150 °C is shown in the inset). (d) Tafel plots of the reference and synthesized materials. (Reproduced with permission from [63]. Copyright 2011 American Chemical Society.)

yields the highest HER efficacy with an overpotential as small as 84 mV at 10 mA cm^{-2} in the basic electrolyte. This overpotential is 120 mV less than the pristine MoS_2 catalyst. The current density at the 200 mV overpotential is 217 mA cm^{-2}, which is twice as large as that of the Pt/C catalyst in alkaline media. This demonstrates that the proposed catalysts have higher performance than most earth-based HER catalysts (figures 4.7(f)–(i)).

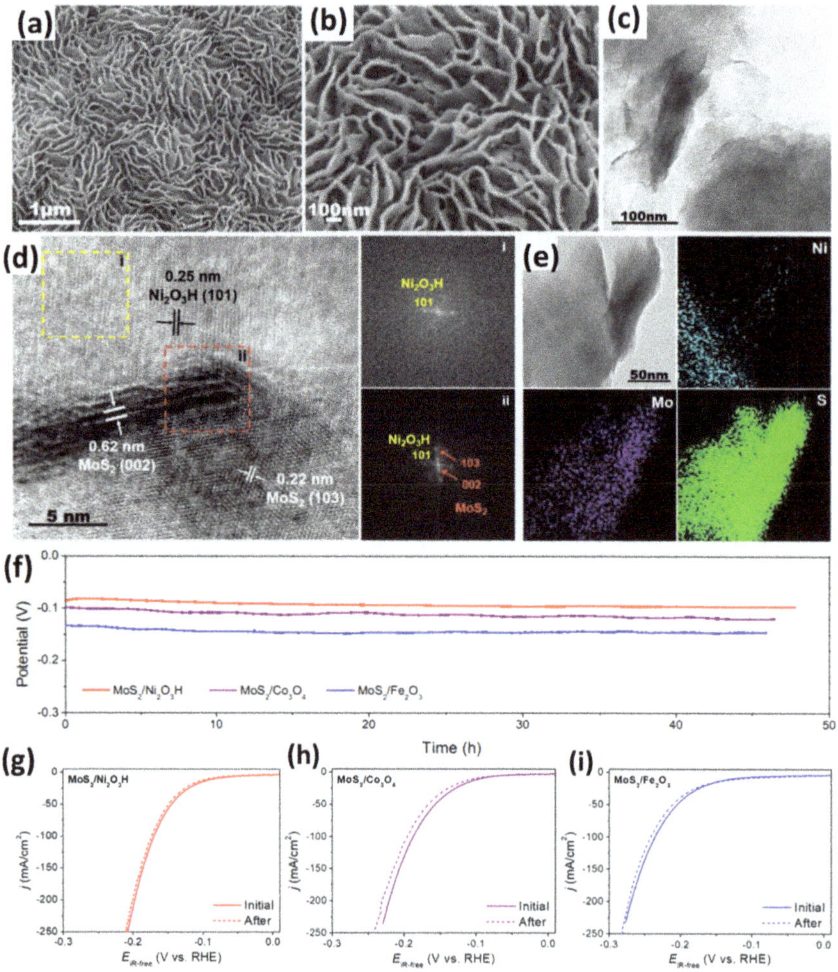

Figure 4.7. Morphological examination of the MoS_2/Ni_2O_3H composite using (a) low- and (b) high-magnification FESEM images. (c) TEM and (d) HR-TEM photographs (FFT patterns are marked by dashed squares indicated by i and ii in the photograph). (e) TEM photograph and corresponding EDS mapping analysis of MoS_2/Fe_2O_3 composite (for S, Mo, and Ni atoms). Electrocatalytic stability of the MoS_2, MoS_2/Ni_2O_3H, MoS_2/Co_3O_4, and MoS_2/Fe_2O_3 composites: (f) chronopotentiometry responses (η–t) noted for the MoS_2/Fe_2O_3 composite (using -10 mA cm^{-2} current densities for more than 46 h). Polarization curves obtained for (g) MoS_2/Ni_2O_3H, (h) MoS_2/Co_3O_4, and (i) MoS_2/Fe_2O_3 composites (5 mV s^{-1} scan rate before and after, represented as solid and dotted curves, respectively, in the chronopotentiometry test). (Reproduced with permission from [64]. Copyright 2020 John Wiley and Sons.)

4.4 2D MXenes for the HER

MXene-based HER catalysts have a promising future, as suggested by current experimental and theoretical measurements. The electrocatalytic performance of MXene is affected by the occurrence of different functional groups in several ways. In their electrochemical HER research, Handoko *et al* [66] employed $Ti_3C_2T_x$ that had been produced using a variety of etchants, including fluorine. The samples with better fluorine coverage on the reference surface had low levels of catalytic activity. This outcome revealed that adding F-terminals over the Ti-based MXene reference plane contributes to HER activity. Jiang *et al* [67] were successful in their efforts to develop and produce a catalyst based on O_2-functionalized Ti_3C_2 MXene ($Ti_3C_2O_x$) for the HER (in acidic media). The as-prepared product demonstrated improved HER progress with a 190 mV overpotential and improved stability through a 10 mA cm^{-2} current density value. The excellent HER activity is due to the highly active O-sites on the basal plane of $Ti_3C_2T_x$ MXenes. Increasing the efficiency of MXene's catalytic activity can be accomplished by the use of an approach that involves modifying the surface functional group. Gao *et al* [68] observed that the ΔG_{H^*} of hydrogen adsorption on the terminal O atom (such as Ti_2CO_2) approximately approached the ideal value (0 eV), indicating that O-terminals played a role in catalytic active sites for an effective HER. Further, the fabrication of dual-TM MXene ($MO_2TiC_2T_x$–V_{Mo}) nanocrystalline sheets was achieved via an electrochemical stripping approach with substantial surface Mo vacancies, as devised by Zhang *et al* (figure 4.8) [69]. The newly created Mo vacancies are used to immobilize single Pt atoms, which increases the catalytic activity of the MXene for the HER. The newly grown catalyst exhibits outstanding catalytic activity, and lower overpotentials (30 and 77 mV) were attained to acquire stability (10 and 100 mA cm^{-2}); further, the mass activity was approximately 40 times greater in comparison to the Pt-on-carbon catalyst. Lastly, the stronger covalent connection present between $MO_2TiC_2T_x$ and Pt atoms restricts coarsening of the catalyst and surface diffusion; that is another factor that contributes to the $MO_2TiC_2T_x$–Pt_{SA} catalyst's good stability for the HER.

The development of a straightforward approach for synthesizing fractional nanorods was revealed by Liu *et al* [70] that includes the fabrication of MoS_2/$Ti_3C_2T_x$ hybrids, that can be achieved by combining the freezing effects of liquid N with an annealing step that comes after it. The quick freezing strategy significantly aided the rolling of $Ti_3C_2T_x$ nanowires and the production of vertically aligned MoS_2 microcrystals. In the electrocatalytic process, MoS_2/$Ti_3C_2T_x$ hybrid fractionation creates supplementary active sites and facilitates charge transfer. Thus it demonstrates an excellent catalytic activity for the HER while having a very low starting overpotential (30 mV). Additionally, the current exchange density is more than 25 times greater compared to MoS_2. The MoS_2/Ti_3C_2 heterostructure was created in a straightforward manner by Huang *et al* [71] using a hydrothermal synthesis technique. It is possible to develop MoS_2 nanosheets in a vertical orientation on top of planar Ti_3C_2 nanosheets. With strong interfacial interaction and an open layer structure, the resultant MoS_2/Ti_3C_2 heterostructures have

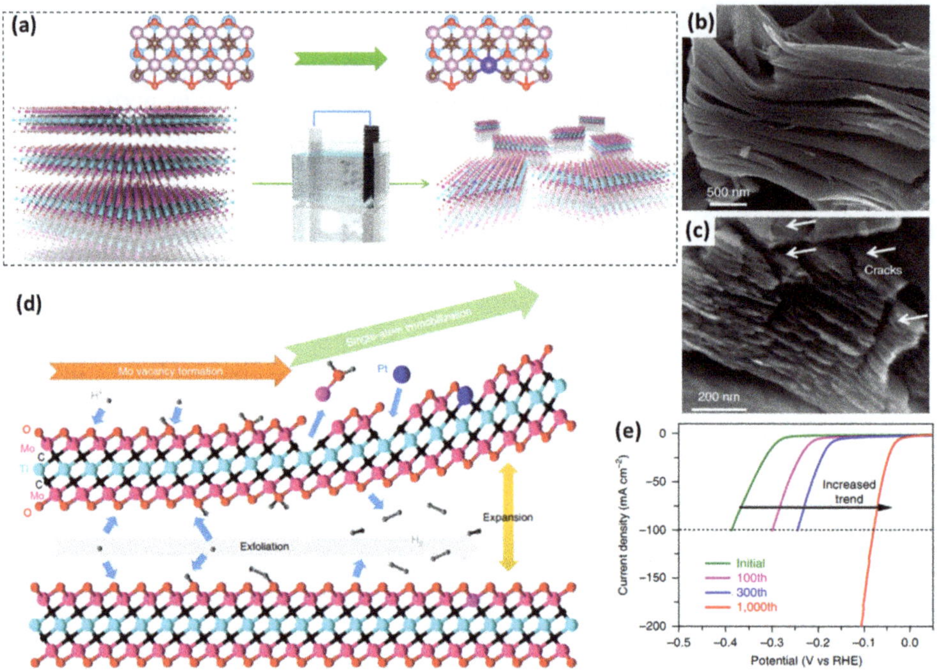

Figure 4.8. (a) Illustration of the electrochemical exfoliation approach for MXene using immobilized single Pt atoms. SEM micrograph for MXene (b) before the exfoliation approach and (c) after. (d) Demonstration of synthesis pathways for $MO_2TiC_2O_2$–Pt_{SA} during the HER procedure. (e) HER polarization curves for $MO_2TiC_2T_x$ with Pt foil (which serves as a counter electrode). (Reproduced with permission from [69]. Copyright 2018 Springer Nature.)

significantly increased HER performance compared to the original MoS_2 nanosheets. Wu *et al* [72] assembled carbon-coated MoS_2 nanoplates on carbon-stable Ti_3C_2 MXene to make nanocrystals with exceptional structural stability, stronger interfacial interaction, and electrical properties. This material is a highly active HER electrocatalyst because it demonstrates outstanding activity and endurance in an acidic medium. The development of N-incorporated carbon encapsulated Mo_2C nanodots on $Ti_3C_2T_x$ MXene was reported by Wang *et al* using Mo precursors on the surface of $Ti_3C_2T_x$ MXene and *in situ* polymerization of dopamine ($Mo_2C/Ti_3C_2T_x$@NC) [73]. During the annealing process, the polydopamine serves various purposes, including the formation of N-doped carbon, the confinement of MoO_4^{2-} ions into extremely tiny Mo_2C nanodots, and the stabilization of MXene flakes against the occurrence of spontaneous oxidation. The hybrid in its as-synthesized state has outstanding HER activity in acidic media, with a 53 mV overpotential at 10 mA cm^{-2} and remarkable stability over a period of thirty hours. Experiments and simulations have both shown that $Ti_3C_2T_x$ MXene aids in charge transport whereas pyridinic N-incorporated carbon-coated Mo_2C nanodots serve as active sites. Both of these factors contribute synergistically to higher HER performance.

An effective electrocatalyst for monolithic water-splitting was proposed by Li *et al* [74] in the form of a 1T/2H-$MoSe_2$ on an MXene nanosheet heterostructure. This structure was designated as 1T/2H-$MoSe_2$/MXene. The 1T/2H-$MoSe_2$/MXene electrocatalyst has significantly enhanced HER activity, with a 95 mV overpotential and 91 mV dec^{-1} Tafel slope, because of the highly conductive support of MXene as well as the outstanding conductivity and plentiful active sites of 1T/2H phase $MoSe_2$. In particular, a voltage of 1.64 V is required for 1T/2H-$MoSe_2$/MXene to achieve a current density of 10 mA cm^{-2} when it is utilized as a bifunctional electrocatalyst for integral water-splitting. The 1T/2H-$MoSe_2$/MXene heterostructure exhibits high endurance in the integrated water-splitting system. One of the MXene-based catalysts for the HER that has received the greatest research attention is molybdenum carbide. The investigation of MXenes as potential HER electrocatalysts was described by Seh *et al* [75]. Because the active phase of the catalytic process is more apparent in Mo_2CT_x than in Ti_2CT_x, the HER activity of Mo_2CT_x is more pronounced.

Because of its significant active sites, the prepared 2D layered Mo_2CT_x (d-Mo_2CT_x) is also preferable to the bulk Mo_2CT_x. Dong *et al* [76] examined a new molybdenum carbide-based electrocatalyst that was enhanced with molybdenum metal. In an acid medium, the electrocatalyst had an initial overpotential of 67 mV and a Tafel slope of $\sim$55 mV dec^{-1} at its lowest point. The electrocatalyst had remarkable cycle stability, which demonstrates that it performs electrocatalysis very well. Intikhab *et al* [77] synthesized two molybdenum-based MXenes, referred to as $Mo_{1.33}CT_z$ and Mo_2CT_z, and evaluated their HER activity and the durability of these compounds. The ordered vacancy on the $Mo_{1.33}CT_z$ substrate caused a considerable drop in HER activity compared to Mo_2CT_z. This is because there is a change in the surface electrical characteristics and a reduction in the fraction of optimum O-terminal HCP sites with the central six-coordinate C atoms. It has been discovered that the stoichiometry of MXene and the atomic surface structure are essential for catalytic activity while having a less significant influence on operational durability. Kuznetsov *et al* [78] successfully synthesized a Mo_2CT_x:Co MXene phase solid solution with a cobalt substituent on the molybdenum site, and the resulting compound displayed high HER activity. Using DFT calculations, researchers could determine that the increase in HER activity following cobalt substitution was due to the thermodynamics of an enhanced hydrogen bonding interaction on the surface of MXene, which promoted the release reaction of H_2. DFT analyses suggest that replacing Mo with Co in the Mo_2CT_x lattice improves hydrogen reaction kinetics. This happens because Co interacts well with the oxygen atoms on the surface of the MXene (figures 4.9(a) and (b)).

An effective method for creating a 2D organ-like Mo_2C MXene matrix produced from MO_2Ga_2C crystals and connected with MoS_2 nanoflowers was proposed by Ren *et al* [79] in order to investigate the HER of molybdenum carbides and disulfides. Compared to pure Mo_2CT_x MXene catalysts, the MoS_2@Mo_2CT_x nanohybrids showed dramatically increased HER activity. These nanohybrids had a low overpotential of 176 mV in alkaline media at a current density of 10 mA cm^{-2} and a very small transfer resistance of 26 Ω. DFT simulations showed that rapid electron transport due to the Mo_2CT_x inherent conductivity and a high number of

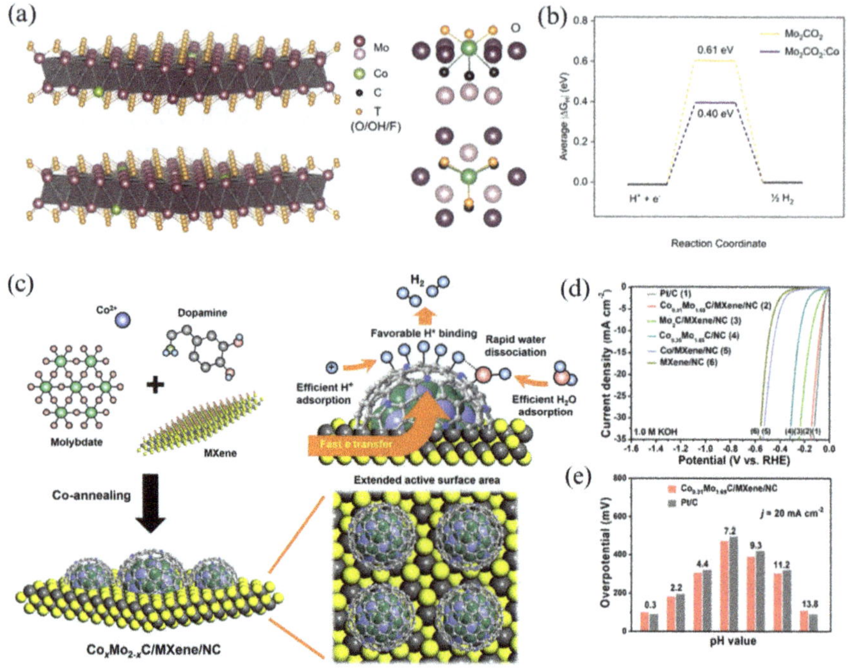

Figure 4.9. (a) Illustration of Mo_2CT_x:Co structure. (b) Reaction coordinate of HER for Mo_2CO_2 and Mo_2CO_2:Co with average ΔG_H. (Reproduced with permission from [78]. Copyright 2019 American Chemical Society.) (c) Demonstration of synthesis approach for the catalyst $Co_xMO_{2-x}C$/MXene/NC. (d) HER polarization curves of the $Co_{0.31}Mo_{1.69}C$/MXene/NC, Mo_2C/MXene/NC, $Co_{0.35}Mo_{1.65}C$/NC, Co/MXene/NC, MXene/NC, and 20% Pt/C at a scan rate of 10 mV s^{-1} in 1.0 m KOH. (e) Assessment between 20% Pt/C and $Co_{0.31}Mo_{1.69}C$/MXene/NC using a pH range of 0.3 to 13.8 at $\eta_j = 20$. (Reproduced with permission from [80]. Copyright 2019 John Wiley and Sons.)

hydrogen adsorption sites from MoS_2 nanoflowers reduced the hydrogen adsorption energy of MoS_2@Mo_2CT_x nanohybrids. In addition, Wu *et al* [80] demonstrated that effective HER could be carried out throughout the whole pH range by meticulously developing a multifunctional collaborative catalytic interface consisting of bimetallic cobalt–molybdenum carbide, hybrid carbon, and MXene. According to the findings, the electrocatalytic performance of the synthesized electrocatalyst ($Co_xMO_{2-x}C$/MXene/NC) competed with the performance of commercial Pt/C catalysts in solutions containing 0.5 m H_2SO_4 or 1.0 m KOH, while outperforming them in solutions with a pH range of 2.2–11.2. When used to carry out a seawater electrolysis hydrogen reaction, this type of catalyst is equal to Pt, with a life expectancy of more than 64 times that of Pt–carbon catalysts and a 98% Faraday efficiency, superior to current catalysts (figures 4.9(c)–(e)). CVD was used to explore the production of high-quality 2D Mo_2C crystals and Mo_2C/graphene heterostructures on liquid gold by Sun *et al* [81]. Atomic force microscopy provided proof of the one-of-a-kind sinking growth phase that Mo_2C exhibits on liquid metal. Simultaneously, hydrogen incorporation resulted in a notable change in the expansion of both Mo_2C and graphene. The heterogeneous structure of graphene/

Mo_2C might be created by modifying the proportion of hydrogen to carbon in the compound. A charge is transferred between the catalyst and electrode more quickly than for pure Mo_2C crystal on the gold foil, which verifies the efficacy and practicability of the produced catalyst for the HER. Based on DFT simulations, Pan *et al* [82] conducted an in-depth investigation of the HER properties of MXene (Mo_2X and W_2X, with X being either carbon or N), and they discovered that the HER performance of these materials is substantially affected by their composition, hydrogen adsorption configuration, and surface functionalization. Compared to other MXenes, the W_2C monolayer has good HER activity and a practically minimal overpotential at high hydrogen density. Furthermore, hydrogenation may improve its catalytic performance across a wide range of hydrogen densities, but oxidation can considerably diminish its activity.

The O_2 must functionalize the HER activity in the Mo_2X monolayer before it may be improved. It is possible to apply MXene to the HER by modifying its composition and functionalization to achieve a level of activity equivalent to that of Pt at a higher hydrogen density. Nguyen *et al* [83] used a hydrothermal technique to make nanoflower W_2C and WS_2. The size of the $W_2C@WS_2$ nanoflowers that were created was in the range of 200–400 nm. The Tafel slope and starting potential of $W_2C@WS_2$ were calculated to be 55.4 mV dec^{-1}. The initial potential was 170 mV. Electrochemical experiments demonstrate that the $W_2C@WS_2$ alloys perform better in the HER than nanosheets and nanoflowers. The incorporation of TM adatoms on the surface of the MXene molecule, in light of the previous research and the subsequent theoretical investigation, is justifiably projected to significantly boost the HER activity of the MXene molecule. High-throughput simulations were utilized by Li *et al* [84] to investigate the HER thermodynamics and kinetics of M_2XO_2 MXene, as well as the mechanism of the boost in HER efficacy brought about by changing various TM atoms. Most initial MXenes have HER activity that is only moderately strong, but incorporating TM atoms into the MXenes will considerably boost their HER performance. According to the study's findings, the introduction of TM atoms to the M_2XO_2 produces a rearrangement of electrons on the outer surface of the MXene. This, in turn, alters the status of charge on O and transfers charge to H, which, in turn, results in a large increase in reactivity. Using first-principles calculations, Ling *et al* [85] chose vanadium carbides with full O_2 termination (V_2CO_2) as the typical material to examine the HER performance. According to the findings, the presence of powerful hydrogen adsorption inhibits the capacity of pure V_2CO_2 to act as an HER catalyst. However, the hydrogen adsorption-free energy may be reduced to zero by inserting an appropriate TM (Fe, Co, and Ni) on the surface. The charge shifting between the O atom on the surface and the promoter atom is the cause of the promotion. In addition, Chen *et al* [86] used DFT simulations to research the HER catalytic characteristics of M_2CO_2 (where $M = Ti$, V, Hf, Zr, and Ta) after TMs were implanted into the compound. Based on the findings, it was determined that the implantation of TM atoms could alter the G and conductivities of MXenes. Through the implantation of TM atoms, it was possible for some MXenes to concurrently attain both high levels of HER catalytic activity and metallic conductivity. There was a correlation between the periodic alteration in the electronegativity of TM elements and the periodic variation in *G*, which occurred at various implantations of different TM elements. After the implantation of TM atoms, it was discovered that a number of M_2CO_2 compounds (M = Ti, Hf, Zr, and Ta) are effective HER catalysts.

A $Ti_3C_2T_x$ MXene catalyst was used by Ramalingam *et al* [87] as a stable support for a ruthenium atom that was N and S coordinated (RuSA). The RuSA-N-S-$Ti_3C_2T_x$ electrocatalyst that was developed is an effective and consistent HER electrocatalyst. It has a small overpotential of 76 mV, making it possible to attain a current density of 10 mA cm^{-2}. The HER efficacy of RuSA-N-S-$Ti_3C_2T_x$ is higher than that of other MXene-based HER catalysts reported in the past and that of the majority of TM-based HER catalytic substances. XAFS and DFT simulations demonstrate that the outstanding HER catalytic activity of RuSA-N-S-$Ti_3C_2T_x$ is primarily due to its active catalytic interface and optimized ΔG_{H*}. The experimental and hypothetical research results have unequivocally demonstrated that the catalytic activity of MXene may be modified by modifying the relation between metals and supports. Zhang *et al* [88] created a type of Pt nanoparticle coated $Ti_3C_2T_x$ MXene catalyst using the ALD method that had a small Pt content (1.70 wt%) and a small size (~2 nm in diameter); this catalyst showed good HER catalytic activity and stability. When ALD cycle approached 40, the electrochemical measurements indicated that the catalysts displayed the ideal HER performance level, with 67.8 mV overpotential approaching the profitable 64.2 mV of the Pt/C catalytic substance. Due to the Pt nanoparticles' uniform dispersion and the 2D $Ti_3C_2T_x$ MXene supports' strong conductivity, the superior behavior may be attributed to both of these factors. The electrocatalytic performance of MXene can be significantly better with the introduction of nitrogen atoms as well as other heteroatoms. Ding *et al* [89] used DFT to conduct a methodical investigation of the HER performance of non-metal heteroatom X (where X = N, B, P, and S) doped M_2C. Compared to the X-incorporated pristine M_2C (M = Mo and Ti), the X-doped M_2CT_2 (M = Mo, Ti; T = O) revealed higher levels of HER catalytic activity. In addition, the ΔG_{H*} value of N–Ti_2CO_2 is relatively low at 0.087 eV (1/9 ML H coverage) and possesses the appropriate electrical conductivity. Moreover, Le *et al* [90] presented a method for improving the HER catalytic activity of $Ti_3C_2T_x$ MXene by modifying it with N heteroatoms via NH_3 heat treatment. This modification would increase the activity of the catalyst during the HER reaction. The DFT calculations demonstrate that a suitable and stable N-doping may be achieved by simply altering the calcination temperature.

Furthermore, the calculations demonstrate that an adequate degree of incorporation of N can be employed to maximize the catalytic activity. The experimental findings show that incorporation of N to $Ti_3C_2T_x$ annealed at 600 °C has better HER electrocatalytic activity, with a 198 mV overpotential and 30 mV starting potential, along with a noticeably reduced 92 mV dec^{-1} Tafel slope. Electrocatalysts with a high level of activity were conceptualized and developed by Zhou *et al* [91] in the form of a single-layer of MXene above a 2D heterostructure of N inclusion in graphene. Calculations based on first principles demonstrated that the graphite sheets on V_2C and Mo_2C MXene have strong HER efficacy, that the ΔG_{H*} of the

HER method is extremely close to zero, and that the slow reaction barrier of Tafel is 1.3 eV. The excellent performance of these heterostructures is due to the significant electron interaction between the MXene and graphite sheet, which modifies graphene's energy band distribution associated with the Fermi stage. Pang *et al* [92] proposed a broad technique to manufacture MXenes based on a route involving electrochemical etching that is thermally assisted (e.g. Ti_2CT_x, Cr_2CT_x, and V_2CT_x). This process effectively overcomes the decades-old difficulty of producing high-concentration hydrofluoric (HF) acid and demonstrates the technology's potential for use as a broad approach to producing MXene.

Moreover, the inclusion of cobalt ions in MXenes shows significantly increased HER efficacy, which is on par with commercial catalysts (figures 4.10(a) and (b)). In their study, Wang *et al* [93] described a straightforward method of $TiOF_2$ intercalation that might be used to make MXene materials more stable. With a starting potential as low as 103 mV, and a 56.2 mV dec^{-1} Tafel slope, the $TiOF_2@Ti_3C_2T_x$ composite produced by this method can successfully stabilize the O_2-consisting terminal on the substrate and achieve remarkable HER catalytic activity and high oxidation resistance. Yu *et al* [94] used a unique strategy to link mesoporous NiFe-LDH nanosheets to a kinetically favorable 3D MXene skeleton using NF. They referred to it as NiFe-LDH/MXene/NF. Consequently, with the incorporation of very hydrophilic MXene, it is now possible to achieve quick, effective activation of water molecules and adsorption at the catalytic edge, which results in a significant increase in the hydrogen development performance of NiFe-LDH. Using the generated NiFe-LDH/MXene/NF as a binder-free electrocatalytic electrode, 500 mA cm^2 at a small HER (205 mV) overpotential in 1.0 m KOH may be attained. The NiFe-LDH/MXene/NF electrode has also achieved 280 h in terms of its endurance, and its performance is superior to that of the expensive metal Pt/C and RuO_2 catalytic substances (figures 4.10(c)–(f)).

Furthermore, Du *et al* [93] successfully obtained a series of novel 0D–2D nanohybrids of $Ni_{1-x}Fe_xPS_3$ nanomosaic decorated on MXene nanosheets (represented as NFPS@MXene) through a simple self-assembly method monitored by a low-temperature solid-state reaction step. When evaluated as electrocatalysts for total water-splitting, the as-synthesized NFPS@MXene nanohybrids display good activity. This is because the ratio of nickel to iron was adjusted. With an initial Ni:Fe ratio of 9:1, $Ni_{0.9}Fe_{0.1}PS_3$@MXene demonstrated a modest HER overpotential (196 mV) in 1 m KOH solution. Catalytic materials for HER that are molybdenum-based MXene show much promise. This is primarily due to the adsorption-free energy being lowest for the H molecule when it is adsorbed on O-terminated HCP sites, as in the middle of the HCP site of Mo_2CT_z is a C atom that has been coordinated six times, improving hydrogen adsorption and increasing the HER rate [77]. Meanwhile, given that the O_2 termination on the base surface functions as an active site of catalysis, certain O_2-terminated MXenes show promise as candidates for HER catalysts. The adsorbing TM on the MXene may reduce the material's adsorption-free energy of hydrogen nearly to zero. This has the potential to increase the catalytic activity of MXene material significantly. Moreover, because $Ti_3C_2T_x$ is an excellent substrate for integrating with other nanomaterials, it gives researchers a

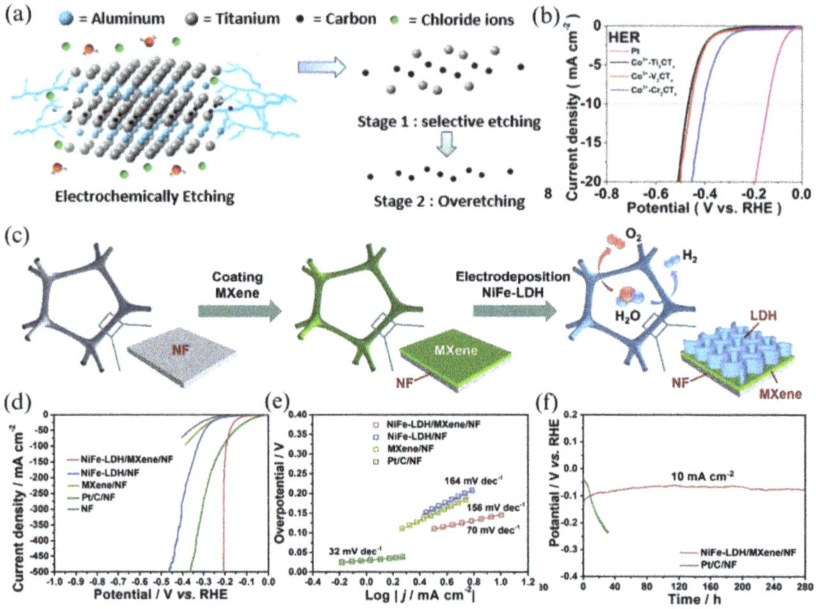

Figure 4.10. (a) E-etching pathways for synthesizing Ti_2CT_x using HCl electrolyte. (b) HER results for Co^{3+}-modified MXenes using 1 M KOH solution (i.e. Co^{3+}–Cr_2CT_x/V_2CT_x/Ti_2CT_x). (Reproduced with permission from [92]. Copyright 2019 American Chemical Society.) (c) Representation of the growth mechanism for a 3D electrocatalytic electrode hierarchically structured on a macroporous MXene/NF frame using an increasing mesoporous system of NiFe-LDH nanosheets. (d) The polarization curves of various catalysts with 90% IR-compensation and a 10 mV s^{-1} scan rate (in 1.0 m KOH solution). (e) Tafel plots for different catalysts. (f) Chronopotentiometric curves for Pt/C/NF and J (current density) = 10 mA cm^{-2} for NiFe-LDH/MXene/NF. (Reproduced with permission from [94]. Copyright 2019 Elsevier.)

stage to develop durable MXene-based hybrid heterostructures for HER. These heterostructures have efficacy in any pH range and adequate operational durability.

4.5 Heterostructure schemes

The theory of heterostructures comes from semiconductor physics. Heterostructures comprise several heterojunctions and interfaces between different constituents and, notably, the heterostructures are semiconductor materials in which the chemical configuration changes with the position. Usually, heterostructures may also be composite structures of interfaces designed differently to solid-state materials, containing semiconductors, insulators, and conductors [16, 95]. Heterostructures are considered a key feature in determining the catalytic activity of 2D materials as they overcome the intrinsic limitations of every material and yield novel characteristics because of interfacial effects. In this regard, the assembly and production of heterostructures, usually based on 0D, 1D, and 2D materials, is an operative strategy to acquire higher electrocatalytic activity. During recent years, advancements have been achieved in synthesizing heterostructures based on 2D materials to yield higher activity electrocatalysts. Thus, in this section we will discuss the production and activity of 2D heterostructures (i.e. 2D/2D, 1D/2D, and 0D/2D heterostructures).

4.5.1 Dimension-based schemes

4.5.1.1 2D/2D heterostructures

Different 2D materials can display valuable catalytic activity because of their exceptional characteristics, but these catalytic yields cannot compete with those catalysts that are based on noble metals due to greater restacking problems. At this time, the appearance of vdW heterostructures could offer innovative approaches to attain the entire response of 2D materials wherein two different 2D materials can be merged to produce 2D/2D heterostructures to compensate for specific weaknesses and reduced interfacial contact resistance, thus producing enhanced catalytic activity [16, 96, 97]. For example, the synthesis of 2D rGO and WS_2 heterostructures was achieved by Yang *et al* [98] using a hydrothermal route that resulted in advanced HER performance caused by enriched charge transfer kinetics owing to intimate contact between two constituents synthesized in solution to slightly regulate structure as well as porosity. Motivated by these outcomes, Tang *et al* [99] constructed a vdW heterostructure composed of nitrogen and graphene-doped MoS_2 using mesoporous magnesia as a template. In this construction process, the porous graphene skeleton was synthesized via CVD, followed by the incorporation of Mo/S/N sources for the growth of nitrogen-doped MoS_2 nanosheets over a graphene skeleton (G@N-MoS_2), as demonstrated in figures 4.11(a) and (b). The plan and synthesis route of the material allowed active control for the hybrid material to have stronger interfacial interactions and the electronic and physical structures of each component. Moreover, the existence of N-MoS_2 was clearly identified using HR-TEM via interlayer spacing (0.62 nm), as shown in figure 4.11 (c). Further, micrographs show the dispersion of N-MoS_2 nanosheets over graphene to produce face-to-face vdW heterostructures, as illustrated in figure 4.11(d). The resulting heterostructure yields efficient multifunctional electrocatalytic activity because of its superior electronic and structural characteristics.

The HER was also investigated in a study that reveals advanced results for N-MoS_2 compared to bare MoS_2 in which the G@N-MoS_2 catalyst offers the best HER activity in an acidic medium with a 10 mA cm^{-2} current density and low overpotential (243 mV), as shown in figure 4.11(e). Further, the onset potential (100 mV) is higher compared to the resultant counterparts in alkaline media, as shown in figure 4.11(f). Additionally, despite limited studies exploring these applications, outstanding ORR and OER performance (figures 4.11(g) and (h), respectively) was shown by the G@N-MoS_2 catalyst using alkaline media. The partial current density was also examined for G@N-MoS_2 catalysts, and indicated that the current density results were quite close to Pt/C for ORR; furthermore, the half-wave potential was relatively reduced compared to other catalysts that have an overpotential (20 mV) of prepared catalyst using current density (10 mA cm^{-2}) that is inferior compared to Ir/C catalysts. In this work, the advanced electrocatalytic activity is due to several aspects. Initially, the electronic structures of MoS_2 can be efficiently controlled using

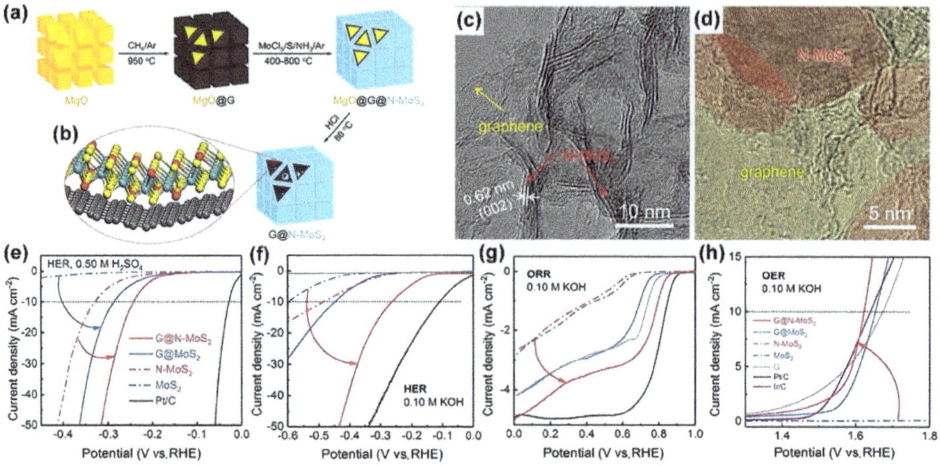

Figure 4.11. 2D/2D heterostructures intended for electrocatalysis. (a) Synthesis of mesoporous 3D G@N-MoS_2 heterostructures. (b) G@N-MoS_2 with vdW heterostructure having N-doping and topological curvature. (c) HR-TEM photograph for G@N-MoS_2. (d) Micrograph representation of the confirmation of the vdW heterostructure of N-MoS_2 and graphene. (e)–(h) Experiments for multifunctional electrocatalysis of mesoporous 3D G@N-MoS_2 heterostructures: HER polarization curves for different contents acquired in (e) H_2SO_4 solution and (f) KOH solutions. Polarization curves acquired for different samples in KOH solution for (g) ORR and (h) OER. (Reproduced with permission from [99]. Copyright 2018 John Wiley and Sons.)

nitrogen doping to offer reduced bandgap energies [100] and superior spin densities [101] that stem from supporting interfacial charge transfer. Second, the interfacial interaction between MoS_2 and graphene can boost adsorption energy and, lastly, the resultant 3D mesoporous structures may enrich proton transport and active site exposure [16]. Despite the valuable tri-functional activity of the G@N-MoS_2 catalyst, the fundamental pathways remain uncertain, and supplementary studies, including experimental and theoretical research work, are required. The 2D/2D heterostructures can also be utilized as self-sustaining electrodes in the direction of direct energy conversion systems rather than 1D/2D heterostructures. For example, the development of a flexible film was done by Duan *et al* [102] using integrating porous C_3N_4 (PCN) nanosheets using nitrogen-doped graphene (PCN@-N-graphene) via adopting a commonly used vacuum filtration route. It includes the generation of prominent active sites via an enormous amount of in- and out-of-plane pores in PCN nanosheets.

Currently, flexible films with a porous structure (hierarchical) are favored for superior mass transport throughout the catalytic experiment; furthermore, the layered structure of graphene and C_3N_4 has more interfaces to enrich charge transfer. Subsequently, remarkable activity was achieved for the self-supporting PCN@-N-graphene electrode with a minor (−0.008 V) onset potential that is close to commercial Pt, an extraordinary (0.43 mA cm^{-2}) exchange current density, as well as exceptional durability (along with slight losses) after 5000 cycles. These fundamental characteristics allow this 2D/2D heterostructure to have greater flexibility, conductivity, and catalytic activity and is an encouraging material for practical electrocatalysis applications. Apart from 2D/2D heterostructures in terms of electronic coupling with substrates, the 2D materials also have these properties directly. For example, electronic coupling between a gold substrate and MoS_2 was reported by Voiry *et al* [103], which causes reduced contact resistance and correspondingly boosts the electron addition to catalyst active sites from a substrate. Herein, for 2H-MoS_2 the basal plane is usually less active than 1T-MoS_2 towards the HER because of its worse charge transfer kinetics and inferior conductivities. Thus, charge transfer facilitation is recognized as an operative strategy to boost the basal planes of 2H-MoS_2 [1]. In addition, Au substrates have an excess of d electrons which can increase the charge transfer in 2H-MoS_2 via coupling between them. Hence, the electrons introduced from Au substrate (towards the 2H-MoS_2 basal plane) result in speeding up of the charge transfer along with the adsorption of hydrogen reactants onto 2H-MoS_2 basal planes, thus improving the electrocatalytic activity of 2H-MoS_2. Generally, this study offers novel understanding of the role of charge transport and contact resistance in the catalytic activity of 2D materials; the outcomes of this section reveal that significantly enhanced catalytic activity of 2D materials can be achieved via constructing 2D/2D heterostructures that provide a base for the growth of 2D materials in electrocatalytic applications.

4.5.1.2 1D/2D heterostructures

The production of superior electrocatalysts includes the development of 1D/2D heterostructures that allow optimal pore size for gas diffusion or mass transfer; further, an extensive variety of 1D/2D heterostructures are also being developed to improve their characteristics as well as microstructures [16, 104, 105]. As an example, the fabrication of CNT/graphene heterostructures was performed by Li *et al* via oxidation of few-walled CNTs [106] that were utilized as a catalyst in the ORR. The exfoliation (under oxidation) strategy was applied to the external walls of CNTs to produce nano-sized graphene. The subsequent graphene, with an enormous number of defects, assisted in the development of ORR catalytic sites after annealing in ammonia. Furthermore, the internal walls of the CNTs remained together, serving as conductors for charge transfer. Consequently, the CNT/graphene catalyst produced an outstanding yield for ORR experiments with a half-wave potential of $\sim$0.76 V and an onset overpotential of $\sim$0.89 V. Likewise, the production of 1D/2D heterostructure was performed by Kim *et al* [107], composed of MoS_2 and CNTs. The direct growth of MoS_2 was achieved over CNTs through the small temperature decomposition of amorphous MoS_x, as shown in figure 4.12(a). HR-TEM photographs in which an aligned structure of CNT forest without any collapse is observed after the deposition of MoS_2 are shown in figure 4.12(b), and an illustration of MoS_2 catalysts grown along CNT strands is shown in figure 4.12(c). This prepared hybrid shows a slight Tafel slope (40 mV dec^{-1}) along with a small overpotential ($\sim$110 mV) at 10 mA cm^{-2}, as demonstrated in figures 4.12(d)–(f). Further investigations regarding the stability of this forest hybrid were also performed using continuous CV, that showed minor falls in cathodic current (after 1000 cycles), which can be recognized as being due to excellent interactions between the support and active materials. Additionally, 1D/2D heterostructures comprised of binary active

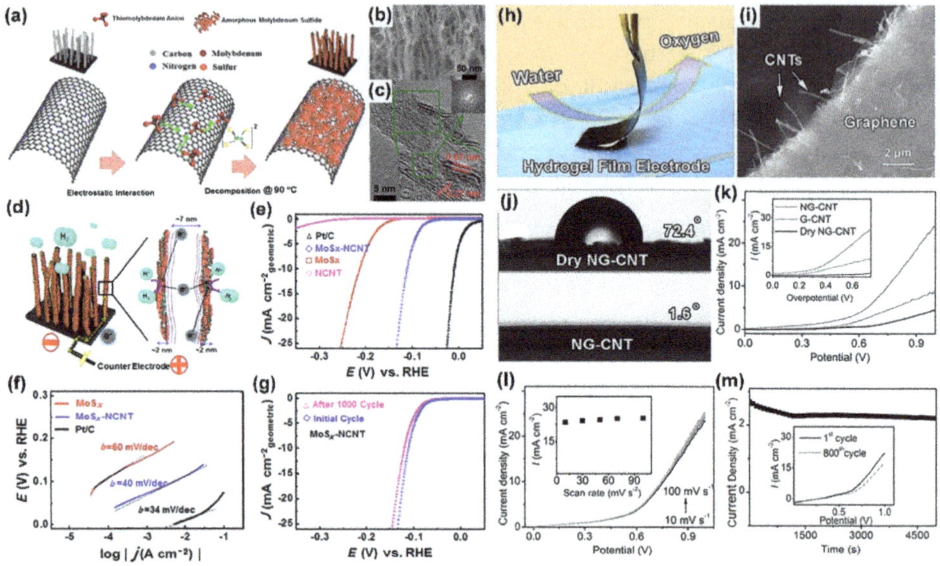

Figure 4.12. Representation of 2D/2D heterostructures for electrocatalysis. (a) The synthesis process for the MoS_x/NCNT (3D) forest-based hybrid catalyst. (b) Corresponding SEM and TEM photographs and respective FFT patterns (inset). (d) The HER of the prepared hybrid catalyst. (e) LSV curves for various catalysts and (f) Tafel plots. (g) Stability analysis of the prepared hybrid catalyst. (Reproduced with permission from [107]. Copyright 2014 American Chemical Society.) (h) Demonstration of a hydrogel film electrode based on NG-CNT. (i) SEM photograph. (j) Contact angles of dry NG-CNT and NG-CNT. (k) LSV curves of dry NG-CNT, and G-CNT. (l) LSV plots. (m) Chronoamperometric response (the inset shows LSV plots for the first and 800th cycles). (Reproduced with permission from [109]. Copyright 2014 John Wiley and Sons.)

materials instead of single ones also significantly impact the results. The synthesis of metal-free NG with N-doped CNT (NG-NCNT) heterostructures, performed by Chen *et al* [108], that consist of four-electron mechanisms for ORR experiments, revealed that entire components of the heterostructured catalyst deliver active sites to intensify electrochemical reactions. In contrast, using CNTs helps to isolate graphene layers that yield pores (gaps) in its structure and boost gas diffusion. Therefore, these features demonstrate that the catalytic progress of 1D/2D heterostructures can increase via synergistic effects. However, these 1D/2D heterostructured catalysts cannot be introduced directly as self-supporting electrodes [16]. Beyond substrate-supported electrodes, the outstanding and more appropriate electrodes for practical applications are self-supporting. To study the optimization of self-supporting electrodes through their electrocatalytic interfacial structure, the growth of N- and O-doped graphene-CNT hydrogel (NG-CNT) was achieved by Chen *et al* [109] via layer-to-layer association of CNTs and graphene, which could be introduced as a self-supporting electrode for electrocatalysis, as shown in figure 4.12(h).

The association of CNTs and graphene follows some kind of order in hydrogel film because of their resilient interactions (figure 4.12(i)), which may assist in charge transport and increase durability throughout catalytic experiments. Additionally, the subsequent NG-CNT was observed to be highly hydrophilic, having a smaller contact angle (~1.6°) than dry NG-CNT (74.2°), as revealed in figure 4.12(j). Hence, NG-CNT film can be utilized as a working electrode directly for OERs that yield higher catalytic activity and small onset potential (315 mV), as shown in figure 4.12 (k). This potential onset value is relatively smaller compared to other samples, including G-CNT (322 mV) and dry NG-CNT (410 mV). Also, surprising results were acquired for this prepared hydrogel compared to some transition metal complex and noble metal oxide (i.e. IrO_2) catalysts [110]. Similar outcomes were gained for catalytic activity with an enriched scan rate (10–100 $mV\,s^{-1}$) for NG-CNT film, suggesting promising catalytic kinetics and extremely effective transport within the electrodes, as shown in figure 4.12(l). Likewise, NG-CNT also showed outstanding stability and durability in performance with little loss (figure 4.12(m)). In conclusion, 1D/2D heterostructures that serve as self-supporting electrodes show outstanding features and excellent activity, so they can be utilized in energy conversion systems directly for enhanced activity between 2D and 1D materials by means of nanostructures, chemical components, and interfaces.

4.5.1.3 0D/2D heterostructures

Before describing 0D/2D heterostructures, let us define the 0D materials. These materials have all their dimensions in the nanometer scale (i.e. ≪100 nm), for example, mono atoms, nanoparticles, clusters, and in particular quantum dots. This class of nanomaterials has been adopted extensively as nanocatalysts because of the higher number of active sites that result in higher catalytic activity. On the other hand, 0D nanomaterials have little electrical conductivity which makes them inappropriate for electrocatalysis. An operational route is to disperse 0D materials on another support with good conductivity and enormous surface area. Thus, for electrocatalysis, the dispersion of single-metal atoms can be done over 2D materials to produce 0D/2D heterostructures. Based on this, the fabrication of a small amount of discrete cobalt atoms dispersed on NG (Co-NG) was performed done by Fei *et al* [111] for an HER electrocatalyst that exhibits decent electrocatalytic progress for the HER along with a small overpotential (~170 mV) at 10 $mA\,cm^{-2}$. Likewise, the production of 0D/2D heterostructures was also reported by Cheng *et al* [14], comprised of distinct Pt atom clusters on N-doped graphene nanosheets (Pt/NGNs). The dispersion and size of the Pt atom clusters were measured precisely using ALD, as shown in figure 4.13(a). To study the adsorption abilities of Pt atoms, both theoretical and experimental approaches were adopted that showed Pt atoms preferably adsorbed over nitrogen sites in NG; also, they have strong interactions with NG that causes promotes electron transfer between them. The results of this study show that a remarkably higher catalytic yield along with decent HER stability were acquired for Pt/NGNs catalysts compared to Co-NG electrocatalysts and, significantly, commercial Pt/C [16].

Other 2D materials, for example TMDs, can also be introduced to advance the catalytic potential of 0D/2D heterostructures. As an example, the development of Rh/MoS_2 heterostructures was performed by Cheng *et al* [112] using MoS_2 nanosheets (as a rapid H_2-desorbing element) morphology as shown in figures 4.13(b)–(f),

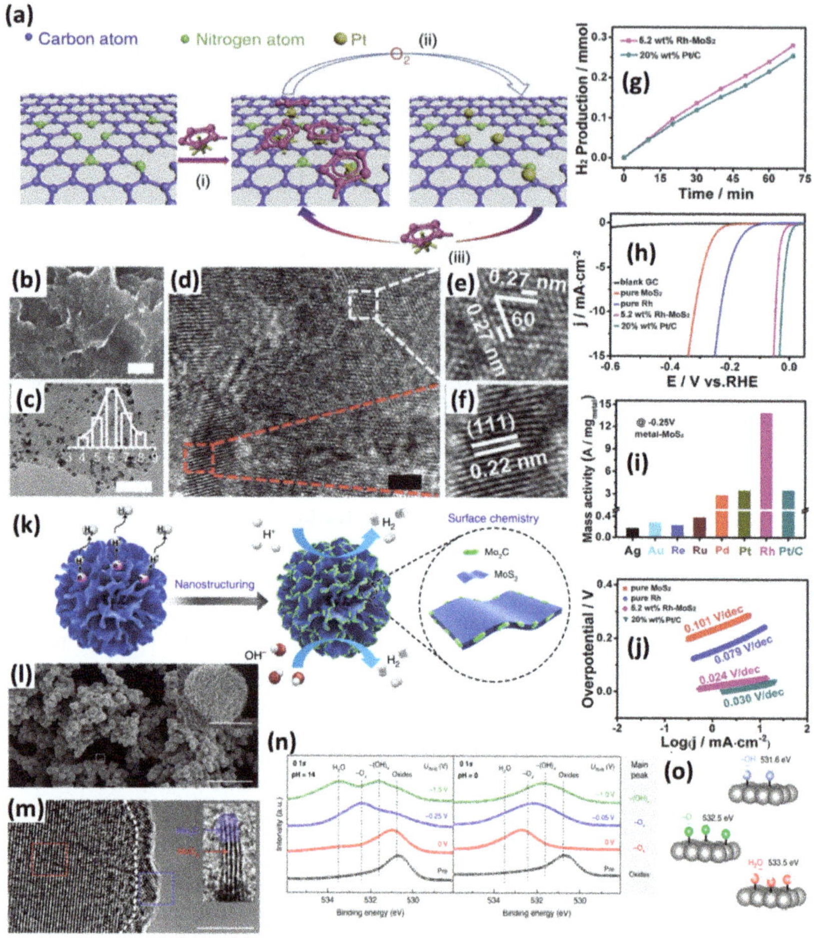

Figure 4.13. (a) Graphic representation of 0D/2D heterostructures for electrocatalysis. ALD progression of Pt on NGNs. (Reproduced with permission from [14]. Copyright 2016 Springer Nature.) (b)–(f) Morphological examination of Rh-MoS_2 nanocomposites for HER. (b) SEM micrograph. (c) TEM photograph of 5.2 wt.% Rh-MoS_2. (d) HR-TEM photograph of 5.2 wt.% Rh-MoS_2. (e), (f) Corresponding enlarged area. (g) The HER at −0.25 V (for 20 wt.% Pt/C and 5.2 wt.% Rh-MoS_2 catalysts). (h) LSV curves for different catalysts. (i) Mass activity of metal–MoS_2 composites and HER for 20 wt.% Pt/C catalysts at −0.25 V. (j) Corresponding Tafel plots and slopes. (Reproduced with permission from [112]. Copyright 2017 John Wiley and Sons.) (k) Visualization of MoS_2/Mo_2C. (l) SEM photographs of the prepared material (20 μm). (m) Corresponding HR-TEM photograph (inset: FFT pattern). (n) XPS spectra of β-Mo_2C HER at different potentials and pH values. (o) Models of oxygenated surface species and binding energies. (Reproduced with permission from [115]. Copyright 2019 Springer Nature.)

and Rh nanoparticles (as a strong H-adsorbing element). The resulting heterostructures revealed a slight Tafel slope (24 mV dec^{-1}) as well as a low overpotential (47 mV) at 10 mA cm^{-2} along with good stability (figures 4.13(g)–(j)). The superior activity of the catalysts was due to the presence of Rh atoms, that show rapid capturing of hydronium ions, as these atoms serve as a vital H-adsorbing element;

thus, the migration of adsorbed H atoms towards the MoS_2 surface was observed as MoS_2 is designated as a rapid H_2-desorbing component. Hence, the 2D MoS_2 nanosheets and 0D Rh nanoparticles can collectively increase the electrocatalytic potential. The significant synthesis of nanoparticle catalysts or single atoms dispersed over 2D materials is challenging. Like metal atom catalysts, the 0D materials can also assist in electrocatalytic reactions via their excellent active sites [113, 114]. The (*in situ*) growth of MoS_2 nanoparticles over rGO was carried out by Li *et al* [61] by adopting a one-step solvothermal scheme. The reports show that MoS_2 nanoparticles possess abundant S-edge sites compared to MoS_2 nanosheets; therefore, the combination of MoS_2 nanoparticles with rGO can considerably stimulate the catalytic potential, leading to a small Tafel slope (41 mV dec^{-1}) and small onset potential (~0.1 V).

MoS_2/rGO catalyst durability was observed for 1000 cycles, resulting in only minor cathodic current losses. A large current density (in addition to small overpotentials and onset potentials) is a significant alternative parameter for the performance of electrocatalysts which is frequently ignored. In this regard, a research group designed and developed an electrocatalyst based on a 0D/2D heterostructure composed of MoS_2 nanosheets and Mo_2C nanoparticles (MoS_2/Mo_2C) using carbonization of MoS_2 (*in situ*), as shown in figures 4.13(k)–(m) [115]. The obtained heterostructure retains rough surfaces and extremely exposed active sites at the micro- and nanoscale. A small overpotential was observed in acidic (227 mV) and alkaline media (220 mV) at a 1000 mA cm^{-2} current density. Moreover, the obtained catalyst offered decent durability in both media during 24 h examination. Surface oxygen grows on Mo_2C and boosts the interfacial mass transfer during the HER experiment, corresponding to outstanding activity of the MoS_2/Mo_2C heterostructures. According to these outcomes, further attention should be paid to the research regarding industrial/practical applications. Generally, 0D/2D heterostructures are considered favorable candidates for electrocatalysis, but the presence of poor durability inhibits its industrial applications; therefore, additional developments are required (figures 4.13(n)–(o)).

4.5.2 Materials-based schemes

4.5.2.1 Transition-metal hydroxide-based heterostructures

Encouraging the water dissociation phase is a profitable scheme for designing emerging catalysts for the alkaline HER. Recently, specific research works revealed that transition metal hydroxides, including $Co(OH)_2$, $Ni(OH)_2$, and $Ni_xFe_y(OH)_2$, are recognized as active promoters for the dissociation of water. To examine the potential of $Ni(OH)_2$ for the HER, Subbaraman *et al* fabricated a Pt-adapted electrode with $Ni(OH)_2$ nanoclusters [116]. The morphological aspects of the Pt (111)- or Pt-island adapted $Ni(OH)_2$ nanoclusters are illustrated in figure 4.14(a), suggesting that the presence of Pt-islands resulted in boosting of active sites. The modification of Pt-islands by $Ni(OH)_2$ (referred to as a Pt(111)/$Ni(OH)_2$ electrode) causes additional enhancement in HER activity (figure 4.14(b)). The water adsorption strategy was believed to be enhanced using simultaneous interaction of H atoms

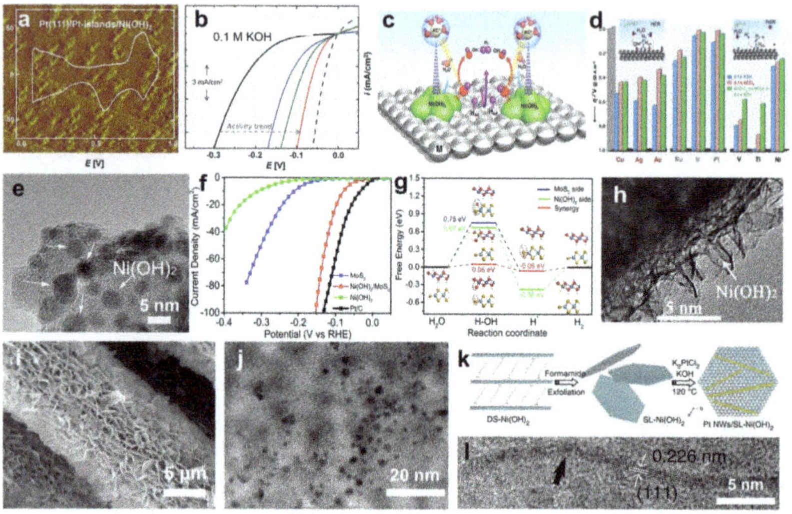

Figure 4.14. (a) STM image (scale bar: 60 × 60 nm^2) and CV trace of the $Ni(OH)_2$/Pt-islands/Pt(111) surface. (b) Confirming HER activity Pt(111) as the substrate. (c) Demonstration of an HER experiment for the Ni $(OH)_2$/Pt electrode surface. (Reproduced with permission from [116]. Copyright 2011 American Association for the Advancement of Science.) (d) Assessment of HER reactions for pristine metal and $Ni(OH)_2$-modified surfaces. (Reproduce with permission from [117]. Copyright 2012 John Wiley and Sons.) (e) TEM photograph of $Ni(OH)_2/MoS_2$ heterostructure. (f) LSV curves for different catalysts. (g) Resultant free energy illustration of the HER for the $Ni(OH)_2/MoS_2$ interface and MoS_2 edge. (Reproduced with permission from [119]. Copyright 2017 Elsevier.) (h) TEM photograph for $Ni(OH)_2/CoS_2$ heterostructure. (Reproduced with permission from [120]. Copyright 2017 The Royal Society of Chemistry.) (i) SEM micrograph for Pt/Co $(OH)_2$/CC electrode. (j) Corresponding TEM photograph. (Reproduced with permission from [127]. Copyright 2017 American Chemical Society.) (k) Synthesis approach for Pt NWs/single-layer $Ni(OH)_2$. (l) TEM photographs of the corresponding prepared product. (Reproduced with permission from [128]. Copyright 2015 Springer Nature.)

with Pt and O atoms with $Ni(OH)_2$ at the boundary between the Pt and $Ni(OH)_2$ domains. Thus, due to the boosted water dissociation using $Ni(OH)_2$ nanoclusters, the hydrogen intermediates produced were successively adsorbed on neighboring Pt surfaces; lastly, they produced recombination with H_2, as shown in figure 4.14(c).

Additionally, the introduction of Li^+ causes boosted production of hydrogen intermediates because of the destabilization effects of Li^+ over the HO–H bond. Later, a systematic study of HER progress for a range of $Ni(OH)_2$ adapted metal electrodes was carried out by Danilovic *et al* [117]. The outcomes of activity yields for $Ni(OH)_2$ that can assist in the HER for all metals, and the connections between them, are shown in figure 4.14(d). This advancement can be recognized as a synergistic effect between the metals and $Ni(OH)_2$; among them, water dissociation was assisted by $Ni(OH)_2$, whereas the produced hydrogens were adsorbed and associated on metal surfaces subsequently. The association between boosted HER activity in alkaline media and $Ni(OH)_2$ structure was studied by Yu *et al* via loading of α-$Ni(OH)_2$ and β-$Ni(OH)_2$ on a Pt electrode [118]. The theoretical and experimental outcomes show that β-$Ni(OH)_2$ is recognized as a stimulating promoter of the HER in alkaline media on Pt because of the better water dissociation capability and stronger interactions between β-$Ni(OH)_2$ and Pt.

Transition metal hydroxides can also enhance HER kinetics in alkaline media over various catalysts, for example, transition metal phosphides, TMDs, and nitrides. In this regard, $Ni(OH)_2/MoS_2$/CC (CC: carbon cloth) 3D-heterostructures with $Ni(OH)_2$ nanoparticles attached above the MoS_2 surface were reported by Zhang *et al*, as shown in figure 4.14(e) [119]. These obtained heterostructures attain a considerably superior HER activity, with an overpotential of 80 mV at 10 mA cm^{-2}, as shown in figure 4.14(f). Furthermore, DFT measurements showed that the water adsorption energy for the as-prepared heterostructures was 0.05 eV compared to MoS_2 (1.17 eV). The free energy photograph of the HER experiment above the MoS_2 and $Ni(OH)_2$ surface and the heterostructure acquired from DFT measurements (figure 4.14(g)), confirm an enormous reduction in the free energy for hydrogen adsorption and water dissociation above the $Ni(OH)_2/MoS_2$ heterostructure surface. Furthermore, Chen *et al* described a $Ni(OH)_2/CoS_2$/CC nanowire array that considerably boosted HER activity (figure 4.14(h)) [120]. A similar reduction in the free energy of the HER experiment of $Ni(OH)_2$ adapted electrodes was confirmed by DFT measurements. Additionally, heterostructures based on transition metal hydroxide were also developed for HER experiments in alkaline media including $Ni(OH)_2$/1T-MoS_2 [121], $Ni(OH)_2$/CuS core–shell heterostructures [122], $Ni(OH)_2$–Fe_2P/TM [123], NiCo–LDH/MoS_2 [124], $Ni(OH)_2/Ni_3N$/TM nanoarrays [125] as well as $Ni(OH)_2$–PtO_2/TM [126] etc. Additionally, transition metal hydroxides are adopted to enhance HER activity in alkaline media as a substrate.

In the same way, Xing *et al* produced a CC adopted $Co(OH)_2$ nanosheet array (by way of a 3D substrate) for electrodepositing ultrafine Pt nanoparticles, as shown in figures 4.14(i) and (j) [127]. Further, the production of ultrathin Pt nanowires on single-layer $Ni(OH)_2$ was described by Yin *et al* (figures 4.14(k) and (l)) [128]. These heterostructures exhibited more remarkable HER progress than commercial Pt/C electrodes because of a fast water dissociation scheme. These outcomes verified the transition metal hydroxides' capability to enhance the water dissociation and adsorption process, which is considered a practical path for fabricating innovative alkaline HER catalysts.

4.5.2.2 Transition metal oxide-based heterostructures

Generally, most of the TMOs (i.e. NiO, MoO_3, TiO_2, and WO_3) are not preferred for HER experiments (using acidic media) due to their fewer adsorption sites for H* [129]. On the other hand, some TMOs may be marked using acidic media (i.e. Fe_2O_3, CuO, and MoO_3), and a limited number of TMO-based heterostructures illustrate potential HER activity using acidic media. For example, vertically oriented MoO_3/MoS_2 nanowires were fabricated using sulfuration in different temperature ranges [63], for which the core–shell structure is represented in figure 4.15(a). Despite the corrosion tendency of MoO_3 (in 0.5 m H_2SO_4), the MoO_3/MoS_2 heterostructure displays an overpotential (~250 mV) at 10 mA cm^{-2} and reveals better stability for 10 000 CV cycles, as shown in figure 4.15(b). The growth of MoS_2 was done on MoO_3 nanowires (a higher aspect ratio substrate) that

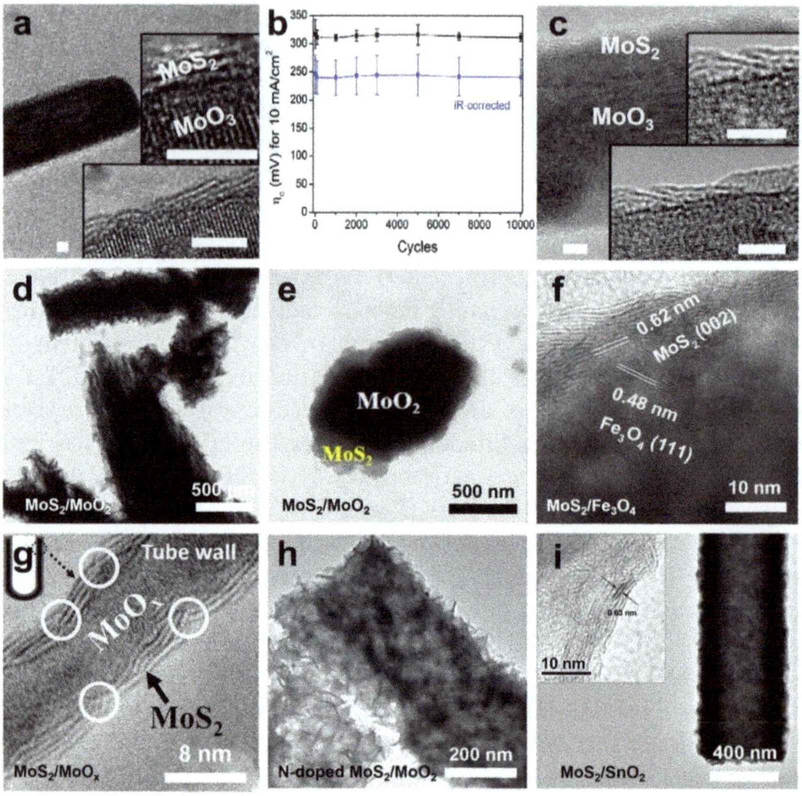

Figure 4.15. (a) TEM photograph of MoS_2/MoO_3 nanowires sulfurized at 200 °C (the inset corresponds to the lattice fringe of MoS_2 and MoO_3 with a 5 nm scale bar). (b) Overpotential stability is necessary to determine 10 mA cm^{-2} without IR correction (black) and with IR correction (blue). (c) TEM photograph of the as-prepared nanowires after 10 000 potential cycles. (Reproduced with permission from [63]. Copyright 2011 American Chemical Society.) (d), (e) TEM photographs of MoS_2/MoO_2 heterostructure. ((d) Reproduced with permission from [130]. Copyright 2015 The Royal Society of Chemistry. (e) Reproduced with permission from [133]. Copyright 2017 Elsevier.) (f) MoS_2/Fe_3O_4 heterostructure. (Reproduced with permission from [132]. Copyright 2017 Elsevier.) (g) MoO_x/MoS_2 core–shell structure. (Reproduced with permission from [136]. Copyright 2016 John Wiley and Sons.) (h) N-doped MoS_2/MoO_2. (Reproduced with permission from [131]. Copyright 2014 The Royal Society of Chemistry.) (i) MoS_2/SnO_2 (the inset shows an HR-TEM photograph of MoS_2). (Reproduced with permission from [62]. Copyright 2014 The Royal Society of Chemistry.)

allowed facile charge transport. The MoS_2 shell prohibited the direct contact of the MoO_3 core with the corrosive electrolyte. The morphological investigation using HR-TEM photographs (figure 4.15(c)), exhibits an excellent core–shell structure, confirming the outstanding stability of the prepared heterostructure using acidic media. In the same way, Yang *et al* described the development of a MoS_2/MoO_2 heterostructure acquired through the sulfuration of highly conductive and porous MoO_2 [130]. This prepared catalyst showed a small Tafel slope (76.1 mV dec^{-1}) and overpotential (240 mV) at a current density of 10 mA cm^{-2}, as well as outstanding durability. The rapid electron transport causes higher electrical conductivity, while the porous structures allow quick mass diffusion. A range of TMO-based heterostructures is described by researchers that serve as electrocatalysts for HER experiments in acidic media [131–133]. Among them, most have core–shell structures with an HER active shell and TMO core (figures 4.15(d)–(i))); the active shell shields the TMO core from corrosion, improving durability [16]. In contrast, several studies ascribe improved HER activity to the higher electrical conductivity values of the TMO core. For example, the general electrical conductivity values for MoS_2 lie in the range of 10^{-4}–10^{-5} S cm^{-1}, whereas for MoO_2, it is 10^2–10^3 S cm^{-1} [134]. The highly conductive MoO_2 core permits fast electron transport in MoS_2/MoO_2 heterostructures, thus improving the HER potential. However, various TMOs, including SnO_2 and MoO_3, are not recognized as decent conductive materials; to date, the higher HER activity of these TMO-based heterostructures is still attributed to a highly conductive TMO core. Furthermore, the morphology and dimensions also correspond to electric conductivity [135]. Therefore, it is indispensable to provide appropriate supporting confirmation when determining electrical conductivity.

4.6 Conclusions, challenges and perspectives

The HER activity of numerous 2D materials could be considerably improved via various techniques to boost active site exposure and intrinsic conductivity. Despite the significant advances made in the application of 2D materials for the HER, there are still several questions that have not been answered. To begin, most electrochemical experimentalists believe that the 1T phase comprises the most active phase in MoS_2. However, other theorists feel that the 1T′ phase is more stable when compared to the 1T phase due to the effects of local strain. Because of this ambiguity, efforts to further increase HER activity by broadly applicable phase engineering are hampered. Second, not many studies investigate the contribution of a number of different components in an organized manner, such as defects, strain, S-vacancy, and phase transition.

The advancement of material engineering technology and theoretical measurements has led to the discovery of a wide range of highly active HER catalysts (alkaline) during the past ten years. These catalysts have been published. In our pursuit of complete comprehension and mastery over alkaline HER, we are still up against a number of obstacles, some of which are listed below.

1. The fundamental question of why alkaline HER has such delayed kinetics remains unanswered. From an experimental point of view, such a mystery can be significantly addressed by gaining knowledge of the interactions between intermediaries and catalysts that occur during reactions on active sites. This can be accomplished by offering supplementary *in situ* characterization techniques (i.e. vibrational spectroscopies, x-ray, TEM, etc) to track the progressing reaction. The local surroundings (for example, the local pH and the electric double layer) and dynamic modification of electrocatalytic surface assembly are significant elements in regulating the HER route (alkaline). In the future, these factors ought to be researched.

2. In addition to concentrating on the reaction mechanism arising from ordinary distinct Pt model catalysts, additional attention should be paid to a comprehensive examination of the electrocatalytic mechanisms occurring for characteristic nanostructured catalysts. These nanostructured catalysts can be utilized in a greater variety of practical applications as they are more active. It is possible that the reaction pathways for nanostructured catalysts, which have more complicated surface structures, will be moderately different to those for different model crystals. The outcome of the original procedure begins with complex surfaces, which can be recognized as a novel guideline for developing innovative catalysts. For example, these nanostructured catalysts comprise transition metal-based bimetallic alloys and compounds.
3. From a computational point of view, more practical operando calculation approaches are required to model the reaction process better. This can be accomplished by considering the local environmental parameters, such as the influence of water, metal cations, and anions. In an ideal scenario, operando spectroscopy and calculation would work together to unveil the fundamental characteristics of the alkaline HER. Notably, this aspect is also one of the most critical concerns regarding additional heterogeneous reduction reactions comprising a few intermediates (i.e. NRR and CO_2RR).

Bibliography

[1] Conway B E and Tilak B V 2002 Interfacial processes involving electrocatalytic evolution and oxidation of H_2, and the role of chemisorbed H *Electrochim. Acta* **47** 3571–94

[2] Lehmann H P, Fuentes-Arderiu X and Bertello L F 1996 Glossary of terms in quantities and units in clinical chemistry (IUPAC-IFCC recommendations 1996) *Pure Appl. Chem.* **68** 957–1000

[3] Parsons R 1958 The rate of electrolytic hydrogen evolution and the heat of adsorption of hydrogen *Trans. Faraday Soc.* **54** 1053–63

[4] Trasatti S 1972 Work function, electronegativity, and electrochemical behaviour of metals: III. Electrolytic hydrogen evolution in acid solutions *J. Electroanal. Chem. Interfacial Electrochem.* **39** 163–84

[5] Nørskov J K, Bligaard T, Logadottir A, Kitchin J R, Chen J G, Pandelov S and Stimming U 2005 Trends in the exchange current for hydrogen evolution *J. Electrochem. Soc.* **152** J23

[6] Hinnemann B, Moses P G, Bonde J, Jørgensen K P, Nielsen J H, Horch S, Chorkendorff I and Nørskov J K 2005 Biomimetic hydrogen evolution: MoS_2 nanoparticles as catalyst for hydrogen evolution *J. Am. Chem. Soc.* **127** 5308–9

[7] Jaramillo T F, Jørgensen K P, Bonde J, Nielsen J H, Horch S and Chorkendorff I 2007 Identification of active edge sites for electrochemical H_2 evolution from MoS_2 nanocatalysts *Science* **317** 100–2

[8] Liao L, Zhu J, Bian X, Zhu L, Scanlon M D, Girault H H and Liu B 2013 MoS_2 formed on mesoporous graphene as a highly active catalyst for hydrogen evolution *Adv. Funct. Mater.* **23** 5326–33

[9] Skúlason E, Tripkovic V, Björketun M E, Gudmundsdóttir S, Karlberg G, Rossmeisl J, Bligaard T, Jónsson H and Nørskov J K 2010 Modeling the electrochemical hydrogen oxidation and evolution reactions on the basis of density functional theory calculations *J. Phys. Chem.* C **114** 18182–97

[10] Seh Z W, Kibsgaard J, Dickens C F, Chorkendorff I, Nørskov J K and Jaramillo T F 2017 Combining theory and experiment in electrocatalysis: insights into materials design *Science* **355** eaad4998
[11] Schmidt T J, Ross P N and Markovic N M 2002 Temperature dependent surface electrochemistry on Pt single crystals in alkaline electrolytes: part 2. The hydrogen evolution/oxidation reaction *J. Electroanal. Chem.* **524–5** 252–60
[12] Danilovic N, Subbaraman R, Strmcnik D, Stamenkovic V and Markovic N M 2013 Electrocatalysis of the HER in acid and alkaline media *J. Serb. Chem. Soc.* **78** 2007–15
[13] Xu K, Ding H, Zhang M, Chen M, Hao Z, Zhang L, Wu C and Xie Y 2017 Regulating water-reduction kinetics in cobalt phosphide for enhancing her catalytic activity in alkaline solution *Adv. Mater.* **29** 1606980
[14] Cheng N *et al* 2016 Platinum single-atom and cluster catalysis of the hydrogen evolution reaction *Nat. Commun.* **7** 13638
[15] Fu Y, Meng F, Rowley M B, Thompson B J, Shearer M J, Ma D, Hamers R J, Wright J C and Jin S 2015 Solution growth of single crystal methylammonium lead halide perovskite nanostructures for optoelectronic and photovoltaic applications *J. Am. Chem. Soc.* **137** 5810–8
[16] Raza A, Rafi A A, Hassan J Z, Rafiq A and Li G 2023 Rational design of 2D heterostructured photo- and electro-catalysts for hydrogen evolution reaction: a review *Appl. Surf. Sci. Adv.* **15** 100402
[17] Qiu H J, Ito Y, Cong W, Tan Y, Liu P, Hirata A, Fujita T, Tang Z and Chen M 2015 Nanoporous graphene with single-atom nickel dopants: an efficient and stable catalyst for electrochemical hydrogen production *Angew. Chem. Int. Ed.* **54** 14031–5
[18] Zheng Y, Jiao Y, Li L H, Xing T, Chen Y, Jaroniec M and Qiao S Z 2014 Toward design of synergistically active carbon-based catalysts for electrocatalytic hydrogen evolution *ACS Nano* **8** 5290–6
[19] Ikram M, Hassan J, Imran M, Haider J, Ul-Hamid A, Shahzadi I, Ikram M, Raza A, Qumar U and Ali S 2020 2D chemically exfoliated hexagonal boron nitride (hBN) nanosheets doped with Ni: synthesis, properties and catalytic application for the treatment of industrial wastewater *Appl. Nanosci.* **10** 3525–8
[20] Raza A, Hassan J Z, Qumar U, Zaheer A, Babar Z U D, Iannotti V and Cassinese A 2024 Strategies for robust electrocatalytic activity of 2D materials: ORR, OER, HER, and CO2RR *Mater. Today Adv.* **22** 100488
[21] Jiao Y, Zheng Y, Davey K and Qiao S-Z 2016 Activity origin and catalyst design principles for electrocatalytic hydrogen evolution on heteroatom-doped graphene *Nat. Energy* **1** 16130
[22] Jiang H, Zhu Y, Su Y, Yao Y, Liu Y, Yang X and Li C 2015 Highly dual-doped multilayer nanoporous graphene: efficient metal-free electrocatalysts for the hydrogen evolution reaction *J. Mater. Chem.* A **3** 12642–5
[23] Waqas U, Farhan A, Haider A, Qumar U and Raza A 2023 Advancements in biofilm formation and control in potable water distribution systems: a comprehensive review and analysis of chloramine decay in water systems *J. Environ. Chem. Eng.* **11** 111377
[24] Hassan J Z, Zaheer A, Raza A and Li G 2023 Au-based heterostructure composites for photo and electro catalytic energy conversions *Sustain. Mater. Technol.* **36** e00609
[25] Ito Y, Cong W, Fujita T, Tang Z and Chen M 2015 High catalytic activity of nitrogen and sulfur co-doped nanoporous graphene in the hydrogen evolution reaction *Angew. Chem. Int. Ed.* **54** 2131–6

[26] de Levie R 1999 The electrolysis of water *J. Electroanal. Chem.* **476** 92–3

[27] Morales-Guio C G, Stern L-A and Hu X 2014 Nanostructured hydrotreating catalysts for electrochemical hydrogen evolution *Chem. Soc. Rev.* **43** 6555–69

[28] Kibsgaard J, Jaramillo T F and Besenbacher F 2014 Building an appropriate active-site motif into a hydrogen-evolution catalyst with thiomolybdate $[Mo_3S_{13}]^{2-}$ clusters *Nat. Chem.* **6** 248–53

[29] Xu Y and Zhang B 2014 Recent advances in porous Pt-based nanostructures: synthesis and electrochemical applications *Chem. Soc. Rev.* **43** 2439–50

[30] Liu P, Rodriguez J A, Asakura T, Gomes J and Nakamura K 2005 Desulfurization reactions on $Ni_2P(001)$ and α-$Mo_2C(001)$ surfaces: complex role of P and C sites *J. Phys. Chem.* B **109** 4575–83

[31] Chianelli R R, Berhault G, Raybaud P, Kasztelan S, Hafner J and Toulhoat H 2002 Periodic trends in hydrodesulfurization: in support of the Sabatier principle *Appl. Catal.* A **227** 83–96

[32] Tributsch H and Bennett J C 1977 Electrochemistry and photochemistry of MoS_2 layer crystals. I *J. Electroanal. Chem. Interfacial Electrochem.* **81** 97–111

[33] Chhowalla M, Shin H S, Eda G, Li L-J, Loh K P and Zhang H 2013 The chemistry of two-dimensional layered transition metal dichalcogenide nanosheets *Nat. Chem.* **5** 263–75

[34] Voiry D, Salehi M, Silva R, Fujita T, Chen M, Asefa T, Shenoy V B, Eda G and Chhowalla M 2013 Conducting MoS_2 nanosheets as catalysts for hydrogen evolution reaction *Nano Lett.* **13** 6222–7

[35] Nørskov J K, Bligaard T, Rossmeisl J and Christensen C H 2009 Towards the computational design of solid catalysts *Nat. Chem.* **1** 37–46

[36] Voiry D, Mohite A and Chhowalla M 2015 Phase engineering of transition metal dichalcogenides *Chem. Soc. Rev.* **44** 2702–12

[37] Voiry D, Goswami A, Kappera R, de Carvalho Castro e Silva C, Kaplan D, Fujita T, Chen M, Asefa T and Chhowalla M 2015 Covalent functionalization of monolayered transition metal dichalcogenides by phase engineering *Nat. Chem.* **7** 45–9

[38] Huang X, Tan C, Yin Z and Zhang H 2014 25th anniversary article: hybrid nanostructures based on two-dimensional nanomaterials *Adv. Mater.* **26** 2185–204

[39] Costi R, Saunders A E and Banin U 2010 Colloidal hybrid nanostructures: a new type of functional materials *Angew. Chem. Int. Ed.* **49** 4878–97

[40] Huang X, Zeng Z, Bao S, Wang M, Qi X, Fan Z and Zhang H 2013 Solution-phase epitaxial growth of noble metal nanostructures on dispersible single-layer molybdenum disulfide nanosheets *Nat. Commun.* **4** 1444

[41] Kajiwara R, Asaumi Y, Nakamura M and Hoshi N 2011 Active sites for the hydrogen oxidation and the hydrogen evolution reactions on the high index planes of Pt *J. Electroanal. Chem.* **657** 61–5

[42] Zeng Z, Yin Z, Huang X, Li H, He Q, Lu G, Boey F and Zhang H 2011 Single-layer semiconducting nanosheets: high-yield preparation and device fabrication *Angew. Chem. Int. Ed.* **50** 11093–7

[43] Zeng Z, Tan C, Huang X, Bao S and Zhang H 2014 Growth of noble metal nanoparticles on single-layer TiS_2 and TaS_2 nanosheets for hydrogen evolution reaction *Energy Environ. Sci.* **7** 797–803

[44] Kang Y *et al* 2014 Plasmonic hot electron induced structural phase transition in a MoS_2 monolayer *Adv. Mater.* **26** 6467–71

[45] Tan Y *et al* 2014 Monolayer MoS_2 films supported by 3D nanoporous metals for high-efficiency electrocatalytic hydrogen production *Adv. Mater.* **26** 8023–8
[46] Tsai C, Abild-Pedersen F and Nørskov J K 2014 Tuning the MoS_2 edge-site activity for hydrogen evolution via support interactions *Nano Lett.* **14** 1381–7
[47] Kim J, Byun S, Smith A J, Yu J and Huang J 2013 Enhanced electrocatalytic properties of transition-metal dichalcogenides sheets by spontaneous gold nanoparticle decoration *J. Phys. Chem. Lett.* **4** 1227–32
[48] Vrubel H, Merki D and Hu X 2012 Hydrogen evolution catalyzed by MoS_3 and MoS_2 particles *Energy Environ. Sci.* **5** 6136–44
[49] Kibsgaard J, Chen Z, Reinecke B N and Jaramillo T F 2012 Engineering the surface structure of MoS_2 to preferentially expose active edge sites for electrocatalysis *Nat. Mater.* **11** 963–9
[50] Wang T, Liu L, Zhu Z, Papakonstantinou P, Hu J, Liu H and Li M 2013 Enhanced electrocatalytic activity for hydrogen evolution reaction from self-assembled monodispersed molybdenum sulfide nanoparticles on an Au electrode *Energy Environ. Sci.* **6** 625–33
[51] Cai Y, Yang X, Liang T, Dai L, Ma L, Huang G, Chen W, Chen H, Su H and Xu M 2014 Easy incorporation of single-walled carbon nanotubes into two-dimensional MoS_2 for high-performance hydrogen evolution *Nanotechnology* **25** 465401
[52] Liu N, Yang L, Wang S, Zhong Z, He S, Yang X, Gao Q and Tang Y 2015 Ultrathin MoS_2 nanosheets growing within an *in situ*-formed template as efficient electrocatalysts for hydrogen evolution *J. Power Sources* **275** 588–94
[53] Zhu H, Du M, Zhang M, Zou M, Yang T, Wang S, Yao J and Guo B 2014 S-rich single-layered MoS_2 nanoplates embedded in N-doped carbon nanofibers: efficient co-electro-catalysts for the hydrogen evolution reaction *Chem. Commun.* **50** 15435–8
[54] Youn D H, Han S, Kim J Y, Kim J Y, Park H, Choi S H and Lee J S 2014 Highly active and stable hydrogen evolution electrocatalysts based on molybdenum compounds on carbon nanotube–graphene hybrid support *ACS Nano* **8** 5164–73
[55] Yuan H, Li J, Yuan C and He Z 2014 Facile synthesis of MoS_2@CNT as an effective catalyst for hydrogen production in microbial electrolysis cells *ChemElectroChem.* **1** 1828–33
[56] Zheng X, Xu J, Yan K, Wang H, Wang Z and Yang S 2014 Space-confined growth of MoS_2 nanosheets within graphite: the layered hybrid of MoS_2 and graphene as an active catalyst for hydrogen evolution reaction *Chem. Mater.* **26** 2344–53
[57] Tan C and Zhang H 2015 Two-dimensional transition metal dichalcogenide nanosheet-based composites *Chem. Soc. Rev.* **44** 2713–31
[58] Firmiano E G S, Cordeiro M A L, Rabelo A C, Dalmaschio C J, Pinheiro A N, Pereira E C and Leite E R 2012 Graphene oxide as a highly selective substrate to synthesize a layered MoS_2 hybrid electrocatalyst *Chem. Commun.* **48** 7687–9
[59] Ma C-B, Qi X, Chen B, Bao S, Yin Z, Wu X-J, Luo Z, Wei J, Zhang H-L and Zhang H 2014 MoS_2 nanoflower-decorated reduced graphene oxide paper for high-performance hydrogen evolution reaction *Nanoscale* **6** 5624–9
[60] Wang H, Lu Z, Kong D, Sun J, Hymel T M and Cui Y 2014 Electrochemical tuning of MoS_2 nanoparticles on three-dimensional substrate for efficient hydrogen evolution *ACS Nano* **8** 4940–7
[61] Li Y, Wang H, Xie L, Liang Y, Hong G and Dai H 2011 MoS_2 nanoparticles grown on graphene: an advanced catalyst for the hydrogen evolution reaction *J. Am. Chem. Soc.* **133** 7296–9

[62] Huang Y, Miao Y-E, Zhang L, Tjiu W W, Pan J and Liu T 2014 Synthesis of few-layered MoS_2 nanosheet-coated electrospun SnO_2 nanotube heterostructures for enhanced hydrogen evolution reaction *Nanoscale* **6** 10673–9
[63] Chen Z, Cummins D, Reinecke B N, Clark E, Sunkara M K and Jaramillo T F 2011 Core–shell MoO_3–MoS_2 nanowires for hydrogen evolution: a functional design for electrocatalytic materials *Nano Lett.* **11** 4168–75
[64] Hu J, Zhang C, Zhang Y, Yang B, Qi Q, Sun M, Zi F, Leung M K H and Huang B 2020 Interface modulation of MoS_2/metal oxide heterostructures for efficient hydrogen evolution electrocatalysis *Small* **16** 2002212
[65] Hu J, Huang B, Zhang C, Wang Z, An Y, Zhou D, Lin H, Leung M K H and Yang S 2017 Engineering stepped edge surface structures of MoS_2 sheet stacks to accelerate the hydrogen evolution reaction *Energy Environ. Sci.* **10** 593–603
[66] Handoko A D, Fredrickson K D, Anasori B, Convey K W, Johnson L R, Gogotsi Y, Vojvodic A and Seh Z W 2018 Tuning the basal plane functionalization of two-dimensional metal carbides (MXenes) to control hydrogen evolution activity *ACS Appl. Energy Mater.* **1** 173–80
[67] Jiang Y, Sun T, Xie X, Jiang W, Li J, Tian B and Su C 2019 Oxygen-functionalized ultrathin $Ti_3C_2T_x$ MXene for enhanced electrocatalytic hydrogen evolution *ChemSusChem.* **12** 1368–73
[68] Gao G, O'Mullane A P and Du A 2017 2D MXenes: a new family of promising catalysts for the hydrogen evolution reaction *ACS Catal.* **7** 494–500
[69] Zhang J, Zhao Y, Guo X, Chen C, Dong C-L, Liu R-S, Han C-P, Li Y, Gogotsi Y and Wang G 2018 Single platinum atoms immobilized on an MXene as an efficient catalyst for the hydrogen evolution reaction *Nat. Catal.* **1** 985–92
[70] Liu J, Liu Y, Xu D, Zhu Y, Peng W, Li Y, Zhang F and Fan X 2019 Hierarchical 'nanoroll' like MoS_2/$Ti_3C_2T_x$ hybrid with high electrocatalytic hydrogen evolution activity *Appl. Catalysis* B **241** 89–94
[71] Huang L, Ai L, Wang M, Jiang J and Wang S 2019 Hierarchical MoS_2 nanosheets integrated Ti_3C_2 MXenes for electrocatalytic hydrogen evolution *Int. J. Hydrogen Energy* **44** 965–76
[72] Wu X, Wang Z, Yu M, Xiu L and Qiu J 2017 Stabilizing the MXenes by carbon nanoplating for developing hierarchical nanohybrids with efficient lithium storage and hydrogen evolution capability *Adv. Mater.* **29** 1607017
[73] Wang H, Lin Y, Liu S, Li J, Bu L, Chen J, Xiao X, Choi J-H, Gao L and Lee J-M 2020 Confined growth of pyridinic N–Mo_2C sites on MXenes for hydrogen evolution *J. Mater. Chem.* A **8** 7109–16
[74] Li N, Zhang Y, Jia M, Lv X, Li X, Li R, Ding X, Zheng Y-Z and Tao X 2019 1T/2H $MoSe_2$-on-MXene heterostructure as bifunctional electrocatalyst for efficient overall water splitting *Electrochim. Acta* **326** 134976
[75] Seh Z W, Fredrickson K D, Anasori B, Kibsgaard J, Strickler A L, Lukatskaya M R, Gogotsi Y, Jaramillo T F and Vojvodic A 2016 Two-dimensional molybdenum carbide (MXene) as an efficient electrocatalyst for hydrogen evolution *ACS Energy Lett.* **1** 589–94
[76] Dong J, Wu Q, Huang C, Yao W and Xu Q 2018 Cost effective Mo rich Mo_2C electrocatalysts for the hydrogen evolution reaction *J. Mater. Chem.* A **6** 10028–35

[77] Intikhab S, Natu V, Li J, Li Y, Tao Q, Rosen J, Barsoum M W and Snyder J 2019 Stoichiometry and surface structure dependence of hydrogen evolution reaction activity and stability of Mo_xC MXenes *J. Catal.* **371** 325–32

[78] Kuznetsov D A, Chen Z, Kumar P V, Tsoukalou A, Kierzkowska A, Abdala P M, Safonova O V, Fedorov A and Müller C R 2019 Single site cobalt substitution in 2D molybdenum carbide (MXene) enhances catalytic activity in the hydrogen evolution reaction *J. Am. Chem. Soc.* **141** 17809–16

[79] Ren J, Zong H, Sun Y, Gong S, Feng Y, Wang Z, Hu L, Yu K and Zhu Z 2020 2D organ-like molybdenum carbide (MXene) coupled with MoS_2 nanoflowers enhances the catalytic activity in the hydrogen evolution reaction *CrystEngComm* **22** 1395–403

[80] Wu X, Zhou S, Wang Z, Liu J, Pei W, Yang P, Zhao J and Qiu J 2019 Engineering multifunctional collaborative catalytic interface enabling efficient hydrogen evolution in all pH range and seawater *Adv. Energy Mater.* **9** 1901333

[81] Sun W, Wang X, Feng J, Li T, Huan Y, Qiao J, He L and Ma D 2019 Controlled synthesis of 2D Mo_2C/graphene heterostructure on liquid Au substrates as enhanced electrocatalytic electrodes *Nanotechnology* **30** 385601

[82] Pan H 2016 Ultra-high electrochemical catalytic activity of MXenes *Sci. Rep.* **6** 32531

[83] Nguyen T P, Kim S Y, Lee T H, Jang H W, Le Q V and Kim I T 2020 Facile synthesis of W_2C@WS_2 alloy nanoflowers and their hydrogen generation performance *Appl. Surf. Sci.* **504** 144389

[84] Li P, Zhu J, Handoko A D, Zhang R, Wang H, Legut D, Wen X, Fu Z, Seh Z W and Zhang Q 2018 High-throughput theoretical optimization of the hydrogen evolution reaction on MXenes by transition metal modification *J. Mater. Chem.* A **6** 4271–8

[85] Ling C, Shi L, Ouyang Y, Chen Q and Wang J 2016 Transition metal-promoted V_2CO_2 (MXenes): a new and highly active catalyst for hydrogen evolution reaction *Adv. Sci.* **3** 1600180

[86] Chen Z, Huang S, Huang B, Wan M and Zhou N 2020 Transition metal atoms implanted into MXenes (M_2CO_2) for enhanced electrocatalytic hydrogen evolution reaction *Appl. Surf. Sci.* **509** 145319

[87] Ramalingam V *et al* 2019 Heteroatom-mediated interactions between ruthenium single atoms and an MXene support for efficient hydrogen evolution *Adv. Mater.* **31** 1903841

[88] Zhang X, Shao B, Sun Z, Gao Z, Qin Y, Zhang C, Cui F and Yang X 2020 Platinum nanoparticle-deposited $Ti_3C_2T_x$ MXene for hydrogen evolution reaction *Ind. Eng. Chem. Res.* **59** 1822–8

[89] Ding B, Ong W-J, Jiang J, Chen X and Li N 2020 Uncovering the electrochemical mechanisms for hydrogen evolution reaction of heteroatom doped M_2C MXene (M = Ti, Mo) *Appl. Surf. Sci.* **500** 143987

[90] Le T A, Bui Q V, Tran N Q, Cho Y, Hong Y, Kawazoe Y and Lee H 2019 Synergistic effects of nitrogen doping on MXene for enhancement of hydrogen evolution reaction *ACS Sustain. Chem. Eng.* **7** 16879–88

[91] Zhou S, Yang X, Pei W, Liu N and Zhao J 2018 Heterostructures of MXenes and N-doped graphene as highly active bifunctional electrocatalysts *Nanoscale* **10** 10876–83

[92] Pang S-Y, Wong Y-T, Yuan S, Liu Y, Tsang M-K, Yang Z, Huang H, Wong W-T and Hao J 2019 Universal strategy for HF-free facile and rapid synthesis of two-dimensional MXenes as multifunctional energy materials *J. Am. Chem. Soc.* **141** 9610–6

[93] Du C-F, Dinh K N, Liang Q, Zheng Y, Luo Y, Zhang J and Yan Q 2018 Self-assemble and *in situ* formation of $Ni_{1-x}Fe_xPS_3$ nanomosaic-decorated MXene hybrids for overall water splitting *Adv. Energy Mater.* **8** 1801127

[94] Yu M, Wang Z, Liu J, Sun F, Yang P and Qiu J 2019 A hierarchically porous and hydrophilic 3D nickel–iron/MXene electrode for accelerating oxygen and hydrogen evolution at high current densities *Nano Energy* **63** 103880

[95] Zhao G, Rui K, Dou S X and Sun W 2018 Heterostructures for electrochemical hydrogen evolution reaction: a review *Adv. Funct. Mater.* **28** 1803291

[96] Lei Y *et al* 2017 Low-temperature synthesis of heterostructures of transition metal dichalcogenide alloys ($W_xMo_{1-x}S_2$) and graphene with superior catalytic performance for hydrogen evolution *ACS Nano* **11** 5103–12

[97] Jia Y *et al* 2017 A heterostructure coupling of exfoliated Ni–Fe hydroxide nanosheet and defective graphene as a bifunctional electrocatalyst for overall water splitting *Adv. Mater.* **29** 1700017

[98] Yang J, Voiry D, Ahn S J, Kang D, Kim A Y, Chhowalla M and Shin H S 2013 Two-dimensional hybrid nanosheets of tungsten disulfide and reduced graphene oxide as catalysts for enhanced hydrogen evolution *Angew. Chem. Int. Ed.* **52** 13751–4

[99] Tang C, Zhong L, Zhang B, Wang H-F and Zhang Q 2018 3D mesoporous van der Waals heterostructures for trifunctional energy electrocatalysis *Adv. Mater.* **30** 1705110

[100] Xiao W, Liu P, Zhang J, Song W, Feng Y P, Gao D and Ding J 2017 Dual-functional N dopants in edges and basal plane of MoS_2 nanosheets toward efficient and durable hydrogen evolution *Adv. Energy Mater.* **7** 1602086

[101] Zhang H, Tian Y, Zhao J, Cai Q and Chen Z 2017 Small dopants make big differences: enhanced electrocatalytic performance of MoS_2 monolayer for oxygen reduction reaction (ORR) by N- and P-doping *Electrochim. Acta* **225** 543–50

[102] Duan J, Chen S, Jaroniec M and Qiao S Z 2015 Porous C_3N_4 nanolayers@N-graphene films as catalyst electrodes for highly efficient hydrogen evolution *ACS Nano* **9** 931–40

[103] Voiry D *et al* 2016 The role of electronic coupling between substrate and 2D MoS_2 nanosheets in electrocatalytic production of hydrogen *Nat. Mater.* **15** 1003–9

[104] Ma T Y, Dai S, Jaroniec M and Qiao S Z 2014 Graphitic carbon nitride nanosheet–carbon nanotube three-dimensional porous composites as high-performance oxygen evolution electrocatalysts *Angew. Chem. Int. Ed.* **53** 7281–5

[105] Ma T Y, Dai S and Qiao S Z 2016 Self-supported electrocatalysts for advanced energy conversion processes *Mater. Today* **19** 265–73

[106] Su M, He C, Zhu L, Sun Z, Shan C, Zhang Q, Shu D, Qiu R and Xiong Y 2012 Enhanced adsorption and photocatalytic activity of BiOI–MWCNT composites towards organic pollutants in aqueous solution *J. Hazard. Mater.* **229–30** 72–82

[107] Li D J, Maiti U N, Lim J, Choi D S, Lee W J, Oh Y, Lee G Y and Kim S O 2014 Molybdenum sulfide/N-doped CNT forest hybrid catalysts for high-performance hydrogen evolution reaction *Nano Lett.* **14** 1228–33

[108] Chen P, Xiao T-Y, Qian Y-H, Li S-S and Yu S-H 2013 A nitrogen-doped graphene/carbon nanotube nanocomposite with synergistically enhanced electrochemical activity *Adv. Mater.* **25** 3192–6

[109] Chen S, Duan J, Jaroniec M and Qiao S-Z 2014 Nitrogen and oxygen dual-doped carbon hydrogel film as a substrate-free electrode for highly efficient oxygen evolution reaction *Adv. Mater.* **26** 2925–30

[110] Cherevko S *et al* 2016 Oxygen and hydrogen evolution reactions on Ru, RuO_2, Ir, and IrO_2 thin film electrodes in acidic and alkaline electrolytes: a comparative study on activity and stability *Catal. Today* **262** 170–80

[111] Fei H *et al* 2015 Atomic cobalt on nitrogen-doped graphene for hydrogen generation *Nat. Commun.* **6** 8668

[112] Cheng Y, Lu S, Liao F, Liu L, Li Y and Shao M 2017 Rh-MoS_2 nanocomposite catalysts with Pt-like activity for hydrogen evolution reaction *Adv. Funct. Mater.* **27** 1700359

[113] Wang X, Wang L, Zhao F, Hu C, Zhao Y, Zhang Z, Chen S, Shi G and Qu L 2015 Monoatomic-thick graphitic carbon nitride dots on graphene sheets as an efficient catalyst in the oxygen reduction reaction *Nanoscale* **7** 3035–42

[114] Dou S, Tao L, Huo J, Wang S and Dai L 2016 Etched and doped Co_9S_8/graphene hybrid for oxygen electrocatalysis *Energy Environ. Sci.* **9** 1320–6

[115] Luo Y, Tang L, Khan U, Yu Q, Cheng H-M, Zou X and Liu B 2019 Morphology and surface chemistry engineering toward pH-universal catalysts for hydrogen evolution at high current density *Nat. Commun.* **10** 269

[116] Subbaraman R, Tripkovic D, Strmcnik D, Chang K-C, Uchimura M, Paulikas A P, Stamenkovic V and Markovic N M 2011 Enhancing hydrogen evolution activity in water splitting by tailoring Li^+-$Ni(OH)_2$-Pt interfaces *Science* **334** 1256–60

[117] Danilovic N, Subbaraman R, Strmcnik D, Chang K-C, Paulikas A P, Stamenkovic V R and Markovic N M 2012 Enhancing the alkaline hydrogen evolution reaction activity through the bifunctionality of $Ni(OH)_2$/metal catalysts *Angew. Chem. Int. Ed.* **51** 12495–8

[118] Yu X, Zhao J, Zheng L-R, Tong Y, Zhang M, Xu G, Li C, Ma J and Shi G 2018 Hydrogen evolution reaction in alkaline media: alpha- or beta-nickel hydroxide on the surface of platinum? *ACS Energy Lett.* **3** 237–44

[119] Zhang B, Liu J, Wang J, Ruan Y, Ji X, Xu K, Chen C, Wan H, Miao L and Jiang J 2017 Interface engineering: the $Ni(OH)_2$/MoS_2 heterostructure for highly efficient alkaline hydrogen evolution *Nano Energy* **37** 74–80

[120] Chen L, Zhang J, Ren X, Ge R, Teng W, Sun X and Li X 2017 A $Ni(OH)_2$–CoS_2 hybrid nanowire array: a superior non-noble-metal catalyst toward the hydrogen evolution reaction in alkaline media *Nanoscale* **9** 16632–7

[121] Zhang X and Liang Y 2018 Nickel hydr(oxy)oxide nanoparticles on metallic MoS_2 nanosheets: a synergistic electrocatalyst for hydrogen evolution reaction *Adv. Sci.* **5** 1700644

[122] Liu S-Q, Wen H-R, Ying G, Zhu Y-W, Fu X-Z, Sun R and Wong C-P 2018 Amorphous Ni $(OH)_2$ encounter with crystalline CuS in hollow spheres: a mesoporous nano-shelled heterostructure for hydrogen evolution electrocatalysis *Nano Energy* **44** 7–14

[123] Zhang X, Zhu S, Xia L, Si C, Qu F and Qu F 2018 $Ni(OH)_2$–Fe_2P hybrid nanoarray for alkaline hydrogen evolution reaction with superior activity *Chem. Commun.* **54** 1201–4

[124] Hu J, Zhang C, Jiang L, Lin H, An Y, Zhou D, Leung M K H and Yang S 2017 Nanohybridization of MoS_2 with layered double hydroxides efficiently synergizes the hydrogen evolution in alkaline media *Joule* **1** 383–93

[125] Gao M, Chen L, Zhang Z, Sun X and Zhang S 2018 Interface engineering of the $Ni(OH)_2$–Ni_3N nanoarray heterostructure for the alkaline hydrogen evolution reaction *J. Mater. Chem.* A **6** 833–6

[126] Xie L, Ren X, Liu Q, Cui G, Ge R, Asiri A M, Sun X, Zhang Q and Chen L 2018 A Ni $(OH)_2$–PtO_2 hybrid nanosheet array with ultralow Pt loading toward efficient and durable alkaline hydrogen evolution *J. Mater. Chem.* A **6** 1967–70

[127] Xing Z, Han C, Wang D, Li Q and Yang X 2017 Ultrafine Pt nanoparticle-decorated Co $(OH)_2$ nanosheet arrays with enhanced catalytic activity toward hydrogen evolution *ACS Catal.* **7** 7131–5

[128] Yin H, Zhao S, Zhao K, Muqsit A, Tang H, Chang L, Zhao H, Gao Y and Tang Z 2015 Ultrathin platinum nanowires grown on single-layered nickel hydroxide with high hydrogen evolution activity *Nat. Commun.* **6** 6430

[129] Quaino P, Juarez F, Santos E and Schmickler W 2014 Volcano plots in hydrogen electrocatalysis—uses and abuses *Beilstein J. Nanotechnol.* **5** 846–54

[130] Yang L, Zhou W, Hou D, Zhou K, Li G, Tang Z, Li L and Chen S 2015 Porous metallic MoO_2-supported MoS_2 nanosheets for enhanced electrocatalytic activity in the hydrogen evolution reaction *Nanoscale* **7** 5203–8

[131] Zhou W, Hou D, Sang Y, Yao S, Zhou J, Li G, Li L, Liu H and Chen S 2014 MoO_2 nanobelts@nitrogen self-doped MoS_2 nanosheets as effective electrocatalysts for hydrogen evolution reaction *J. Mater. Chem.* A **2** 11358–64

[132] Zhang X, Ding P, Sun Y, Wang Y, Wu Y and Guo J 2017 Shell–core MoS_2 nanosheets@Fe_3O_4 sphere heterostructure with exposed active edges for efficient electrocatalytic hydrogen production *J. Alloys Compd.* **715** 53–9

[133] Wu C-L, Huang P-C, Brahma S, Huang J-L and Wang S-C 2017 MoS_2-MoO_2 composite electrocatalysts by hot-injection method for hydrogen evolution reaction *Ceram. Int.* **43** S621–7

[134] Pu E, Liu D, Ren P, Zhou W, Tang D, Xiang B, Wang Y and Miao J 2017 Ultrathin MoO_2 nanosheets with good thermal stability and high conductivity *AIP Adv.* **7** 025015

[135] Opitz M, Go D, Lott P, Müller S, Stollenwerk J, Kuehne A J C and Roling B 2017 On the interplay of morphology and electronic conductivity of rotationally spun carbon fiber mats *J. Appl. Phys.* **122** 105104

[136] Jin B, Zhou X, Huang L, Licklederer M, Yang M and Schmuki P 2016 Aligned MoO_x/MoS_2 core–shell nanotubular structures with a high density of reactive sites based on self-ordered anodic molybdenum oxide nanotubes *Angew. Chem. Int. Ed.* **55** 12252–6

IOP Publishing

2D Materials as Electrocatalysts

Muhammad Ikram, Ali Raza, Jahan Zeb Hassan and Salamat Ali

Chapter 5

The electrochemical oxygen reduction reaction

Parts of this chapter have been reprinted with permission from [91]. Copyright 2023 Elsevier.

Oxygen electrocatalysis is a widely explored area in electrochemistry and catalysis owing to its implications for electrochemical energy conversion applications. The ORR shows additional inactive kinetics to the anodic ORR. For device activity, the ORR is reported to be kinetically restrictive. Thus, developing an operative ORR electrocatalyst is a vital objective for the development of fuel cell technology. This chapter presents electrochemical ORRs for various 2D materials (i.e. graphene-based electrocatalysts, TMDs, 2D MXenes, etc). Essentially, in aqueous solutions, the ORR strategy is highly irreversible and includes various reaction stages, including O-containing species (i.e. OOH*, OH*, and O*). Furthermore, kinetic energy blockades between two stages and the thermodynamic adsorption energetics of these intermediates regulate the apparent rates of the ORR. From the mechanistic and practical viewpoints, Pt metal, including its alloys, is considered a well-researched ORR electrocatalyst. Thus, numerous outstanding pieces of research on the rational design and expansion of these varieties of electrocatalysts exist. Thus far, a wide variety of 2D materials have shown favorable ORR activity, comprising graphene and its derivatives and other 2D materials. Here, we mainly focus on the most feasible and emerging 2D materials for ORR electrocatalysts because of their uniform physicochemical structures. Finally, a wide-ranging understanding of the challenges for 2D electrocatalysts for the ORR is provided.

5.1 Reaction kinetics and mechanisms

The ORR can follow a two-step $2e^-$ mechanism, producing H_2O_2 in acidic media or HO_2 in alkaline media as intermediate species. This can also be done more efficiently via a $4e^-$ route, where the acidic medium-based H_2O or alkaline medium-based OH are produced directly [1]. The adsorption energies of these intermediates and the

doi:10.1088/978-0-7503-5291-8ch5

reaction barriers on an electrode catalyst's surface are the main determinants of the selectivity in the direction of the $4e^-$ or $2e^-$ route. The direct $4e^-$ mechanism offers an increased current efficiency, thus it is favorable for fuel cell routes, whereas industrial application can be achieved using the $2e^-$ reduction pathway to acquire H_2O_2. Two potential molecular mechanisms (associative and dissociative pathways) comprise the numerous elementary processes for the $4e^-$ pathway. For alkaline solutions, these are compiled in table 5.1 [2]. The primary distinction between these two pathways is the type of intermediates used (associative: O*, OOH*, and OH*; dissociative: O* and OH*), which changes how the free energy route is built. It has been discovered that the initial O_2 dissociation energy barrier significantly affects whether a reaction proceeds along an associative or dissociative pathway [2]. According to DFT computational simulations, the dissociation barrier is relatively higher on carbon surfaces and thus unfavorable for $4e^-$ dissociative mechanisms [3]. As a result, practically all carbon materials have an empirically measured electron transfer number that is always less than four and frequently precisely two [4].

In contrast, the intense initial O_2 adsorption causes the ORR on metal surfaces (such as Pt) to often follow a dissociative pathway. Pt-based materials invariably illustrate selectivity for the $4e^-$ pathway, which supports this theoretical finding, and is also supported by experimental data. The OOH* intermediate's adsorption energy is another consideration; a $2e^-$ pathway is made possible by reasonably weak OOH* adsorption energy; on the other hand, the $4e^-$ pathway is made possible by an enormous OOH* adsorption energy [5].

The traditional Butler–Volmer equation can be used to explain the kinetic current–overpotential correlation for the ORR, which occurs on the surfaces of solid materials. The overpotential (η) in actual ORR situations is quite large. As a result, the reverse reaction is minimal, transforming the Butler–Volmer equation as

$$j_k = j_0 e^{n\alpha F\eta/RT}, \tag{5.1}$$

Table 5.1. Mechanisms of the ORR in alkaline conditions.

Mechanism	Reactions
Dissociative ($4e^-$)	$O_2 + 2^* \rightarrow 2O^*$
	$2O^* + 2e^- + 2H_2O \rightarrow 2OH^* + 2OH^-$
	$2OH^* + 2e^- \rightarrow 2OH^- \rightarrow 2^*$
Associative ($4e^-$)	$O_2 + {}^* \rightarrow O_2^*$
	$O_2^* + H_2O + e^- \rightarrow OOH^* + OH^-$
	$OOH^* + e^- \rightarrow O^* + OH^-$
	$O^* + H_2O + e^- \rightarrow OH^* + OH^-$
	$OH^* + e^- \rightarrow OH^- + {}^*$
Associative ($2e^-$)	$O_2 + {}^* \rightarrow O_2^*$
	$O_2^* + H_2O + e^- \rightarrow OOH^* + OH^-$
	$OOH^* + e^- \rightarrow OOH^- + {}^*$

where n represents the total electrons migrated in RDS, F corresponds to the Faraday constant, α indicates the electron migration coefficient whose value varies from 0.5–1 for ORR, T represents temperature (in kelvins), and finally, R is the gas constant. The J_o and kinetic ORR current (j_k) density are the current density of the forward reaction that must be equivalent to the backward reaction at equilibrium. The Tafel slope, η versus $\log(j_k)$, which is based on equation (5.1), reveals a linear connection having a $2.303RT/\alpha nF$ slope. At a definite value of η, high cathodic j_k can be achieved by taking a considerable value of J_o and/or a smaller Tafel slope. At this point, J_o is associated with the adsorption energetics for chains of reaction intermediates, as stated by a simple microkinetic model in electrochemistry industrialized by Nørskov *et al* [6]:

$$J_0 = nFk^0 C_{\text{total}}[(1 - \theta)^{1-\alpha}\theta^{\alpha}], \tag{5.2}$$

where k^0 is the standard rate constant, C_{total} is the total number of active sites, and the coverage of surface adsorbates θ is a quantity related to the highest free energy change of the whole reaction as

$$\theta = \frac{C_{\text{R}}}{C_{\text{total}}} = \frac{K}{1 + K}, \tag{5.3}$$

where the equilibrium constant K is related to the free energy change (ΔG) for the RDS as

$$K = \exp\left(-\frac{\Delta G}{k_{\text{B}}T}\right). \tag{5.4}$$

In the above equations (5.2–5.4), k_{B} is the Boltzmann constant. According to the electrode material and applied potential, the ORR typically has two values for the respective Tafel slopes, for instance, 60 and 120 mV dec^{-1}. The first number shows that RDS constant is a pseudo-two-electron reaction, whereas the latter value suggests that the RDS constant is the first-electron reduction of oxygen [7].

5.2 Graphene-based electrocatalysts

Many efforts have been undertaken towards developing graphene-based nanomaterials with improved activity for ORR depending on the unique characteristics of graphene. However, experimental and theoretical studies reveal improved chemical reactivity and electronic characteristics for heteroatom-doped (S, B and N) graphene nanomaterials and innovative functions have been developed. These materials are promising categories of (non-precious metal) electrocatalysts for the ORR. An outstanding illustration is N-doped graphene, which shows higher electrocatalytic activity and CO tolerance than traditional Pt catalysts for the ORR. This material thus shows tremendous promise for replacing noble metal catalysts in the direction of supplementary electrochemical devices and fuel cells [8, 9].

However, during fuel cell operating, metal oxide/metal catalysts regularly experience agglomeration, dissolution, and sintering, leading to catalyst degradation. Since the electron movement in the electrodes might be severely impeded during the course of the ORR, low electrical conductivity of the electrocatalysts often represents a considerable barrier to their usage. Given their superior thermal and electrical conductivities and previously noted high specific surface area, graphene-based materials have become novel catalytic supports.

The addition of graphene, a crucial element in these functional nanocomposites, could effectively boost the catalyst's surface area (i.e. electroactive area). Additionally, it improves conductivities that boost the catalytic activity along with durability. Significantly, GO, which is recognized as graphene's precursor, has strong solvent solubility and offers favorable conditions for developing hybrid nanocomposites based on graphene and their future use in advanced electrocatalysts for the ORR. For example, graphene can be used as an innovative catalytic support for the ORR and a high-performing metal-free electrocatalyst.

5.2.1 Pt-supported graphene catalysts

Ideal Pt catalyst supports offer several characteristics, including electrochemical stability, large surface areas, uniform nanoparticle distribution, and excellent electrical conductivity. All of these requirements can be satisfied by graphene, and its derivatives have shown promise as substrates for Pt electrocatalysts. The support of Pt catalysts directly on graphene-based systems or their attachment using surface functionalization techniques will be covered first in this section. The difference of functionalized graphene involves directly incorporating heteroatoms into graphitic planes. Next, we will address the graphene-based supports doped with N or S (heteroatom species). It is challenging to deposit evenly distributed nanoparticles of uniform size and this frequently leads to forming Pt agglomerates or restricting nanoparticle deposition to edge sites. The similar surface interactions and prominent features of CNTs can be comparable to this, and functionalization techniques are frequently used to help catalyst particles disperse evenly [10]. By functionalizing the surface of graphene nanoplatelets [11], also known as GO [12], with poly (diallyldimethylammonium chloride-PDDA), many researchers have a made variety of efforts towards the facilitation as well as strengthening of Pt nanoparticle deposition. Each approach verified the admirable dispersion of particles (2–3 nm), consistent with decent electrochemical stability using half-cell examinations. Prominently Pt surfaces are liable to contamination by polymers, surfactants, and tiny adsorbing molecules that limit the utility of this strategy because of the obstruction of active sites, which unavoidably takes place with trace amounts of remnant species [9, 13]. The development of an innovative strategy for functionalization was reported by Zeng *et al* [14] that utilizes GO functionalized with anion exchange ionomer (AEI) prepared in-house, which is commonly known as quaternary ammonia polymer (2,6-dimethyl-1,4-phenylene oxide). Since this AEI belongs to the same class of ionomers that must be added to alkaline-based fuel cells electrodes so that they can conduct hydroxide species grown at the anode, they deliberately chose to use AEI (comparable to the Nafion-ionomers utilized in polymer electrolyte fuel cells).

Furthermore, functionalization of GO with AEI enabled the production of catalysts with reduced-sized ordinary Pt nanoparticles (i.e. c. ~2.46 nm) compared to those grown without AEI (c. 3.95 nm), as shown in figure 5.1(a). Enhanced ORR activity can be acquired using Pt/AEI/rGO, consistent with half-cell electrochemical examinations in 0.1 M KOH, probably because of the higher ECSA, which arises from the smaller Pt nanoparticles. In addition, the alkaline exchange membrane (AEM) process has been used for the spray-coating of prepared catalysts before they were tested in AEM fuel cells. Compared to commercial Pt/C or AEI-free Pt/rGO catalysts, Pt/AEI/rGO shows increased activity, demonstrating the advantages of the smaller Pt particle size, as shown in figure 5.1(b). Further, AEI should not be removed; by adding more AEI to the electrode during preparation, even more performance enhancements were made. With additional AEI, the Pt/AEI/rGO-based membrane electrode assembly (MEA) was able to reach it is extreme (264.8 mW cm^{-2}) power density (figure 5.1(c)). He *et al* [15] reported a proton conducting ionomer often employed in polymer electrolyte fuel cells based on acidic conditions; further, perfluorosulfonic acid was functionalized on rGO to bind discrete Pt nanoparticles uniformly (1–4 nm).

Other research groups have employed GO during the deposition stage to produce excellent Pt nanoparticle dispersion. It is easy for GO to be dispersed in various solvents [16] due to the highly concentrated surface functional materials, as this allows appropriate connections with the catalytic precursor material. These surface species were observed to have poor electronic conductivity and disrupted graphitic structure [17]. However, with this method, solvothermal [18] or chemical procedures (such as $NaBH_4$) [19] can be used to reduce Pt nanoparticles and GO simultaneously. Interestingly, concurrently exfoliating and reducing GO [20] with heat will

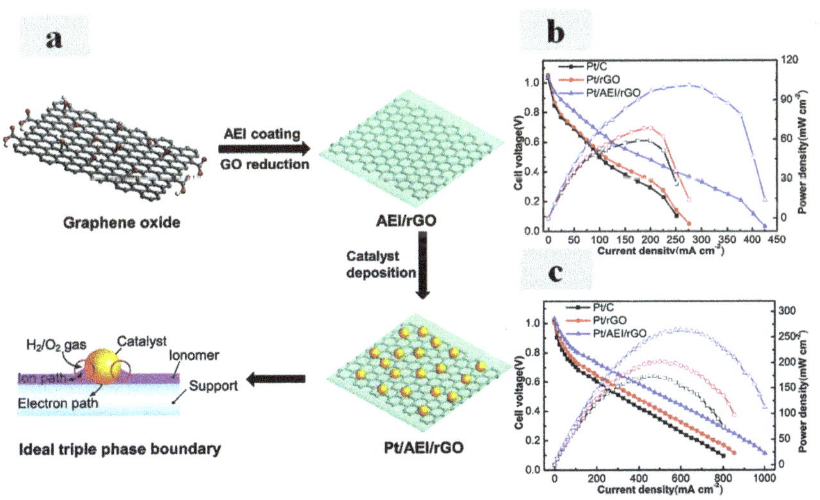

Figure 5.1. (a) The synthesis strategy for the Pt/AEI/rGO catalyst (based on the ideal triple level border). (b) Analysis of polarization (solid symbols) and power density (hollow symbols) curves for a H_2/O_2 AEM fuel cell deprived of an extra ionomer, and (c) catalyst layers consistent with an ionomer (a total of 20 wt.%). (Reproduced with permission from [14]. Copyright 2015 Elsevier.)

produce rGO with surface characteristics that favor the deposition of evenly distributed Pt nanoparticles [21]. With the help of XPS, Higgins *et al* [21] discovered that rGO produced using this method had an oxygen surface concentration of 7.49 wt. %, that the authors hypothesized to be the nucleation spots for well-isolated particles of approximately 2.25 nm in size. Tan *et al* [22] used commercially available graphene-based materials that had been functionalized with carboxylic species to deposit Pt and Pt–Ni nanoparticles on the surface. The authors then used an ionic liquid (IL) ([MTBD][bmsi]) to impregnate their graphene nanosheet–Pt catalysts. This IL is less methanol-philic than water/aqueous electrolytes and is protic through higher oxygen solubility. By using this method, the generated graphene–Pt–IL showed a unique Pt mass-based experiment (320 mA mg_{Pt}^{-1}) and more remarkable activity (870 mA mg_{Pt}^{-1}) for graphene–Pt_3Ni–IL at 0.9 V against RHE using 0.1 M perchloric acid ($HClO_4$). According to the authors, 2D structures, hydrophobic connections between the two materials, and graphene-based supports allow them to trap the IL. When the IL is added to the catalyst, there are notable increases in ORR activity and alleviation of the consequences of methanol poisoning (since methanol tends to stay in the aqueous electrolyte phase). These findings are probably comparable to Snyder *et al* [23], who reported that nanoporous Pt–Ni catalysts impregnated with IL [MTBD][beti] exhibited ORR activity that was on par with the finest single-crystal catalysts available. The scaffold configuration created by 2D graphene nanosheets coated with Pt nanoparticles will demonstrate an IL confinement comparable to nanoporous metallic catalysts [9]. Thus, graphene-based materials with various arrangements of IL have the potential for additional activity enhancement, and it is likely possible to use three-dimensional scaffold structures to create a confinement effect. It is necessary to conduct more research in these areas.

It can be challenging to fully understand the support effects in graphene-based catalysts since the morphology and size of the nanoparticles are greatly influenced by the particular synthesis procedures used. These factors, which are well recognized to regulate ORR activity [24], make it difficult to compare the ORR activity of various catalytic supports directly. Guo *et al* [25] created distinct polyhedral Pt–iron nanoparticles with a ~7 nm diameter regulated in size and morphology to prevent these effects. The nanoparticles were then applied to carbon black or rGO via a solution phase self-assembly (rGO is produced by reducing GO by refluxing in DMF). Compared to carbon-supported particles created using reference and commercial Pt/C, the Pt–iron particles dispersed on rGO demonstrated improved ORR activity using 0.1 M $HClO_4$ (from the fuel cell store). The rGO-assisted catalyst demonstrated significantly better double-layer capacitance in the CV profile after 10 000 cycles, showing outstanding electrochemical stability (at 100 mV s^{-1}; value ranges from 0.4 to 0.8 V versus Ag/AgCl). Therefore, graphene can offer a greater electrochemically available surface area than traditional supports and this is frequently reported when employing graphene-based catalyst supports [26]. The tendency of graphene materials to ‘restack’ due to the potent π–π surface interactions is a well-known problem, particularly when drying samples. The result is an agglomeration structure that, even with Pt nanoparticles on the surface, is impervious to the mass movement of reactive species [27]. As shown by Xin *et al* [27], their Pt/graphene catalyst could be quickly re-dispersed in catalytic inks after being collected by freeze-drying. However, preparing fuel cell electrodes, which calls for the deposition of catalyst ink over a gas diffusion layer/electrolyte membrane monitored via solvent elimination, may not be viable using freeze-drying. In this case, restacking graphene-based catalysts will cause problems with product water removal and oxygen inaccessibility throughout the entire catalytic layer. When preparing a ring disc electrode (RDE), Li *et al* [28] looked into this phenomenon and devised a method to stop graphene clumping. They demonstrated enhanced ORR activity (using 0.1 M $HClO_4$) by combining carbon black with Pt/rGO catalyst. In particular, Pt/rGO did not exhibit the characteristic mass transport limitation of high surface area Pt catalysts at electrode potentials lower than 0.8 V versus RHE. An intriguing aspect of this work is that the authors discovered that adding carbon black to the electrode structure might significantly reduce ECSA losses caused by electrochemical cycling (i.e. 20 000 cycles) at 100 mV s^{-1} that ranges from 0.6 to 1.1 V versus RHE in air. The authors assert that carbon black can operate using a trap, which could nucleate and anchor every dissolved Pt particle that would move into the bulk electrolyte solution based on post-characterization of the catalyst. It has become clear from studies of this kind that adding graphene-based catalysts to an electrode configuration will probably cause mass transport issues.

Therefore, there is much room to develop creative electrocatalyst and electrode engineering approaches to deal with this issue. It might be significant if the stability and activity gained by adding carbon black to catalyst ink through half-cell analysis may be applied to the MEA process. It would undoubtedly offer a captivating way to utilize graphene's unique qualities in real-world fuel cell scenarios. The usage of sun-exfoliated graphene–CNT composites as Pt nanoparticle supports was examined by Aravind *et al* (figure 5.2(a)) [29]. Based on the ECSA ratio to the estimated specific surface area, they discovered through half-cell testing that composite support offered better exploitation of Pt at ~82%, as opposed to employing only CNT (70%) or graphene-based (49%) supports. TEM photographs are shown in figures 5.2(b) and (c). The Pt/GCNT composite offers a maximum power density (675 mW cm^{-2}) using the particular cell H_2/O_2 MEA, as opposed to the Pt/graphene catalyst's 355 mW cm^{-2}, despite having somewhat larger average sized Pt nanoparticles (i.e. 2.81 nm versus 2.10 nm) (figure 5.2(d)). Extremely conductive CNTs, which serve using spacers (rapid diffusion pathway) between graphene-based materials, were thought to be responsible for the improved electrical and mass transport within the electrode 781 mW cm^{-2}; more power density could be achieved by applying back pressure (1 atm).

Although several research reports, as described above, have developed and analysed Pt/graphene catalysts, the primary research aim is to determine in what way and to what degree the surface and structural characteristics of graphene-based materials affect the stability and activity of Pt catalysts. Unknown variables include the number of layers, nanosheet size, and degree of reduction. Comparing graphene-supported catalysts with other nanostructures and graphitic carbon supports (i.e. CNTs) might be beneficial. Understanding the (intrinsic) electrochemical stability of graphene supports compared to other carbon-based supports is also essential. Since

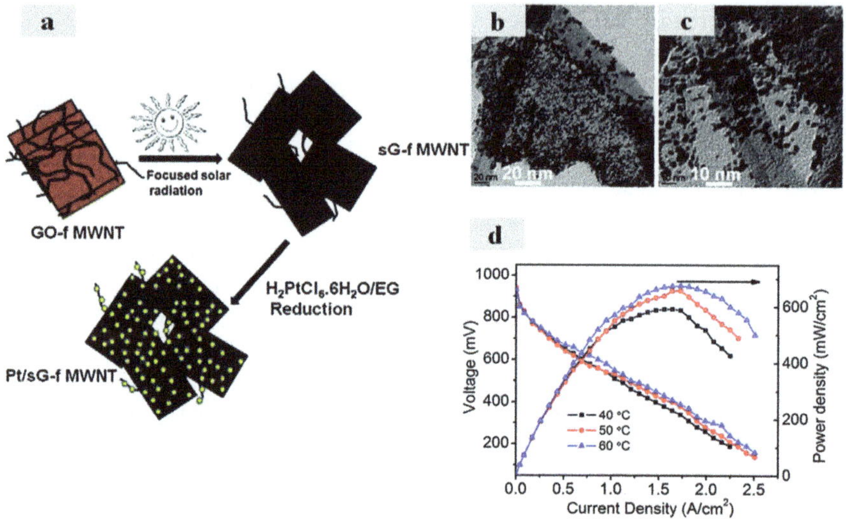

Figure 5.2. (a) Synthesizing Pt/sG-f (solar exfoliated graphene) multi-walled nanotubes. TEM photographs for (b) low and (c) high magnification of Pt/graphene–CNTs. (d) Test data of Pt/ graphene–CNT for H_2/O_2 MEA at different temperatures (without back pressure). (Reproduced with permission from [29]. Copyright 2011 The Royal Society of Chemistry.)

Pt nanoparticles are well recognized to accelerate carbon support corrosion rates, investigations on pristine graphene supports could be combined with Pt–graphene composites [9]. Engineering techniques can be used to create novel catalysts with exceptional stability if reduced carbon corrosion kinetics can be achieved on graphene support systems and we can understand how the features of the support affect these rates.

5.2.2 Pt-supported heteroatom-doped graphene catalysts

The significant impact that support materials have on the activity, stability, and characteristics of nanoparticle catalysts has, in turn, a significant impact in electrocatalysis. Incorporating heteroatom dopants in the graphitic carbon support is one way to modify the underlying catalyst–support interactions. When used properly, these dopant species have a threefold positive impact [9, 30]. First, the heteroatom material acts as nucleation spots, eliminating the need for surface functionalization techniques and enabling the addition of evenly sized Pt nanoparticles. Second, the dopant heteroatoms can advantageously alter the Pt nanoparticles' electrical structure. Weakening interactions with adsorbed surface species modify the Pt's surface characteristics and can increase ORR activity (i.e. OH_{ads}). Lastly, enhanced contact between the Pt nanoparticles and the doped carbon supports can result in a potent 'tethering' influence, which successfully mitigates Ostwald ripening and dissolution, two systematic processes of ECSA loss [21, 31]. The method that has been the most thoroughly studied is nitrogen doping of nanostructured carbons. Numerous studies have shown the advantages of employing N-CNTs through catalyst supports [32, 33]. Recently, scientists have focused their research on N-doped graphene-based species (NG), which have shown higher ORR stability and improved MEA activity compared to their pristine material. By heating a mixture of hydrogen and polypyrrole at 800 °C to exfoliate rGO, Vinayan *et al* [34] produced NG, which they then used to deposit Pt_3Co/Pt nanoparticles via refluxing the mixture in ethylene glycol or commonly in water. Nanoparticles with approximate sizes of 2.3 nm for Pt_3Co and 2.6 nm for Pt were acquired using XPS measurements that also gave a 5.9 at% nitrogen surface concentration. The catalyst was retained on a gas diffusion layer to grow the cathode structure using Pt loading (0.4 mg cm^{-2}). Improved activity for Pt/NG was achieved compared to commercial (E-TEK) Pt/C (512 mW cm^{-2} versus 241 mW cm^{-2}), and also for Pt_3Co/NG in contrast to Pt_3Co/C, with no back pressure and MEA evaluation by H_2/O_2 (805 mW cm^{-2} versus 379 mW cm^{-2}). Their advanced research [35] reveals that by using N-doped CNT/graphene composites as Pt_3Co nanoparticle supports, the high power density value might be elevated to 935 mW cm^{-2}. At 60 °C, using a cell voltage of 0.5 V, preliminary stability analyses of the catalyst layers in both reports [35] revealed that after 100 h, the current density remained unchanged for the NG-based catalyst support. Good stability was revealed that could be credited to boosted Pt support binding energies predicted via theoretical measurements for N-dopant systems. Furthermore, stability experiments can offer supplementary understanding of real stability aptitudes and the nature of any enrichment.

In contrast to standard zero-dimensional nanoparticles, Zhu *et al* [31] aimed to combine the advantages of NG with the stability and activity frequently reported for 1D Pt species (i.e. nanotubes/nanowires) [31, 36]. These improvements are due to lower surface energies and a higher number of surface Pt atoms with fewer connections to neighboring atoms. This reduces the binding strength with molecules that interfere with the ORR and helps prevent the breakdown or clumping of the catalyst. In parallel with the reduction and N-doping of GO, tellurium nanowire templates were produced at 180 °C via the hydrothermal strategy for four hours [31]. Following washing, an auxiliary reaction between the tellurium nanowires and Pt^{+4} grows Pt nanotubes (PtNTs) over the NG surface, as shown in figures 5.3(a)–(c). A higher mass activity of Pt-based ORR was acquired (350 mA mg_{pt}^{-1}) through PtNTs/NG in 0.1 M $HClO_4$ based on half-cell electrochemical examinations (figure 5.3(d)), which is an improvement for PtNTs and NG that were combined (PtNTs/NG-PM), as well as commercial Pt/C. The durability investigation regimen, which comprises 10 000 cycles at 50 mV s^{-1}, with between 0.6 and 1.2 V versus RHE in O-saturated electrolytes for the PtNTs/NG, revealed outstanding electrochemical stability. This method resulted in PtNTs/NG showing just a 7 mV drop in half-wave potential and retaining >90% of the initial ECSA (figures 5.3(f)–(g)).

Sulfur and Pt interact quite strongly, often leading to surface pollution or poisoning problems [37]. However, it is hypothesized that these interactions could be utilized to create favorable catalyst–support interactions if sulfur could be trapped inside the graphitic framework of the support material [9]. Using this as a foundation, we suggest sulfur-doped graphene-based materials (SG) as a fresh approach to ORR catalytic support. A mixture of phenyl disulfide and GO

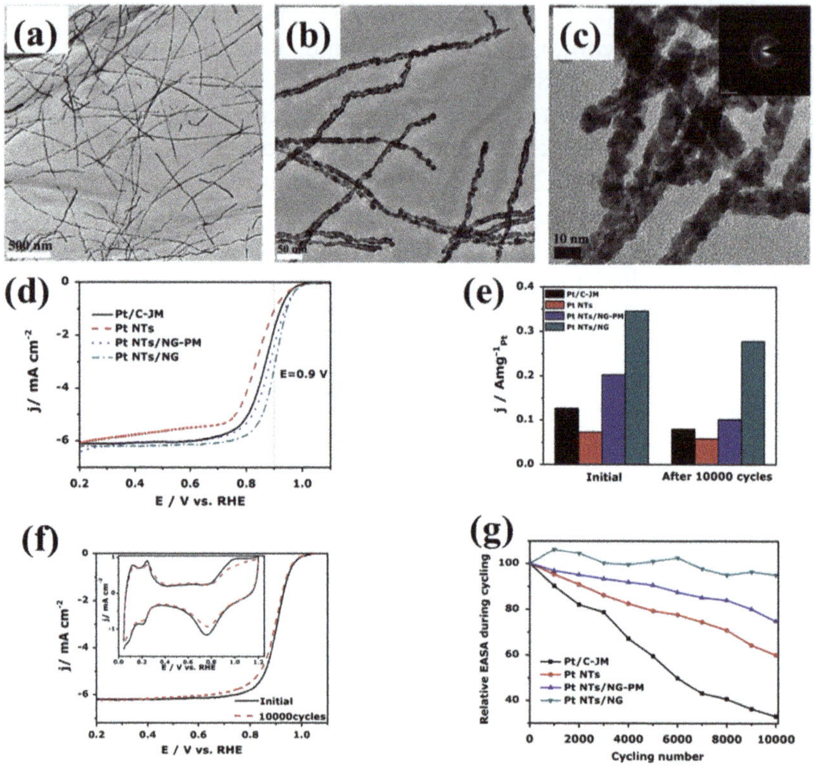

Figure 5.3. (a)–(c) TEM photographs for PtNTs/NG. (d) ORR polarization curves for several catalysts. (e) Mass activities before and after durability investigation for altered catalysts. (f) ORR polarization curves. Inset: CV curves before and after durability examination. (g) Relative ESCA residual for altered catalysts throughout durability examination. (Reproduced with permission from [31]. Copyright 2015 Elsevier.)

produced using a modified Hummer's technique was heated to 1000 °C to create SG [21]. The same method was used to prepare undoped rGO, and Pt nanoparticles were deposited using an ethylene glycol-based deposition process on both materials. On the surface of SG (Pt/SG, figure 5.4(a)) and rGO (denoted as Pt/graphene, not shown), uniformly sized nanoparticles were successfully produced, with average diameters based on TEM imaging of 2.10 and 2.25 nm, respectively. At a surface concentration of around 2.32 at%, sulfur was successfully incorporated into the structure of SG, where it principally occurs in the form of the five-membered ring thiophene, according to XPS (figure 5.4(b)). It is hypothesized that the sulfur dopant species' advantages enable the Pt nanoparticles' decreased size and efficient dispersion. Wang *et al* produced Pt nanowires (2–5 nm in diameter) on the surface of SG and rGO in an aqueous formic acid solution [37]. The entire SG surface could be covered with Pt nanowires of highly uniform size when utilizing modest Pt loadings and a diluted precursor solution; however, growth on rGO was restricted to small islands.

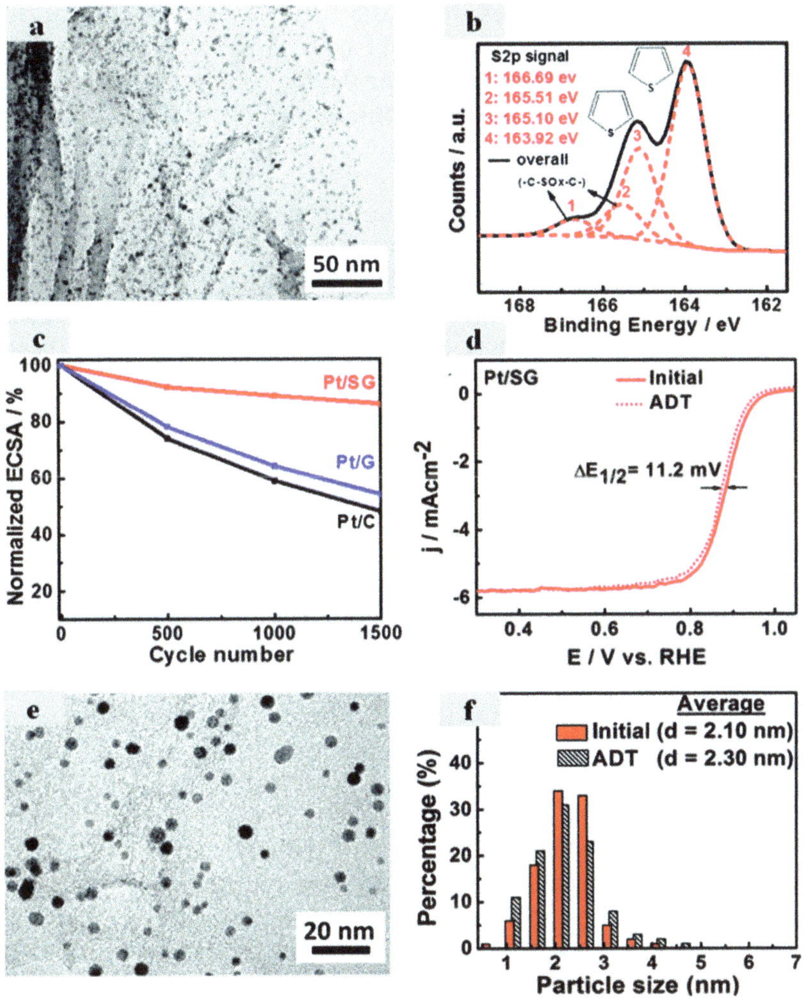

Figure 5.4. (a) TEM image of Pt/SG and (b) S2p spectra of SG. (c) Normalized ECSA results for durability testing. (d) ORR polarization curves initially and after durability testing for Pt/SG. (e) TEM image after durability testing and (f) particle size distribution initially and after ADT for Pt/SG. (Reproduced with permission from [21]. Copyright 2014 John Wiley and Sons.)

Using half-cell electrochemical testing in 0.1 M $HClO_4$, the ORR activity and electrochemical stability of Pt/SG, Pt/graphene, and commercial Pt/C were assessed. When compared to Pt/ graphene (101 mA mg_{pt}^{-1}) and commercial Pt/C, the Pt-based mass activity of the Pt/SG was 139 mA mg_{pt}^{-1} at 0.9 V versus RHE (121 mA mg_{pt}^{-1}). Although this can be related to variations in Pt nanoparticle size, Pt/SG showed the highest specific activity among the tested catalysts. Even though the Pt nanoparticles' average size was the smallest possible, it is frequently associated with a decline in specific activity due to the increased percentage of step and kink sites [24]. In nitrogen-saturated conditions, Pt/SG showed remarkable ECSA and ORR activity retention after 1500 cycles from 0.05 to 1.3 V versus RHE at 50 mV s^{-1} (figures 5.4(c) and (d)). The development and aggregation of Pt nanoparticles on the surface of Pt were significantly reduced after the durability test, with an increase in average size from 2.10 nm to only 2.30 nm, according to TEM characterization (figures 5.4(e) and (f)). Compared to Pt/graphene and commercial Pt/C, which both showed an increase in average nanoparticle size from 2.25 to 3.80 nm and from 2.15 to 5.25 nm, respectively, this was a significant improvement. To shed light on these intriguing results, computational simulations study the interactions between a model sulfur-doped graphene surface structure and individual Pt clusters of 13 atoms. Interestingly, higher cohesive and adsorptive energies between the Pt atoms and clusters and the SG were seen when undoped graphene was used as a benchmark.

In an extension of this work, one-dimensional Pt nanowires (PtNWs) solvothermally doped with sulfur have been synthesized to achieve the advantages of sulfur doping. On the surface of the SG, nanowires of micrometer length made of many single crystallites aligned in the ⟨111⟩ direction were produced successfully. The Pt-based mass activity of the as-prepared PtNW/SG was higher than that of PtNWs grown directly on the surface of rGO (PtNW/G 167 mA mg mg_{pt}^{-1} at 0.9 V versus RHE in 0.1 M $HClO_4$). These catalysts were put through an accelerated durability test that included 3000 cycles at 50 mV s^{-1} between 0.05 and 1.5 V versus RHE. These highly demanding conditions cover the potential range in which Pt catalyst degradation and support corrosion will occur in appreciable quantities [38]. PtNW/SG only had a 22 mV decrease in ORR half-wave potential, compared to an 86 mV loss for PtNW/G and a nearly total performance loss for commercial Pt/C. These findings demonstrate that the sulfur doping of graphene-based substrates can favorably influence catalytic activity and stability, independent of Pt morphology (i.e. nanoparticles or nanowires). The cause of these effects in the PtNW/SG still needs to be determined because it is unclear whether the sulfur species interacts specifically with the catalyst or if it contributes to the nucleation and growth of Pt, giving nanowires their favorable surface structure and morphology for ORR electrocatalysis. This uncertainty also applies to the PtNTs/NG catalysts that were discussed previously [31]. Therefore, well-planned studies that can demonstrate and pinpoint the precise impact dopant species have on ORR performance will be essential for learning information that can help in the innovative design of 1D Pt/doped-graphene catalysts with unusual activity and stability.

Pt catalysts supported by graphene doped with heteroatoms (N and S) have shown tremendous potential thus far. Despite this, there are still only a few research teams using this strategy, and there is still much to add to the body of existing studies and to advance scientific knowledge [9]. For example, no systematic research has been done to determine how the concentrations of various nitrogen functionalities in graphene affect the ORR activity and stability of deposited Pt nanoparticles. The foundation for such investigations is provided by the ease with which nitrogen speciation and contents can be modified using various synthetic procedures and pyrolysis temperatures [39]. The same is true for sulfur doping, where it is difficult to determine precisely how the presence of sulfur affects graphene characteristics and subsequent catalyst–support interactions. Although it is well known that the initiation of carbon corrosion frequently occurs in oxygen functional groups, it is yet unknown whether this problem may be solved by using other dopants. Finally, fabrication into functional electrode structures and their integration into MEAs for performance validation are required in order to translate any remarkable properties into useful accomplishments, regardless of the success in designing and developing heteroatom-doped graphene-supported Pt catalysts in terms of activity and stability in half-cell electrochemical tests.

5.2.3 Non-Pt-based precious metal catalysts on graphene

Pd-based nanoparticles have been investigated as potential Pt-free metal-based catalysts for electrochemical oxygen reduction, primarily in alkaline conditions, due to their equivalent catalytic activity to Pt but lower cost [40, 41]. Therefore, their coupling with graphene substrates is anticipated to provide inexpensive electrocatalysts with higher activity and endurance. To date, only a few investigations have concentrated on ORR graphene-supported Pd-based electrocatalysts. Investigated in this context were the electrocatalytic properties of graphene-supported CuPd alloyed nanoparticles [42]. The electronic structure of these catalysts, as well as their catalytic efficacy, phase structure, and size, were significantly influenced by the Cu: Pd ratio. Electrochemical impedance spectroscopy (EIS) was used to investigate these hybrids' interfacial electronic transfer resistance, and the results showed a significant association with the Cu: Pd ratio. The Cu_3Pd/graphene hybrid is a promising candidate for fuel cell applications because the 5.3 nm Cu_3Pd nanoparticles supported on graphene had the lowest interfacial resistance and the best ORR activity in alkaline electrolyte, close to that of the commercial Pt/C catalyst but with superior durability. In the same context, 6.8 nm PdCu nanoparticle–GO hybrids were created [43] using a one-pot solvothermal approach. It was discovered that their mass activity toward the ORR was 2.5 times more than that of the Pd black reference. The catalyst showed consistent performance after 10 000 cycles in an alkaline solution since there was little change in the half-wave potential. The PdCu/rGO catalyst was also demonstrated to be methanol-tolerant and selectively responsive to ORR in an alkaline medium. It has also been reported that monodisperse AuPd bimetallic alloyed nanoparticles (6–10 nm in diameter) were synthesized through solvothermal synthesis and used as ORR electrocatalysts [44]. Compared to the Pd black catalyst, LSV studies on RDE in an alkaline environment showed a 5.2-fold increase in mass activity toward ORR, better durability, and methanol tolerance. This enhancement is directly attributed to synergistic interaction of Au and Pd atoms and the rGO support, which facilitates electron transmission. Another illustration involves the investigation of graphene-supported PdAg nanorings as prospective cathode alkaline electrocatalysts with excellent methanol tolerance [45]. More specifically, a two-step synthesis procedure that involved the hydrothermal production of silver (Ag) nanoparticles on graphene nanosheets and a galvanic replacement reaction between palladium ions and Ag nanoparticles produced the hybrid. The consistent dispersion of 27.5 nm PdAg nanorings, with an average ring wall thickness of 5.5 nm, onto the graphene support was confirmed by HR-TEM images (figure 5.5(a)). CV tests in an O_2-saturated alkaline environment showed that the hybrid had better ORR activity than the Pd/C catalyst since a 24 times higher reduction in current density was reported. The hybrid is the optimal electrocatalyst for use in direct methanol fuel cells because, in contrast to the commercial catalyst, the current reduction density remained nearly unaltered

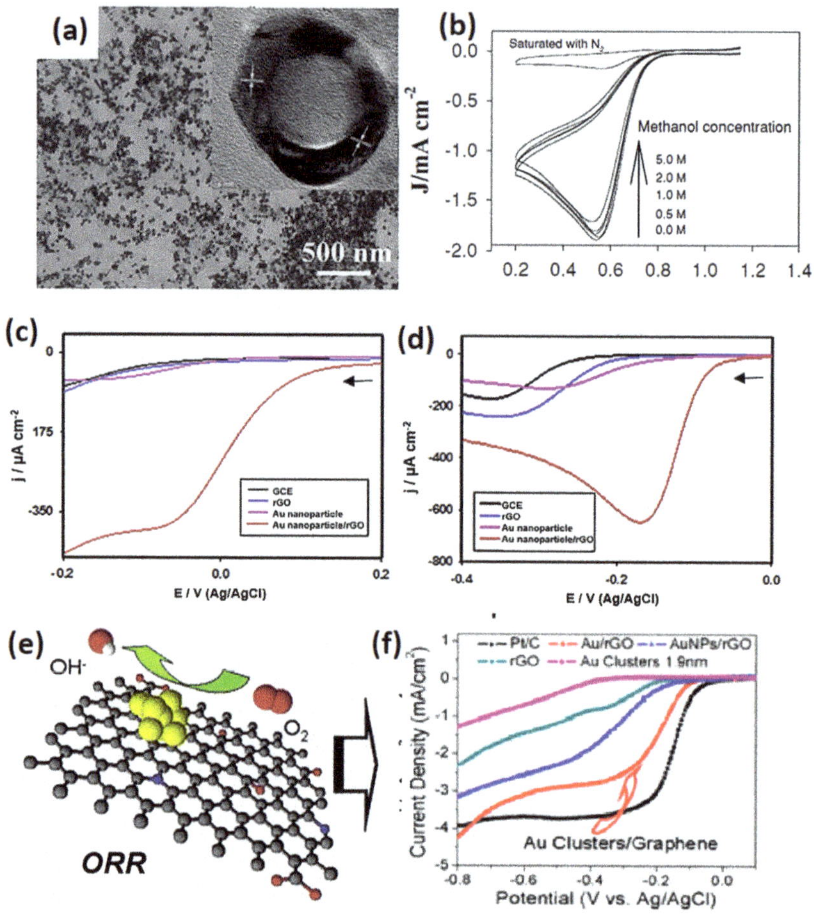

Figure 5.5. (a) HR-TEM photographs for PdAg nanorings on a graphene substrate and (b) CVs of PdAg/graphene hybrids (using a saturated N_2- or O_2 0.1 M KOH medium containing different quantities of methanol (0.0, 0.5, 1.0, 2.0, and 5.0 M)). (Reproduced with permission from [45]. Copyright 2013 John Wiley and Sons.) LSV curves for the pristine glassy carbon (GC) electrode in black, GC-adapted rGO electrode in blue, Au nanoparticle electrode in pink and Au nanoparticle/rGO electrode in red, in (c) O_2-saturated 0.1 M H_2SO_4 and (d) KOH. (Reproduced with permission from [48]. Copyright 2015 Elsevier.) (e) Schematics of surfactant-free Au cluster/graphene hybrids for high-performance ORR. (f) RDE curves of commercial Pt/C, Au/rGO hybrids, Au NP/rGO hybrids, rGO sheets, and Au clusters in O_2-saturated 0.1 M KOH at a scanning rate of 50 mV s^{-1} at 1600 rpm. (Reproduced with permission from [47]. Copyright 2012 American Chemical Society.)

with the addition of large methanol concentrations, up to 5.0 M (figure 5.5(b)). The ring-shaped nanoparticles' hollow structure offers a remarkably high surface area, and synergistic interaction between Ag and Pd atoms is responsible for the hybrid's overall increased ORR performance and methanol tolerance. Although bulk gold is inert, Au nanoparticles and nanoclusters have caught the attention of scientists as promising non-Pt-based electrocatalysts [46].

However, several obstacles must be overcome during their synthesis since the capping agents used to prevent NP dissolution and aggregation may hinder electron transmission, negatively impacting their electrocatalytic efficiency. Without capping agents, a simple technique for synthesizing Au nanoclusters supported on rGO sheets was achieved in this frame (figures 5.5(e) and (f)) [47]. Comparing the Au cluster/rGO hybrids to the thiol-capped Au clusters, Au NP/rGO hybrids, and rGO sheets, the Au cluster/rGO hybrids outperformed them in terms of electrocatalytic performance toward ORR. Additionally, they demonstrated an onset potential and ORR current density that were equivalent to the 20% Pt/C catalyst, but had better methanol tolerance and improved electrocatalytic stability. Their small size, lack of capping agent, and presence of the graphene support are all attributed to this property. The four-electron reduction pathway is facilitated by their tiny size, promoting O_2 dissociation. Additionally, the absence of the capping agent and the presence of the graphene support led to a more effective interfacial charge transfer. Recently, 6.8 nm Au nanoparticles dispersed on rGO sheets were made by electrochemically reducing GO and Au^{3+} [48]. In comparison to unsupported Au nanoparticles in acidic and alkaline solutions, it was shown that the Au/rGO hybrid had a tremendous positive reduction peak potential and 8 and 3.5 times higher mass activity (figures 5.5(c) and (d)).

It is crucial to emphasize that no binding agent, such as Nafion® ionomer, was used while applying the rGO decorated with Au nanoparticles directly to the electrode surface. Electrochemical impedance spectroscopy found negligible charge transfer resistance at the electrode/electrolyte interface for fuel cell applications. LSV studies on RDE in an oxygen-saturated alkaline solution revealed that the hybrid with the smallest Au nanoparticles appeared to have followed a four-electron reduction process, as opposed to the larger ones, which had followed a two-electron reduction pathway. This suggests that electrocatalytic behavior of nanoparticles can be regulated by varying their density and, as a result, their particle size. The difference in behavior is due to the smaller nanoparticles' higher number density on the rGO film. In addition to Pd and Au, Ir-based nanoparticles have been studied as potential electrocatalysts for ORR. Iridium–vanadium binary alloyed nanoclusters with a diameter of 2 nm were subsequently made, and the ORR performance of these nanoclusters was evaluated in an alkaline medium [49]. The data demonstrate that their ORR activity is closely related to their composition, with the Ir_2V/rGO exhibiting the best electrocatalytic behavior. For instance, compared to the Pt/C and Ir/rGO catalysts, the ORR current density of the Ir_2V/rGO catalyst was three and five times greater, respectively. The synthesized catalyst also showed higher methanol tolerance and good electrochemical stability in an alkaline medium.

5.2.4 Non-precious metal catalysts on graphene

Several studies have focused on creating graphene-supported non-precious metal-based ORR catalysts to lower the electrocatalysts' cost further. The most popular alternative ORR electrocatalysts are TMOs or chalcogenides. The fundamental problem with their use in fuel cells is that these materials typically experience agglomeration and disintegration under the demanding operating conditions of fuel cells, which leads to catalyst deterioration. By producing hybrids with higher electrical conductivity, their immobilization on a graphene support may boost their long-term stability and durability and improve their ORR activities [9, 50]. In this respect, several cobalt-based nanostructures placed on graphene substrates, including Co_3O_4 [51], Co/CoO [52], $Co_{1-x}S$ [52], and $MnCO_2O_4$ [53], are promising ORR electrocatalysts. For example, compared to conduction band-supported electrocatalysts, graphene-supported Co/CoO core–shell nanoparticles with an 8 nm Co core and a 1 nm CoO shell showed higher ORR activity in alkaline solution, which was comparable to that of commercial Pt/C catalyst and had improved long-term durability [54]. Their electrocatalytic performance was due to the Co core and CoO shell dimensions' tailoring and interactions with the graphene support. The Co_3O_4/rGO hybrid demonstrated significantly increased durability compared to both unsupported Co_3O_4 nanorods and the commercial Pt/C catalyst, and the same conclusions were drawn for Co_3O_4 nanorods uniformly scattered on rGO sheets [51]. Another non-precious metal-based electrocatalyst example is copper oxide (Cu_2O) nanoparticles. They are typically ineffective ORR catalysts due to their poor electrical conductivity but show promise when supported on a graphene substrate. It was shown [55] that Cu_2O/graphene hybrids showed enhanced stability and extraordinary tolerance to methanol and CO but with lesser catalytic activity towards ORR compared to that of the Pt/C catalyst.

5.3 2D-MXenes for ORR

In the past few years, fuel cells have presented prospective and sustainable alternative solutions to tackle increasingly critical energy and environmental challenges. This is owing to the benefits of fuel cells, which include their ecologically favorable features and their high level of reliability [56, 57]. The ORR that occurs at the cathode of a fuel cell is recognized as a rate-controlling period that can be governed by Pt-based materials that serve as electrocatalysts. These materials are scarce and expensive [58, 59]. On the other hand, Pt-based electrocatalysts have CO deactivation and time-dependent drift [60] that significantly impede the scaling up of fuel cell applications and their ability to remain stable. As a result, one of the most active areas of study is dedicated to using non-precious metal catalysts as ORR catalysts [61]. In recent years, research has been conducted on various MXene-based catalysts to boost the ORR. The development of the principle (metal-free) nano-catalyst was reported by Lin *et al* [62], that consists of 2D Ti_3C_2 MXene nanosheets obtained through bulk stiff ceramics using a merging TPAOH intercalation (disintegration) liquid exfoliation strategy and HF etching (delamination). This resulted in the nanosheets being able to stand independently (MAX phase). The

MXene-based Ti_3C_2 catalyst demonstrates stability and activity of the ORR in alkaline conditions, both of which are desired. The sandwich-like structure of the catalyst, which consists of atoms of titanium at the surface layer and atoms of carbon at the internal layer, makes it an ideal structure in for comprehending the ORR active sites of the 2D layered catalysts being discussed here. The effect of surface termination on the ORR strategy using MXene was described by Liu *et al* [63] via modeling Pt/v-$Ti_{n+1}C_nT_2$ and $Ti_{n+1}C_nT_2$ (T = F or O; n = 1–3,) surfaces. Their findings showed a substantial impact of surface termination towards the ORR. Based on these outcomes, it is expected that F-terminated surfaces will demonstrate a higher performance in the ORR than O-terminated surfaces.

On the other hand, due to the weaker chemical bonds, the F termination can have lower stability. The results provide a semi-quantitative study over Pt/v-$Ti_{n+1}C_nT_2$ or $Ti_{n+1}C_nT_2$ surfaces with different O/F termination ratios. This analysis may encourage experimentalists to build viable MXene systems to enhance ORR activity. Additionally, the development of an iron nitrogen carbon (Fe–N–C) catalyst that is highly stable and an active non-precious metal catalyst shows promise for use in the ORR. In this regard, the fabrication of an innovative Fe–N–C/MXene hybrid nanosheet was conveyed by Wen *et al* [64] by engaging dispersed MXene nanosheets as a carrier in their synthesis of the material. Here, the function of catalyst support was performed by MXene, which offers improved electrical channels and an exclusive 2D platform, while the robust Fe–N–C coating intimately bound to the MXene surface provides applicable ORR active centers, yielding an efficient electrocatalyst. Furthermore, electrochemical experiments indicate that the prepared hybrid nanosheet had a remarkable electrocatalytic activity as it displays a 25 mV higher half-wave potential than its Pt/C counterpart (0.814 V versus RHE). Additionally, it had excellent durability (2.6% decay) after 20 000 s cycles. This is a significant improvement over the 15.8% deterioration that is typical for commercial Pt/C catalysts.

Moreover, the fabrication of 2D/2D Fe–N–C/MXene-superlattice-like heterostructure was informed by Jiang *et al* [65], who established that metal clusters, as visualized in figure 5.6(a), such as Fe clusters (5.2 wt.%), were capable of assisting its development. The produced heterostructure could be engaged for an electrocatalytic ORR that had a positive potential (0.92 V), strong durability (20 h using alkaline media) and four-electron transfer channels (figures 5.6(b)–(e)). In the LSVs curve on a rotating disk electrode at 1600 rpm, the produced heterostructure demonstrated a small (0.92 V) onset potential (versus RHE, figure 5.6(b)) and increased its current density, which are common explanations for the ORR strategy. Lastly, 0.84 V half-wave potentials were acquired for an as-prepared product equivalent to Pt/C (0.9 V), overtaking various supplementary counterparts (figure 5.6(c)). Using $Ti_3C_2T_x$ for the first time as a support, Li *et al* [66] achieved a two-fold increase in the intrinsic ORR activity of bare iron phthalocyanine (FePc) catalyst. The coupling of FePc with $Ti_3C_2T_x$ MXene dramatically alters the spin configuration and delocalization of the Fe 3d electrons. As a result of these strong interactions, active FeN_4 sites can quickly absorb the reaction species that are taking part, which boosts ORR catalysis. The optimized catalyst has a particular ORR activity that is two and five times

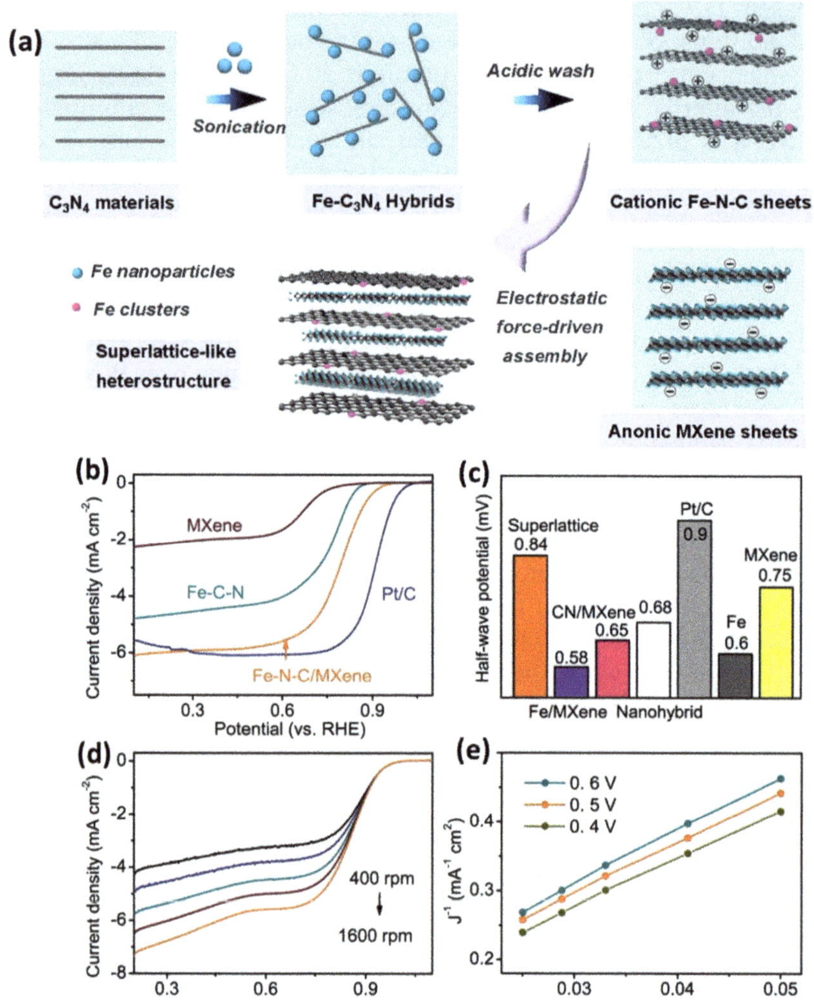

Figure 5.6. (a) The synthesis process for the Fe cluster is directed towards 2D/2D Fe–N–C/MXene-superlattice-like heterostructures. (b) LSV plots for O_2-saturated 0.1 M KOH electrolyte (5 mV s^{-1} scan rate). (c) Half-wave potential histogram. (d) Visualization of LSVs at different rotating speeds. (e) Plots for *K–L* at different applied potentials. (Reproduced with permission from [65]. Copyright 2020 American Chemical Society.)

greater than for pure FePc and Pt/C. It surpasses most of the Fe–N–C catalysts that have ever been reported under identical experimental circumstances.

Cobalt-based carbon nanotube (CoCNT) Ti_3C_2 nanosheet composites (CoCNT/Ti_3C_2) produced from ZIF-67/Ti_3C_2 were reported by Chen *et al* [67] for the electrochemical ORR. The Ti_3C_2 nanosheets act as a 2D conductive support and lead to a balanced trade-off between carbon graphitization and surface area toward forming CoCNTs. The elevated CoCNT/Ti_3C_2 had similar ORR activity compared to commercial Pt/C in addition to a more stable catalyst with 5.55 mA cm^{-2} current density (diffusion-limiting) and a 0.82 V half-wave potential (figures 5.6(d) and (e)). Yu *et al* [68] produced a straightforward strategy to alter g-C_3N_4 nanosheets through Ti_3C_2 nanoparticles. By changing the existing amount of Ti_3C_2 nanoparticles, numerous different g-C_3N_4/Ti_3C_2 heterostructures were produced. Compared with pure g-C_3N_4 nanosheets, the ORR of the new g-C_3N_4/Ti_3C_2 heterostructures exhibits a substantial improvement. The improvement might be recognized as higher conductivity of the highly dispersed Ti_3C_2 MXene nanoparticles and the increased O_2 adsorption brought about by electron coupling outcomes between Ti_3C_2 and g-C_3N_4. Direct reduction of $AgNO_3$ in an alkalization intercalated MXene solution, called alk-MXene, comprising polyvinylpyrrolidone, was used by Zhang *et al* [69] to generate a novel urchin-like MXene $Ag_{0.9}Ti_{0.1}$ nanowire composite (MXene/NW-$Ag_{0.9}Ti_{0.1}$), as shown in figure 5.7(a). The one-of-a-kind bimetallic nanowires make it easier to complete the four-electron migration strategy and display an extraordinary current density with a high degree of stability. They do this by providing numerous O-adsorption sites and reducing the distance that adsorbed oxygen must travel before it is released. The experiments' findings demonstrate that the half-wave and onset potentials of MXene/NW-$Ag_{0.9}Ti_{0.1}$ are greater than formerly described for Ag/C and Ag-based catalysts (20 wt.%), revealing remarkable electrocatalytic ORR activity (figures 5.7(b)–(e)). Xie *et al* proposed an extensively dispersed Pt nanoparticle scaffold on 2D titanium dioxide ($Ti_3C_2X_2$) nanosheets [70]. The stability measurements for a Pt catalyst supported by

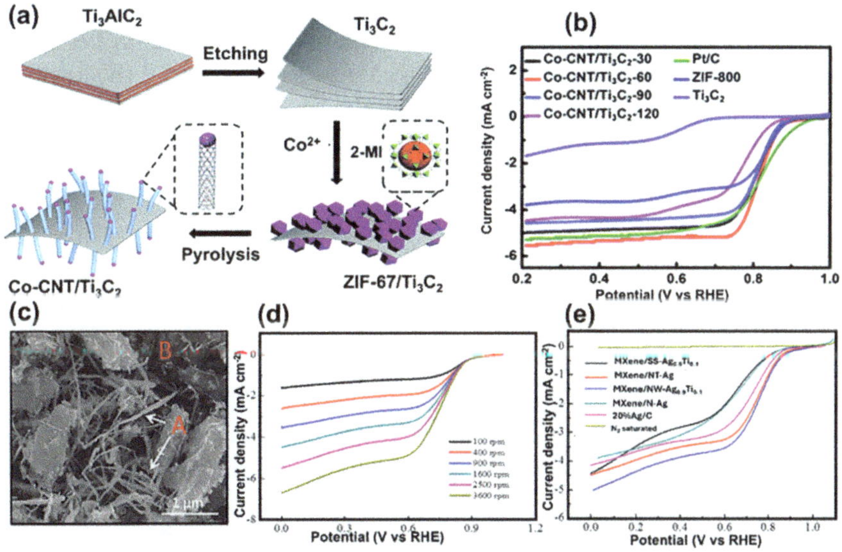

Figure 5.7. (a) Schematic diagram of the CoCNT/Ti_3C_2 composite synthesis strategy. (b) ORR activities via rotating disk electrodes for different catalysts. (Reproduced with permission from [67]. Copyright 2019 The Royal Society of Chemistry.) (c) FESEM photograph of MXene/NW-$Ag_{0.9}Ti_{0.1}$. Inset: A corresponds to the nanowire and B the nanosized particle. (d) LSV curves for the MXene/NW-$Ag_{0.9}Ti_{0.1}$ electrode at various rotation rates (5 mV s^{-1} sweep rate). (e) ORR polarization curves. (Reproduced with permission from [69]. Copyright 2016 American Chemical Society.)

$Ti_3C_2X_2$ are superior compared to commercial Pt/C. The half-wave potential for the ORR and Pt particle size did not exhibit any substantial decrease after accelerated durability testing, even when subjected to heavy cyclic loading [9]. This was the case even though the accelerated durability tests were performed under extreme conditions. Cheng *et al* [71] were motivated to study the ORR activity of MXene-based monolayer metal-modified Mo_2C (i.e. M_{ML}/Mo_2C, M = Ag, Au Pt, and Cu, Pd) via DFT. They were encouraged by the outstanding MXene characteristics as a substrate, and recent investigations towards the deposition of metal nanoparticles over MXenes. According to the findings, the Au monolayer-modified Mo_2C was the most effective ORR catalyst among the M_{ML}/Mo_2C (M = Cu, Pd, Pt, and Ag) samples. It also exhibited high stability and long-term performance. The ORR on Au_{ML}/Mo_2C is accomplished using a four-electron reduction route along with thermodynamic developments. Kinetics are recognized as similar or superior to the Pt/C, Pt (111), and Pt (100) catalysts used in commercial applications. The intense interaction between the metal and the carrier generated a massive electronic disturbance over the Au monolayer when subjected to Mo_2C. This led to increases in O-containing adsorption for compounds that caused improvement in its catalytic activity. Therefore, the expensive Pt/C catalyst is anticipated to be replaced with the less expensive Au_{ML}/Mo_2C as a prospective catalyst for ORR. It is recommended that research be conducted on nanocomposite catalyst scaffolds via $Ti_3C_2T_x$ MXene to gain considerably enhanced electrocatalytic activity for ORR. It is a substance that is high in titanium, and because of this it can withstand mechanical, acidic, and oxidative stress well. In addition to this, it possesses an excellent electrically conductive channel, which allows it to efficiently evade the collective carrier corrosion that occurs in noble metal/carbon catalysts. In addition, the advantageous characteristics of $Ti_3C_2T_x$ are not exclusive to the MXene family as a whole. We are looking forward to expanding our study into different MXenes since this will open up other doors for developing higher-activity ORR electrocatalysts.

5.4 Transition metal dichalcogenides

The 2D TMDs exhibit suitable bandgap energies controlled via layer stacking and coordination bonds between X and M atoms. As a result, various surface characteristics, catalytic activity, and electronic structures can be achieved [72]. Among the sixty TMDs that are now known, 2D MoS_2 has received much attention as an ORR catalyst. The atoms with low coordination at their edges and corners are the active sites. Due to their unsaturated coordinated environment (sulfide-rich), dangling bonds, and their location at the margins, Mo atoms passivated by S atoms are the most active locations [73]. However, pure MoS_2 has few active sites and weak electron transport characteristics. Numerous efforts have been made to increase conductivity and reveal more active edge sites. The activation of basal plane atoms can be done, for example, by heteroatom doping. O, B, P, and N were added to MoS_2 in a technique akin to the doping of graphene [74]. Additionally, it has been demonstrated that metallic atoms such as Fe, Co, and Ni help to enhance ORR activity [75, 76]. To augment the catalysts' ability to transport electrons, nanomaterials such as graphene are used as the support. The N-MoS_2/C composite materials were developed through thermal processing of thiourea, ammonium, melamine, molybdate, and pluronic F127 mixture [77]. The abundance of active sites produced by the rich defects on the N-doped MoS_2/carbon materials and higher electron conductivity increased the ORR activity.

5.5 Transition metal oxides

Owing to their excellent stability, low cost, and customizable activity, 2D TMOs have much potential in electrocatalysis. Layered and non-layered TMOs are the two primary kinds of 2D TMOs. The metal trioxides (such as WO_3, TaO_3, and MoO_3) that belong to the layered class have a layered assembly comparable to graphene and can be exfoliated conventionally using liquid exfoliation to generate 2D nanosheets [78, 79]. Non-layered TMOs frequently feature 3D crystal structures that are difficult to synthesize using top-down methods because they have strong chemical interactions connecting the various crystal layers. As a result, techniques such as salt-template approaches and ultrathin 2D precursor-based topotactic alteration approaches have been devised to synthesize these 2D non-layered nanomaterials [9, 80, 81]. For example, utilizing salt crystals as a template, Xiao *et al* devised a universal technique to synthesize a variety of 2D TMO nanosheets (WO_3, MnO, MoO_2, and MoO_3). Lattice matching and salt crystal geometry both influence and direct the adjacent development of TMOs on salt surfaces. Precursor depletion restricts any development that is perpendicular to the salt surface. The process even works with materials with non-layered crystallographic assemblies and produces TMO nanosheets with a nanometer thickness [81]. So far, investigations have shown that 2D TMOs are capable of a wide range of electrocatalytic processes in the carbon and water cycles. Most 2D TMOs can be employed directly as electrocatalysts without further alterations. However, the poor electron transport capabilities of 2D TMOs severely restrict their application. Therefore, maximizing their electrical conductivity is the best strategy to enhance their catalytic effectiveness. As an illustration, plasma-etched Co_3O_4 nanosheets demonstrated adequate OER activity because of their large exposed surface area, improved electrical conductivity, and active sites created by vacancies [82]. Defect engineering can also enhance the 2D TMOs' catalytic performance. For example, CO oxidation occurred more quickly in 2D CeO_2 nanosheets with surface-confined pits than in the material's bulk form [83].

5.6 Transition metal hydroxides

Layered transition-metal hydroxides (LMHs), which include LDHs and layered single metal hydroxides (LSHs), are a class of layered materials with metal hydroxyl host layers that may or may not have intercalated anions (figure 5.8) [84]. Investigations reveal that various materials with 2D assembly have distinctive characteristics, which are different to the respective bulk counterparts, which is a result of the rising interest in graphene. LDHs, in particular, consist of layers that resemble brucite and are positively charged, as well as an interlayer area that

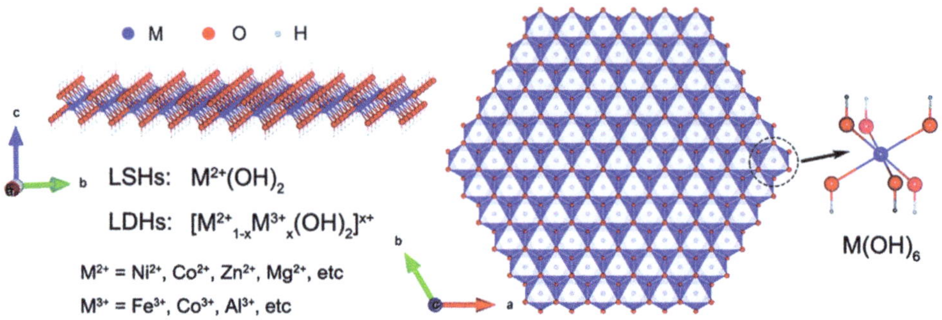

Figure 5.8. Illustration of a single layer (brucite or brucite-like) in layered metal hydroxides comprising LDHs and LSHs. (Reproduced with permission from [84]. Copyright 2016 The Royal Society of Chemistry.)

contains anions that balance charges and solvation molecules. Incorporating anions and metal cations also produces large interlayer spacing and distinct redox properties. LDHs have significant electrochemical performance as a result of all of these characteristics [85]. Several pieces of research have been done on 2D LMHs as favorable possibilities for electrocatalytic applications because of their unique combination of physiochemical characteristics and ultrathin nanostructures [86, 87]. Various 2D LMHs have been prepared recently, and most of them perform effectively electrocatalytically. For example, Song *et al* demonstrated that 2D single-layer LDHs perform better than bulk LDHs in OER in alkaline conditions [88]. The excellent electrical conductivity and amplified active site density of the 2D LDHs were credited with this improved OER performance.

Additionally, an electrode that combined an ultrathin layer of $Ni(OH)_2$ with a Pt substrate showed synergistic results that significantly enhanced the HER (using alkaline conditions) [89]. It was discovered that in this hybrid system, $Ni(OH)_2$ accelerated water dissociation and adsorption, whereas the Pt surface accordingly enhanced hydrogen recombination and desorption. There are very few investigations on LMH-based electrocatalysts for the carbon cycle. However, research has demonstrated that LDHs have CO_2 photocatalytic activity, opening up possibilities for their use in the ECR process [90].

5.7 Conclusion, challenges and perspectives

It has been demonstrated that the electrocatalytic material 2D MoS_2 has significant application possibilities for the HER; however, its development in ORRs has come to a standstill. Because most 2D MoS_2 materials have limited electron transport properties and substantial areas of unusable inert basal plane, their potential applications are severely constrained. This chapter offers a summary of the research progress that has been made in 2D MoS_2 in ORRs over the past few years, concentrating on detailed works that have been done on 2D MoS_2, ranging from synthesis strategies to ORR activity optimization, and recognized the primary challenges appearing for 2D MoS_2 that must be overcome in ORRs. The improvement of ORR activity for these materials is required to overcome the fundamental difficulties of electron transport along with growth in active sites. This article also discusses a number of potential solutions to these issues, including increasing the amount of heteroatom doping, designing composite materials, and constructing heterostructures, exposed edges, and stable phase transitions to improve electron transport, amongst other potential solutions. Because of these methodologies, developing higher-activity 2D MoS_2-based catalysts for ORR is now feasible. However, despite the many positive results that have been achieved, there remain plenty of problems that need to be solved:

1. The current bottom-up synthesis approach requires rather extreme conditions, while the aggregation of nanosheets produced through a top-down synthesis strategy can reduce the quality of the MoS_2. The fundamental necessity is to continue working toward producing vast quantities of high-quality 2D MoS_2 materials.
2. There is no general agreement about the precise pathways for pure MoS_2 and MoS_2-based catalysts for the ORR, particularly regarding the contributions of other elements (i.e. P, N, and O). For this reason, it is vital to carry out more profound research into the link between the composition of the structure and its performance to direct future development.
3. An efficient use of the enormous area basal plane while maintaining reactive inertness is needed. Most of the attention paid to this aspect concentrates on material recombination and heteroatom doping. However, precise doping control using heteroatoms has not been done successfully. The primary focus of future research will be on elucidating both the internal mechanism of improved ORR activity and the synergy between MoS_2 and other active materials in the ORR process.
4. The innate characteristics of MoS_2 may be brought out by engineering stable phase transitions and strain in the material. Therefore, in order to mitigate the detrimental impact that spontaneous phase change has on catalytic performance, we have to look into more practical techniques for synthesizing 1T-phase MoS_2 nanosheets that are stable over the long term.

Bibliography

[1] Dai L, Xue Y, Qu L, Choi H-J and Baek J-B 2015 Metal-free catalysts for oxygen reduction reaction *Chem. Rev.* **115** 4823–92

[2] Yu L, Pan X, Cao X, Hu P and Bao X 2011 Oxygen reduction reaction mechanism on nitrogen-doped graphene: a density functional theory study *J. Catal.* **282** 183–90

[3] Yan H J, Xu B, Shi S Q and Ouyang C Y 2012 First-principles study of the oxygen adsorption and dissociation on graphene and nitrogen doped graphene for Li–air batteries *J. Appl. Phys.* **112** 104316

[4] Chai G-L, Hou Z, Ikeda T and Terakura K 2017 Two-electron oxygen reduction on carbon materials catalysts: mechanisms and active sites *J. Phys. Chem.* C **121** 14524–33

[5] Jiao Y, Zheng Y, Jaroniec M and Qiao S Z 2014 Origin of the electrocatalytic oxygen reduction activity of graphene-based catalysts: a roadmap to achieve the best performance *J. Am. Chem. Soc.* **136** 4394–403

[6] Nørskov J K, Rossmeisl J, Logadottir A, Lindqvist L, Kitchin J R, Bligaard T and Jónsson H 2004 Origin of the overpotential for oxygen reduction at a fuel-cell cathode *J. Phys. Chem.* B **108** 17886–92

[7] Ge X, Sumboja A, Wuu D, An T, Li B, Goh F W T, Hor T S A, Zong Y and Liu Z 2015 Oxygen reduction in alkaline media: from mechanisms to recent advances of catalysts *ACS Catal.* **5** 4643–67

[8] Peterson A A, Abild-Pedersen F, Studt F, Rossmeisl J and Nørskov J K 2010 How copper catalyzes the electroreduction of carbon dioxide into hydrocarbon fuels *Energy Environ. Sci.* **3** 1311–5

[9] Perivoliotis D K and Tagmatarchis N 2017 Recent advancements in metal-based hybrid electrocatalysts supported on graphene and related 2D materials for the oxygen reduction reaction *Carbon* **118** 493–510

[10] Higgins D C, Meza D and Chen Z 2010 Nitrogen-doped carbon nanotubes as platinum catalyst supports for oxygen reduction reaction in proton exchange membrane fuel cells *J. Phys. Chem.* C **114** 21982–8

[11] Shao Y, Zhang S, Wang C, Nie Z, Liu J, Wang Y and Lin Y 2010 Highly durable graphene nanoplatelets supported Pt nanocatalysts for oxygen reduction *J. Power Sources* **195** 4600–5

[12] He W, Jiang H, Zhou Y, Yang S, Xue X, Zou Z, Zhang X, Akins D L and Yang H 2012 An efficient reduction route for the production of Pd–Pt nanoparticles anchored on graphene nanosheets for use as durable oxygen reduction electrocatalysts *Carbon* **50** 265–74

[13] Borup R *et al* 2007 Scientific aspects of polymer electrolyte fuel cell durability and degradation *Chem. Rev.* **107** 3904–51

[14] Zeng L, Zhao T S, An L, Zhao G, Yan X H and Jung C Y 2015 Graphene-supported platinum catalyst prepared with ionomer as surfactant for anion exchange membrane fuel cells *J. Power Sources* **275** 506–15

[15] He D, Cheng K, Li H, Peng T, Xu F, Mu S and Pan M 2012 Highly active platinum nanoparticles on graphene nanosheets with a significant improvement in stability and CO tolerance *Langmuir* **28** 3979–86

[16] Paredes J I, Villar-Rodil S, Martínez-Alonso A and Tascón J M D 2008 Graphene oxide dispersions in organic solvents *Langmuir* **24** 10560–4

[17] Dreyer D R, Park S, Bielawski C W and Ruoff R S 2010 The chemistry of graphene oxide *Chem. Soc. Rev.* **39** 228–40

[18] Chen D, Zhao X, Chen S, Li H, Fu X, Wu Q, Li S, Li Y, Su B-L and Ruoff R S 2014 One-pot fabrication of FePt/reduced graphene oxide composites as highly active and stable electrocatalysts for the oxygen reduction reaction *Carbon* **68** 755–62

[19] Seger B and Kamat P V 2009 Electrocatalytically active graphene-platinum nanocomposites. Role of 2-D carbon support in PEM fuel cells *J. Phys. Chem.* C **113** 7990–5

[20] Lee D U, Park H W, Higgins D, Nazar L and Chen Z 2013 Highly active graphene nanosheets prepared via extremely rapid heating as efficient zinc-air battery electrode material *J. Electrochem. Soc.* **160** F910–5

[21] Higgins D, Hoque M A, Seo M H, Wang R, Hassan F, Choi J-Y, Pritzker M, Yu A, Zhang J and Chen Z 2014 Development and simulation of sulfur-doped graphene supported platinum with exemplary stability and activity towards oxygen reduction *Adv. Funct. Mater.* **24** 4325–36

[22] Tan Y, Xu C, Chen G, Zheng N and Xie Q 2012 A graphene–platinum nanoparticles–ionic liquid composite catalyst for methanol-tolerant oxygen reduction reaction *Energy Environ. Sci.* **5** 6923–7

[23] Snyder J, Fujita T, Chen M W and Erlebacher J 2010 Oxygen reduction in nanoporous metal–ionic liquid composite electrocatalysts *Nat. Mater.* **9** 904–7

[24] Shao M, Peles A and Shoemaker K 2011 Electrocatalysis on platinum nanoparticles: particle size effect on oxygen reduction reaction activity *Nano Lett.* **11** 3714–9

[25] Guo S and Sun S 2012 FePt nanoparticles assembled on graphene as enhanced catalyst for oxygen reduction reaction *J. Am. Chem. Soc.* **134** 2492–5

[26] Hoque M A, Hassan F M, Higgins D, Choi J-Y, Pritzker M, Knights S, Ye S and Chen Z 2015 Multigrain platinum nanowires consisting of oriented nanoparticles anchored on sulfur-doped graphene as a highly active and durable oxygen reduction electrocatalyst *Adv. Mater.* **27** 1229–34

[27] Xin Y, Liu J-g, Zhou Y, Liu W, Gao J, Xie Y, Yin Y and Zou Z 2011 Preparation and characterization of Pt supported on graphene with enhanced electrocatalytic activity in fuel cell *J. Power Sources* **196** 1012–8

[28] Li Y, Li Y, Zhu E, McLouth T, Chiu C-Y, Huang X and Huang Y 2012 Stabilization of high-performance oxygen reduction reaction Pt electrocatalyst supported on reduced graphene oxide/carbon black composite *J. Am. Chem. Soc.* **134** 12326–9

[29] Jyothirmayee Aravind S S, Imran Jafri R, Rajalakshmi N and Ramaprabhu S 2011 Solar exfoliated graphene–carbon nanotube hybrid nano composites as efficient catalyst supports for proton exchange membrane fuel cells *J. Mater. Chem.* **21** 18199–204

[30] Zhou Y, Neyerlin K, Olson T S, Pylypenko S, Bult J, Dinh H N, Gennett T, Shao Z and O'Hayre R 2010 Enhancement of Pt and Pt-alloy fuel cell catalyst activity and durability via nitrogen-modified carbon supports *Energy Environ. Sci.* **3** 1437–46

[31] Zhu J, Xiao M, Zhao X, Liu C, Ge J and Xing W 2015 Strongly coupled Pt nanotubes/N-doped graphene as highly active and durable electrocatalysts for oxygen reduction reaction *Nano Energy* **13** 318–26

[32] Chen Y, Wang J, Liu H, Banis M N, Li R, Sun X, Sham T-K, Ye S and Knights S 2011 Nitrogen doping effects on carbon nanotubes and the origin of the enhanced electrocatalytic activity of supported Pt for proton-exchange membrane fuel cells *J. Phys. Chem.* C **115** 3769–76

[33] Chen Y, Wang J, Liu H, Li R, Sun X, Ye S and Knights S 2009 Enhanced stability of Pt electrocatalysts by nitrogen doping in CNTs for PEM fuel cells *Electrochem. Commun.* **11** 2071–6

[34] Vinayan B P, Nagar R, Rajalakshmi N and Ramaprabhu S 2012 Novel platinum cobalt alloy nanoparticles dispersed on nitrogen-doped graphene as a cathode electrocatalyst for PEMFC applications *Adv. Funct. Mater.* **22** 3519–26

[35] Vinayan B P and Ramaprabhu S 2013 Platinum–TM (TM = Fe, Co) alloy nanoparticles dispersed nitrogen doped (reduced graphene oxide-multiwalled carbon nanotube) hybrid structure cathode electrocatalysts for high performance PEMFC applications *Nanoscale* **5** 5109–18

[36] Higgins D C, Wang R, Hoque M A, Zamani P, Abureden S and Chen Z 2014 Morphology and composition controlled platinum–cobalt alloy nanowires prepared by electrospinning as oxygen reduction catalyst *Nano Energy* **10** 135–43

[37] Wang R, Higgins D C, Hoque M A, Lee D, Hassan F and Chen Z 2013 Controlled growth of platinum nanowire arrays on sulfur doped graphene as high performance electrocatalyst *Sci. Rep.* **3** 2431

[38] Higgins D C, Ye S, Knights S and Chen Z 2012 Highly durable platinum-cobalt nanowires by microwave irradiation as oxygen reduction catalyst for PEM fuel cell *Electrochem. Solid-State Lett.* **15** B83
[39] Wang H, Maiyalagan T and Wang X 2012 Review on recent progress in nitrogen-doped graphene: synthesis, characterization, and its potential applications *ACS Catal.* **2** 781–94
[40] Choi S-I, Shao M, Lu N, Ruditskiy A, Peng H-C, Park J, Guerrero S, Wang J, Kim M J and Xia Y 2014 Synthesis and characterization of Pd@Pt–Ni core–shell octahedra with high activity toward oxygen reduction *ACS Nano* **8** 10363–71
[41] Jiang G, Zhu H, Zhang X, Shen B, Wu L, Zhang S, Lu G, Wu Z and Sun S 2015 Core/shell face-centered tetragonal FePd/Pd nanoparticles as an efficient non-Pt catalyst for the oxygen reduction reaction *ACS Nano* **9** 11014–22
[42] Zheng Y, Zhao S, Liu S, Yin H, Chen Y-Y, Bao J, Han M and Dai Z 2015 Component-controlled synthesis and assembly of Cu–Pd nanocrystals on graphene for oxygen reduction reaction *ACS Appl. Mater. Interfaces* **7** 5347–57
[43] Lv J-J, Li S-S, Wang A-J, Mei L-P, Feng J-J, Chen J-R and Chen Z 2014 One-pot synthesis of monodisperse palladium–copper nanocrystals supported on reduced graphene oxide nanosheets with improved catalytic activity and methanol tolerance for oxygen reduction reaction *J. Power Sources* **269** 104–10
[44] Lv J-J, Li S-S, Wang A-J, Mei L-P, Chen J-R and Feng J-J 2014 Monodisperse Au-Pd bimetallic alloyed nanoparticles supported on reduced graphene oxide with enhanced electrocatalytic activity towards oxygen reduction reaction *Electrochim. Acta* **136** 521–8
[45] Liu M, Lu Y and Chen W 2013 PdAg nanorings supported on graphene nanosheets: highly methanol-tolerant cathode electrocatalyst for alkaline fuel cells *Adv. Funct. Mater.* **23** 1289–96
[46] Fujigaya T, Kim C, Hamasaki Y and Nakashima N 2016 Growth and deposition of Au nanoclusters on polymer-wrapped graphene and their oxygen reduction activity *Sci. Rep.* **6** 21314
[47] Yin H, Tang H, Wang D, Gao Y and Tang Z 2012 Facile synthesis of surfactant-free Au cluster/graphene hybrids for high-performance oxygen reduction reaction *ACS Nano* **6** 8288–97
[48] Govindhan M and Chen A 2015 Simultaneous synthesis of gold nanoparticle/graphene nanocomposite for enhanced oxygen reduction reaction *J. Power Sources* **274** 928–36
[49] Zhang R and Chen W 2013 Non-precious Ir–V bimetallic nanoclusters assembled on reduced graphene nanosheets as catalysts for the oxygen reduction reaction *J. Mater. Chem.* A **1** 11457–64
[50] Zhu C and Dong S 2013 Recent progress in graphene-based nanomaterials as advanced electrocatalysts towards oxygen reduction reaction *Nanoscale* **5** 1753–67
[51] Wang M, Huang J, Wang M, Zhang D, Zhang W, Li W and Chen J 2013 Co_3O_4 nanorods decorated reduced graphene oxide composite for oxygen reduction reaction in alkaline electrolyte *Electrochem. Commun.* **34** 299–303
[52] Wang H, Liang Y, Li Y and Dai H 2011 $Co_{1-x}S$–graphene hybrid: a high-performance metal chalcogenide electrocatalyst for oxygen reduction *Angew. Chem. Int. Ed.* **50** 10969–72
[53] Liang Y, Wang H, Zhou J, Li Y, Wang J, Regier T and Dai H 2012 Covalent hybrid of spinel manganese–cobalt oxide and graphene as advanced oxygen reduction electrocatalysts *J. Am. Chem. Soc.* **134** 3517–23
[54] Guo S, Zhang S, Wu L and Sun S 2012 Co/CoO nanoparticles assembled on graphene for electrochemical reduction of oxygen *Angew. Chem. Int. Ed.* **51** 11770–3

[55] Yan X-Y, Tong X-L, Zhang Y-F, Han X-D, Wang Y-Y, Jin G-Q, Qin Y and Guo X-Y 2012 Cuprous oxide nanoparticles dispersed on reduced graphene oxide as an efficient electrocatalyst for oxygen reduction reaction *Chem. Commun.* **48** 1892–4

[56] Yang L, Zeng X, Wang W and Cao D 2018 Recent progress in MOF-derived, heteroatom-doped porous carbons as highly efficient electrocatalysts for oxygen reduction reaction in fuel cells *Adv. Funct. Mater.* **28** 1704537

[57] Zhao D, Zhang S, Yin G, Du C, Wang Z and Wei J 2012 Effect of Se in Co-based selenides towards oxygen reduction electrocatalytic activity *J. Power Sources* **206** 103–7

[58] Sa Y J, Woo J and Joo S H 2018 Strategies for enhancing the electrocatalytic activity of M–N/C catalysts for the oxygen reduction reaction *Top. Catal.* **61** 1077–100

[59] Wu J *et al* 2019 Surface confinement assisted synthesis of nitrogen-rich hollow carbon cages with Co nanoparticles as breathable electrodes for Zn-air batteries *Appl. Catalysis* B **254** 55–65

[60] Liu C, Wang J, Li J, Luo R, Sun X, Shen J, Han W and Wang L 2017 Fe/N decorated mulberry-like hollow mesoporous carbon fibers as efficient electrocatalysts for oxygen reduction reaction *Carbon* **114** 706–16

[61] Chen Z, Higgins D, Yu A, Zhang L and Zhang J 2011 A review on non-precious metal electrocatalysts for PEM fuel cells *Energy Environ. Sci.* **4** 3167–92

[62] Lin H, Chen L, Lu X, Yao H, Chen Y and Shi J 2019 Two-dimensional titanium carbide MXenes as efficient non-noble metal electrocatalysts for oxygen reduction reaction *Sci. China Mater.* **62** 662–70

[63] Liu C-Y and Li E Y 2019 Termination effects of Pt/v-$Ti_{n+1}C_nT_2$ MXene surfaces for oxygen reduction reaction catalysis *ACS Appl. Mater. Interfaces* **11** 1638–44

[64] Wen Y, Ma C, Wei Z, Zhu X and Li Z 2019 FeNC/MXene hybrid nanosheet as an efficient electrocatalyst for oxygen reduction reaction *RSC Adv.* **9** 13424–30

[65] Jiang L, Duan J, Zhu J, Chen S and Antonietti M 2020 Iron-cluster-directed synthesis of 2D/2D Fe–N–C/MXene superlattice-like heterostructure with enhanced oxygen reduction electrocatalysis *ACS Nano* **14** 2436–44

[66] Li Z *et al* 2018 The marriage of the FeN_4 moiety and MXene boosts oxygen reduction catalysis: Fe 3d electron delocalization matters *Adv. Mater.* **30** 1803220

[67] Chen J, Yuan X, Lyu F, Zhong Q, Hu H, Pan Q and Zhang Q 2019 Integrating MXene nanosheets with cobalt-tipped carbon nanotubes for an efficient oxygen reduction reaction *J. Mater. Chem.* A **7** 1281–6

[68] Yu X, Yin W, Wang T and Zhang Y 2019 Decorating g-C_3N_4 nanosheets with Ti_3C_2 MXene nanoparticles for efficient oxygen reduction reaction *Langmuir* **35** 2909–16

[69] Zhang Z, Li H, Zou G, Fernandez C, Liu B, Zhang Q, Hu J and Peng Q 2016 Self-reduction synthesis of new MXene/Ag composites with unexpected electrocatalytic activity *ACS Sustain. Chem. Eng.* **4** 6763–71

[70] Xie X, Chen S, Ding W, Nie Y and Wei Z 2013 An extraordinarily stable catalyst: Pt NPs supported on two-dimensional $Ti_3C_2X_2$ (X = OH, F) nanosheets for oxygen reduction reaction *Chem. Commun.* **49** 10112–4

[71] Cheng C, Zhang X, Fu Z and Yang Z 2018 Strong metal–support interactions impart activity in the oxygen reduction reaction: Au monolayer on Mo_2C (MXene) *J. Phys. Condens. Matter* **30** 475201

[72] Splendiani A, Sun L, Zhang Y, Li T, Kim J, Chim C-Y, Galli G and Wang F 2010 Emerging photoluminescence in monolayer MoS_2 *Nano Lett.* **10** 1271–5

[73] Jayabal S, Saranya G, Wu J, Liu Y, Geng D and Meng X 2017 Understanding the high-electrocatalytic performance of two-dimensional MoS_2 nanosheets and their composite materials *J. Mater. Chem.* A **5** 24540–63

[74] Sadighi Z, Liu J, Zhao L, Ciucci F and Kim J-K 2018 Metallic MoS_2 nanosheets: multifunctional electrocatalyst for the ORR, OER and Li–O_2 batteries *Nanoscale* **10** 22549–59

[75] Kwon I S, Kwak I H, Kim J Y, Abbas H G, Debela T T, Seo J, Cho M K, Ahn J-P, Park J and Kang H S 2019 Two-dimensional MoS_2/Fe-phthalocyanine hybrid nanostructures as excellent electrocatalysts for hydrogen evolution and oxygen reduction reactions *Nanoscale* **11** 14266–75

[76] Mao J, Liu P, Du C, Liang D, Yan J and Song W 2019 Tailoring 2D MoS_2 heterointerfaces for promising oxygen reduction reaction electrocatalysis *J. Mater. Chem.* A **7** 8785–9

[77] Hao L, Yu J, Xu X, Yang L, Xing Z, Dai Y, Sun Y and Zou J 2017 Nitrogen-doped MoS_2/carbon as highly oxygen-permeable and stable catalysts for oxygen reduction reaction in microbial fuel cells *J. Power Sources* **339** 68–79

[78] Zhang H, Tian Y, Zhao J, Cai Q and Chen Z 2017 Small dopants make big differences: enhanced electrocatalytic performance of MoS_2 monolayer for oxygen reduction reaction (ORR) by N- and P-doping *Electrochim. Acta* **225** 543–50

[79] Kalantar-zadeh K, Ou J Z, Daeneke T, Mitchell A, Sasaki T and Fuhrer M S 2016 Two dimensional and layered transition metal oxides *Appl. Mater. Today* **5** 73–89

[80] Yang W, Zhang X and Xie Y 2016 Advances and challenges in chemistry of two-dimensional nanosheets *Nano Today* **11** 793–816

[81] Xiao X *et al* 2016 Scalable salt-templated synthesis of two-dimensional transition metal oxides *Nat. Commun.* **7** 11296

[82] Xu L, Jiang Q, Xiao Z, Li X, Huo J, Wang S and Dai L 2016 Plasma-engraved Co_3O_4 nanosheets with oxygen vacancies and high surface area for the oxygen evolution reaction *Angew. Chem. Int. Ed.* **55** 5277–81

[83] Sun Y, Liu Q, Gao S, Cheng H, Lei F, Sun Z, Jiang Y, Su H, Wei S and Xie Y 2013 Pits confined in ultrathin cerium(IV) oxide for studying catalytic centers in carbon monoxide oxidation *Nat. Commun.* **4** 2899

[84] Yin H and Tang Z 2016 Ultrathin two-dimensional layered metal hydroxides: an emerging platform for advanced catalysis, energy conversion and storage *Chem. Soc. Rev.* **45** 4873–91

[85] Qian L, Lu Z, Xu T, Wu X, Tian Y, Li Y, Huo Z, Sun X and Duan X 2015 Trinary layered double hydroxides as high-performance bifunctional materials for oxygen electrocatalysis *Adv. Energy Mater.* **5** 1500245

[86] Tang C and Zhang Q 2017 Nanocarbon for oxygen reduction electrocatalysis: dopants, edges, and defects *Adv. Mater.* **29** 1604103

[87] Jiang J, Zhang A, Li L and Ai L 2015 Nickel–cobalt layered double hydroxide nanosheets as high-performance electrocatalyst for oxygen evolution reaction *J. Power Sources* **278** 445–51

[88] Song F and Hu X 2014 Exfoliation of layered double hydroxides for enhanced oxygen evolution catalysis *Nat. Commun.* **5** 4477

[89] Subbaraman R, Tripkovic D, Strmcnik D, Chang K-C, Uchimura M, Paulikas A P, Stamenkovic V and Markovic N M 2011 Enhancing hydrogen evolution activity in water splitting by tailoring Li^+–$Ni(OH)_2$–Pt interfaces *Science* **334** 1256–60

[90] Teramura K, Iguchi S, Mizuno Y, Shishido T and Tanaka T 2012 Photocatalytic conversion of CO_2 in water over layered double hydroxides *Angew. Chem. Int. Ed.* **51** 8008–11
[91] Raza A, Rafi A A, Hassan J Z, Rafiq A and Li G 2023 Rational design of 2D heterostructured photo- and electro-catalysts for hydrogen evolution reaction: a review *Appl. Surf. Sci. Adv.* **15** 100402

Muhammad Ikram, Ali Raza, Jahan Zeb Hassan and Salamat Ali

Chapter 6

Electrochemical CO_2 reduction and beyond

The increasing scarcity of fossil fuels and continued release of CO_2 need to be addressed, and using ECR combined with renewable energy as the foundation of fuel and chemical assembly provides a highly appropriate solution for future energy sustainability. Even if ECR is feasible from a thermodynamics viewpoint, the CO_2 molecule is highly stable, which causes sluggish reaction kinetics. Hence, effectual electrocatalysts are required to achieve ECR. In recent decades, wide-ranging efforts have been dedicated to developing effective catalysts for CO_2 electrolysis. This chapter describes the ECR for various 2D catalysts, makes a comparison with other reactions, and examines its industrial significance for clean energy. Traditionally, bulk metals are categorized into three groups depending on the binding energy strength for numerous intermediates and final products. Despite their use in ECR experiments, bulk metals have highly limited active reaction sites; further, they are subject to deactivation through different firmly bound intermediates or impurities in the form of poisoning. In recent years, numerous nanostructured materials have been produced that serve as innovative ECR catalysts. Additionally, 2D materials (specifically atomically thin nanosheets) are promising as they present plentiful unsaturated surface atoms, therefore offering an enormous amount of active reaction sites. In this chapter, we primarily describe emerging 2D materials (i.e. TMDs, TMOs, transition metals, and carbon-based materials) produced for advanced ECR activity while serving as efficient catalysts. Moreover, a summary of 2D materials with their applications is also discussed based on a critical analysis of their suitability for the nitrogen cycle and oxidation of carbon fuels with significance on an industrial scale. Lastly, challenges for 2D electrocatalysts for the ECR are also described.

6.1 Reaction kinetics and mechanisms

The ECR is composed of three significant steps that are very similar to those found in other electrocatalytic methods: (i) using the active site as the cathode catalyst for

doi:10.1088/978-0-7503-5291-8ch6

Table 6.1. Standard redox potentials of CO_2 reduction to different products.

Electron transfer	Reaction	E° (V versus SHE)
e^-	$CO_2 + e^- \rightarrow CO_2^-$	−1.9
$2e^-$	$CO_2 + 2H^+ + 2e^- \rightarrow CO + H_2O$	−0.53
	$CO_2 + 2H^+ + 2e^- \rightarrow HCOOH$	−0.61
	$2CO_2 + 2H^+ + 2e^- \rightarrow H_2C_2O_4$	−0.913
$4e^-$	$CO_2 + 4H^+ + 4e^- \rightarrow HCHO + H_2O$	−0.48
$6e^-$	$CO_2 + 6H^+ + 6e^- \rightarrow CH_3OH + H_2O$	−0.38
$8e^-$	$CO_2 + 8H^+ + 8e^- \rightarrow CH_4 + 2H_2O$	−0.24
$12e^-$	$2CO_2 + 12H^+ + 12e^- \rightarrow C_2H_4 + 4H_2O$	−0.349
	$2CO_2 + 12H^+ + 12e^- \rightarrow C_2H_5OH + 3H_2O$	−0.329
$14e^-$	$2CO_2 + 14H^+ + 14e^- \rightarrow C_2H_6 + 4H_2O$	−0.27
$18e^-$	$3CO_2 + 18H^+ + 18e^- \rightarrow C_3H_7OH + H_2O$	−0.31

chemical adsorption of CO_2 molecules; (ii) migration of electrons/protons to split C–O bonds to produce C–H bonds; and (iii) rearrangement of the final products as monitored through desorption phenomena take place on the electrode surface followed by discharge to the electrolyte [1]. Furthermore, the subsequent phase of ECR is quite a challenging route since it is recognized as a multi-step (proton-coupled) reaction including 2-, 4-, 6-, 8-, 12-, 14-, or 18-electron reaction paths to specific products; this stage also involves the same reaction that has various products (table 6.1) [2]. It is more desirable for ECR (from an energy efficiency viewpoint) to continue by multi-electron (proton-coupled) transfer than any other mechanism. However, the exact reaction pathways are highly dependent on various circumstances, some of which will be covered here.

6.1.1 Measurement criteria

According to the Nernst equation, the ECR is thermodynamically advantageous for the reaction mechanisms with progressive positive standard redox potentials (E°). However, the actual ECR kinetics depends on numerous elements, including the electrocatalyst, electrolyte, temperature, and pressure; this is the case regardless of where the thermodynamic equilibrium is located. The electrocatalyst, in particular, contributes a highly vital part to the mechanism and kinetics of ECR. For example, the bulk counterparts of transition metals are classified into one of three categories based on their capacity to bind certain chemical intermediates. In addition to the catalyst's chemical makeup, a substantial amount of research indicates that factors such as its size, shape, crystallographic configuration, and chemical conditions play an essential role in determining its reaction kinetics [3, 4]. The electrolyte is yet another significant component that contributes to ECR kinetics. The majority of example have aqueous solutions that include inorganic salts (i.e. $KHCO_3$ and $NaHCO_3$) as the electrolyte; hence, the HER acts as an unwanted viable reaction

(E^{o} = −0.42 V versus SHE). This is because of the difference in potential between the two solutions [5]. Although non-aqueous organic electrolytes can efficiently prevent the limiting HER and therefore advance CO_2 solubility, the comparatively low viscosity of these electrolytes makes electrolyte diffusion difficult. In addition, the high cost of these electrolytes and their potential toxicity further limit their application in practical settings [6, 7]. The pressure and temperature both alter CO_2 solubility in an electrolyte through influencing the kinetics of ECR. Enhanced CO_2 solubility, low temperature, and an advanced CO_2 partial pressure are generally considered advantageous [6]. Higher CO_2 concentrations do, to some extent, result in better reaction rates [8]. However, additional energy is required to generate circumstances with these characteristics. Consequently, the vast majority of ECR investigations has been carried out in CO_2-saturated aqueous electrolytes under normal environmental circumstances. In addition to the onset potential, two additional parameters, j (energy efficiency) and η (faradaic efficiency) are sensitive and vital for assessing ECR activity using electrocatalysts because of the variety of products designed in the strategy. Thus, the evaluation of faradaic efficiency for an actual product can be given by the equation

$$\mathrm{FE} = \frac{\alpha n F}{Q}. \tag{6.1}$$

In the above equation, α represents the total number of electrons migrated in mechanisms that yield certain products. In addition, n corresponds to the number of moles produced by that specific product, and Q indicates the total charge transmitted during the electrolysis route. An estimation of the energy efficiency may be made using the formula

$$\mathrm{EE} = \frac{E^0}{E^0 + \eta} \times \mathrm{FE}. \tag{6.2}$$

6.2 Graphene-based electrocatalysts

For CO_2 electroreduction, both homogeneous and heterogeneous catalysts have been utilized successfully. Homogeneous catalysis involves molecular catalysts and enzymes, providing several benefits, including product specificity and productivity, which may be achieved by accurately switching distinct active centers [9]. However, this type comprises various limitations, including high costs, limited stability, and costly post-separation processes, which prevent scalable and practical uses. Immobilization on conductive supports such as graphene is a promising new approach that might be used to alleviate these problems [10]. A wide range of heterogeneous electrocatalysts, such as transition (bi-)metals that have active d electrons and vacant orbits (such as Pd, Rh, Cu, Ni, Au and Pt, and Ag) [11, 12], p-block (bi-)metals (such as Bi, In, Sn and Pb) [13], oxides (such as RuO_2 [14], IrO_2 [14], TiO_2) have been used. Among these, carbon-based materials have shown significant potential in ECR [15, 16]. These materials offer various distinct advantages, including greater quantity, environmental friendliness, lower costs, resilient structural strength, excellent conductivity, and the potential to be renewed. Recently, graphene, which is an innovative material, has become the desirable carbon allotrope because of the distinctive structural character and the numerous excellent qualities it possesses. The enormous specific surface area calculated using theoretical measurements (~2630 $m^2\ g^{-1}$), which affords a higher density of active surface sites, is an attractive property of this material. It also boasts impressive electrical conductivity, superior stability, and great mechanical strength, all of which are advantageous when applied in the form of catalysts or catalytic supports. The surface of graphene can be doped, with both nonmetallic and metallic species holding equal importance. Because of this, degrees of freedom are available toward the modification of associated catalytic activity [17]. Due to the minor capability of neutral carbon to activate CO_2, pristine graphene and GO do not affect the electron capture reaction. Doping graphene with heteroatoms including P, N, and B (i.e. single-, dual-, and multiple-doping of heteroatoms) could considerably change the electrical structural characteristics of graphene that, in turn, significantly boost the material's activity in the electron capture reaction. Developing heterostructures based on graphene can improve charge transfer and generate synergy, contributing to decreasing CO_2 [18].

6.2.1 Alteration of graphene to improve CO_2 reduction

Graphene/GO surfaces containing a higher degree of exposed atoms could be changed using nonmetallic or metallic species.

6.2.1.1 Nonmetallic doping

Pure graphene cannot efficiently catalyze ECR because it has a minimal capacity to activate and adsorb CO_2 molecules and is an extraordinary free energy obstacle for the fundamental stage of *COOH production. Doping of graphene with single/multi heteroatoms (i.e. Br, P, N, S, Cl, F, B, and I (p-block elements) visualized in figure 6.1) [19] causes its structure to be modified along with the fundamental

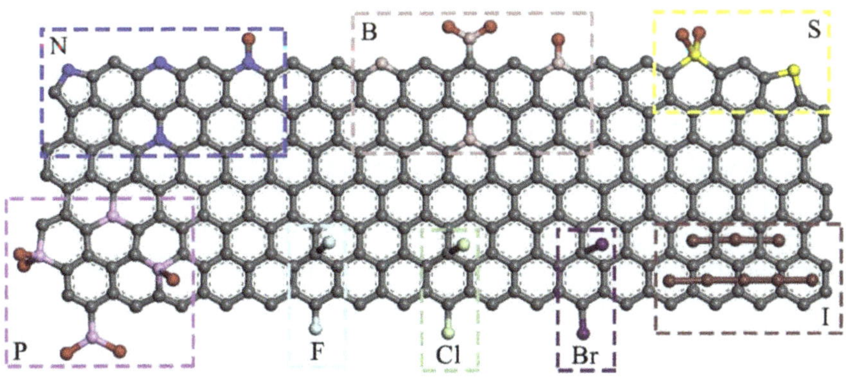

Figure 6.1. Visualization of heteroatom-doped graphene (Br, P, N, S, Cl, F, B, and I). (Reproduced with permission from [24]. Copyright 2019 Elsevier.)

characteristics (edge strain, spin densities, electronic characteristics, and hydrophilicity) of graphene, resulting in boosted ECR. Furthermore, certain heteroatom doping in graphene can increase the surface wettability of pristine graphene [20], making transferring hydrated CO_2 to active areas easier. Boron atoms ($2s^22p^1$) have one less valence electron than carbon atoms ($2s^22p^2$), which causes p-type doping and charge polarization in the graphene matrix. Boron is most commonly found in B-doped graphene in the forms of B_4C doping (out-of-plane), BC_3 doping (in-plane), borinic BCO_2 and boronic BC_2O. Boron doping causes charge polarization in the carbon framework, stabilizing the CO_2 atoms that are negatively polarized [21]. As a result, the chemisorption of carbon dioxide onto the surface of carbon increases. It was hypothesized that the positively charged atoms of B and C assisted as catalytically active sites for the reduction of CO_2, which in turn promoted the chemisorption of COOH to create formate [22]. In contrast to boron doping, nitrogen generates polarization in the carbon network because nitrogen has a greater electronegativity (3.04) than carbon. There have been reports of four primary bonding topologies in NG: edge pyrrolic N, basal plane quaternary N (central graphitic), and nitrile N [23]. A material with n-type characteristics is produced by graphitic nitrogen, whereas a material with p-type characteristics is produced by pyrrolic-, pyridinic-, or nitrile nitrogen.

It has been shown that the presence of an N dopant (at the pyridinic N site) can modify the electronic configuration of carbon atoms and encourage a 200–400 meV density of states below the Fermi level [25]. This can potentially offer sites for CO_2 adsorption. Results acquired from various DFT measurements reveal that the energy barrier to the *COOH intermediate is lowered when nitrogen is doped into a material. Additionally, the adsorption energy (ΔE_{CO}) of CO is reduced, while that for COOH (ΔE_{COOH}) is increased. Furthermore, pyridinic N can produce a Lewis base phase due to a lone pair of electrons toward CO_2 adsorption [26]. As a direct result of this, the electroreduction was strengthened for CO_2–CO [27] or formate (i.e. formic acid) [28]. The presence of active sites in NG is not understood well, and these are most important for CO_2 reduction. Further, pyridinic N (triple- and single-pyridinic N) [27] and a C atom adjacent to pyridinic N [29] were indicated as being the most active [30]. Pyridinic N at the edge location for graphene quantum dots may stimulate C–C bond development compared to that of the basal plane, which leads to an increased yield of products (C_2 and C_3) [31]. It was hypothesized that pyrrolic N would initiate ECR towards HCOOH with the lowest probable overpotential (0.24 V) [32]. In addition, for planar graphene, the graphitic N adjacent to the edge was revealed to be the leading active site, with a CO_2 activation barrier of just 0.58 eV [33].

Similarly, it was suggested that the effectiveness in boosting CO_2 conversion to CO follows the trend graphitic N > pyridinic N > pyrrolic N [34]. These disparities in the roles that N dopants play in ECR may be explained by the fact that the amounts of dopant found in each species can vary widely, as can the degree of effort involved in forming a single configuration. The electronegativity of sulfur (2.58) is closer to carbon (2.55) than phosphorous (2.19), and it is lower than for nitrogen (3.04). Therefore, the charge density distribution is less affected by the presence of

P and S atoms (slight polarization for S-doping). Moreover, doping with P and S atoms is likely to arise at the edges of graphene due to the fact that the atomic radii for S (1.04 Å) and P (1.10 Å) are more significant compared to the atomic radius of C (0.77 Å). Moreover, interaction between 3p orbitals (i.e. for P) and 2p orbitals (for C) can offer sp^3 hybridization of carbon atoms. This results in a distorted structure resembling a distorted tetrahedron containing three proximal carbon atoms. Catalytic activity is attributed to graphene's non-uniform scattering of spin density, which is made possible by the mismatch between the outmost orbitals of C and S atoms. C–S–C (also known as thiophene) and C–SO_x–C (where $x = 2$, 3, or 4) are two primary products of S-doping. It was claimed that S-doping would result in little ECR activity compared to N doping. This poor activity was due to fewer positively charged carbon atoms adjacent to the S atoms [29]. An inferior (1.01 eV) Gibbs free energy blockade for *COOH arises at pyridinic N adjacent carbon-bonded S atoms [35] contrary to pristine pyridinic N atoms that involve an advanced Gibbs free energy (1.34 eV) to overcome the blockade of *COOH adsorption. This is because pristine pyridinic N atoms involve an advanced Gibbs free energy to adsorb *COOH. Doping a material with either fluorine or chlorine yields a p-type material with sp^3 bonding between the carbon atoms. Other dopants such as I and Br interrelate with graphene through the development of charge transfer complexes or physisorption, so neither disturbs the sp^2 carbon network. It was determined that doping material with fluorine would activate neighboring carbon atoms by producing asymmetrical charge supplies, leading to greater polarization. Asymmetrical spin scattering was encouraged, which promoted COOH* interacting with the activated carbon [36]. Moreover, these carbon atoms with an advanced asymmetrical spin density and positive charge revealed a diminished energy barrier for COOH* formation. As a result, they were speculated to be active sites for ECR.

6.2.1.2 Construction of graphene composites

Graphene is frequently utilized as a substrate for active phases (nanoparticles or nanosheets) because of its high conductivity and large, accessible surface areas. This combination can potentially improve charge migration and yield structural and electronic coupling effects [37] that will speed up CO_2 conversion. At the interface, a synergy can be generated, which improves CO_2 adsorption and intermediate binding and stability, which is to the benefit of ECR. The immobilization of copper [38] or silver [39] on the surface of N-doped graphene might change stable carbon monoxide (through the presence of N-doped carbon) into *OC–COH, hence increasing the production of ethanol or n-propanol [39]. Pyridinic N has been reported to serve as an absorber of both CO_2 and protons, making it easier for hydrogenation and C–C coupling processes on copper to take place, which leads to the formation of ethylene [18]. The CuMn, Cu_2, and CuNi dimers maintained on graphene with neighboring vacancies were anticipated to be suitable electrocatalysts for the generation of CO, CH_4, and CH_3OH, respectively, based on the results of large-scale screening DFT and microkinetics modeling [40].

6.3 Transition metals

It is important to state that the morphological aspects are recognized as the main features that affect the catalytic performance of a specific catalyst [41]. However, little research has been dedicated to this aspect of ECR electrocatalysts. In recent times, the impact of nanostructured morphologies aimed at developing higher-activity ECR catalysts was demonstrated by Kim *et al* using three shape-controlled Bi nanostructures, namely, nanoflakes, nanodots, and dendrites [42]. Using a similar chemical composition for ECR, the Bi nanostructures demonstrated altered selectivities and efficiencies. Bi nanoflakes attained a $HCOO^-$ faradaic efficiency (FE_{HCOO-}) of 79.5 % at −0.4 V versus RHE, whereas nothing was acquired for formate employing nanodendrites or nanodots using a similar potential. A simulation study was used to study the electric field distribution, and shows that the increased edge and corner site ratios of the 2D nanostructure resulted in the production of robust local electric fields at small overpotential values. Researchers rely on the fact that intense localized fields assist the ECR and improve selectivity for Bi nanoflakes. In particular, the simulation outcomes suggest that the stronger electric fields associated with the edge as well as corner sites are related to the lower thickness of Bi nanoflakes. Morphological aspects may offer an excellent base for the establishment of rational design schemes for higher-activity ECR electrocatalysts. In recent times, the fabrication of ultrathin $Cu/Ni(OH)_2$ nanosheets (consisting of $Ni(OH)_2$ nanosheets dispersed in atomically thick Cu nanosheets) was described by Dai *et al* and they demonstrated interesting ECR characteristics [43]. The production of nanosheets was done using a solvothermal route in DMF that involves Cu^{2+}, Ni^{2+}, and $HCOO^-$, as shown in figure 6.2(a). During synthesis, Ni $(OH)_2$ nanosheets were produced first. Then the occurrence of ion exchange between Cu^{2+} and Ni^{2+} is established over its surface (as the solubility product constant for $Ni(OH)_2$ is more significant than for $Cu(OH)_2$). The adsorption of $HCOO^-$ was carried out over recently produced $Cu(OH)_x$. The reaction indicated the steady reduction of deposited Cu precursor into Cu nanosheets, as demonstrated in figure 6.2(b). Electrochemical analysis revealed that the $Cu/Ni(OH)_2$ nanosheets were highly selective towards the reduction process at low η (CO_2 to CO), that yielded an FE_{CO} of 92% (maximum) at −0.5 V, as shown in figures 6.2(c) and (d). An additional study was carried out in the same work by adopting atomically thick Cu nanosheets produced via etching the basic $Ni(OH)_2$ that resulted in an FE_{CO} of up to 89%, indicating a high intrinsic ECR reaction using ultrathin Cu nanosheets. Consequently, the protection of transition metal 2D nanosheet surfaces with capping molecules and decreasing their thickness can offer innovative schemes for the development of durable and efficient ECR electrocatalysts.

6.4 Transition metal-oxides

The innovation of an initial class of oxide-derived metals for ECR electrocatalysts has led to research interest in TMOs for ECR applications [44, 45], as described by Chen *et al.* The pioneering work was reported by Gao *et al*, which focused on the enhancement of the selectivity and electrocatalytic activity of TMOs via reducing

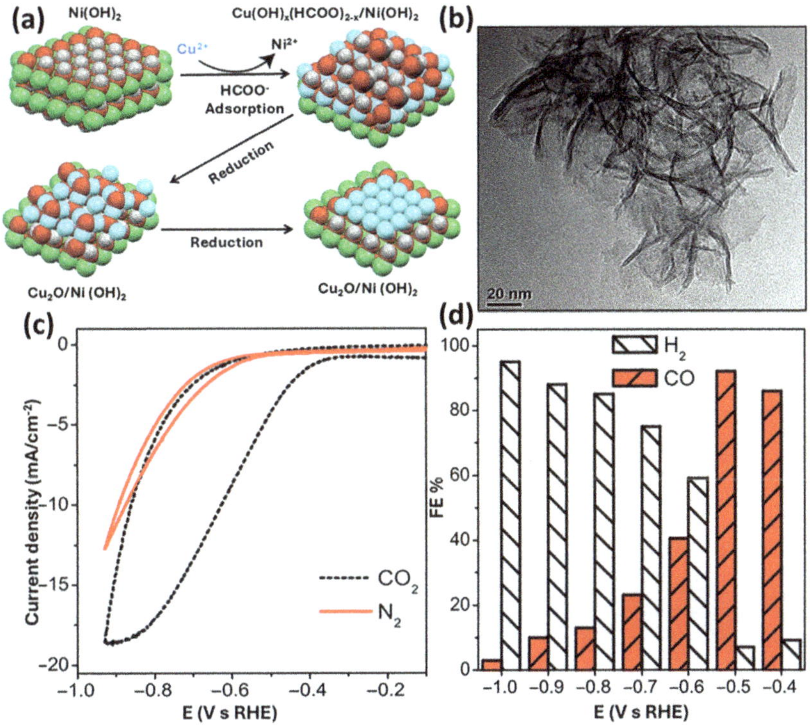

Figure 6.2. (a) Representation of synthesis pathways for Cu/Ni$(OH)_2$ nanosheets. (b) TEM photograph of the prepared product. (c) CV curves of prepared nanosheets in CO_2-saturated and N_2-saturated 0.5 M $NaHCO_3$. (d) Percentage FE for H_2 and CO at different potentials. (Reproduced with permission from [43]. Copyright 2017 American Association for the Advancement of Science.)

their thickness to the atomic level, as shown in figures 6.3(a)–(d) [46]. A thickness of 2 nm for the TMOs causes the majority of the transition metal atoms to be exposed above the surface. As these transition metal atoms are inferior in coordination number to the interior atoms, therefore they can act as active sites for efficient CO_2 adsorption. Moreover, DFT measurements reveal that higher electrical conductivity values significantly affect atomic thickness. For example, an ECR current density value of 0.68 mA cm^{-2} was observed for 1.72 nm thick Co_3O_4 layers at −0.88 V versus SCE, which is 1.5 nm and found to be 20 times higher in comparison to bulk Co_3O_4 samples and 3.51 nm thick layers (figure 6.3(c)).

Furthermore, an FE_{HCOO-} of 64.3% (maximum) was attained for 1.72 nm Co_3O_4 layers at −0.88 V versus SCE, whereas bulk Co_3O_4 and 3.51 nm layers correspond to 18.5% and 51.2%, respectively, as indicated in figure 6.3(d). Adsorption isotherms and ECSA studies verified that ultrathin Co_3O_4 layers might support a higher portion of low coordinated surface atoms that serve as active sites, resulting in enriched intrinsic ECR activity. Further investigation was done by Gao *et al* by adopting a solvothermal strategy to fabricate oxidized Co layers having thickness measurements of only a few atoms, presented in figure 6.3(e) [47]. These atomically

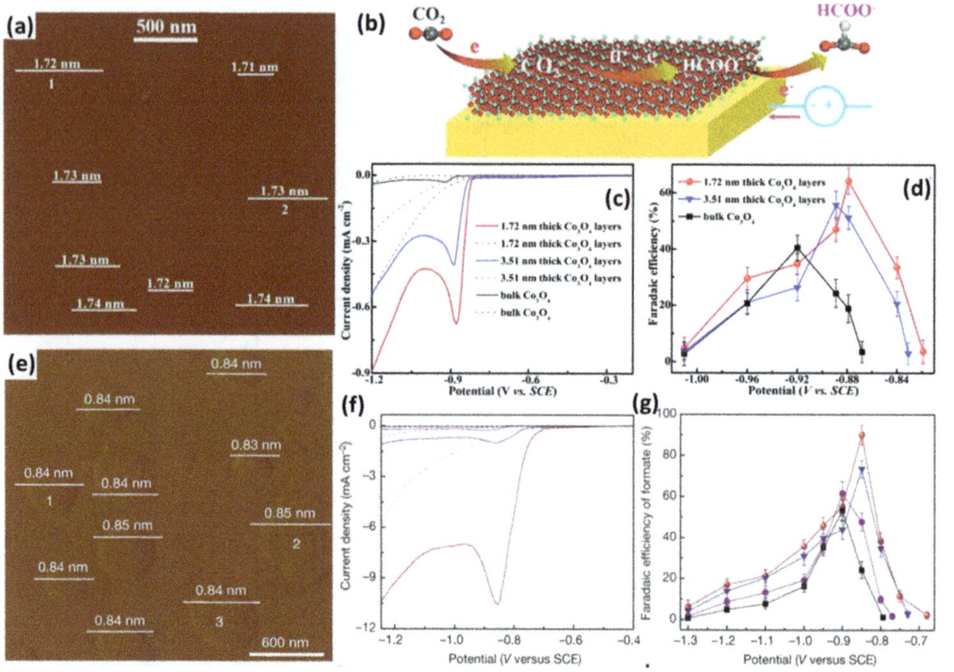

Figure 6.3. (a) AFM photograph for Co_3O_4 atomic layers (1.72 nm average thickness). (b) Schematic view of CO_2 electroreduction into formate over Co_3O_4 atomic layers. (c) LSV curves of the described electrocatalysts in N_2-saturated (dashed lines) and CO_2-saturated (solid lines) 0.1 M $KHCO_3$ aqueous solution. (d) FEs of formate for the described electrocatalysts using different applied potentials after 4 h. (Reproduced with permission from [46]. Copyright 2016 John Wiley and Sons.) (e) AFM photograph of partially oxidized Co (four-atom-thick layers). (f) LSV curves of the described electrocatalysts in N_2-saturated (dashed lines) and CO_2-saturated (solid lines) 0.1 M Na_2SO_4 aqueous solution. (g) FEs of formate for different electrocatalysts using different applied potentials after 4 h. (Reproduced with permission from [47]. Copyright 2016 Copyright Springer Nature.)

thick 2D Co-based layers could present plentiful unsaturated surface atoms allowing additional improvement of ECR activity. For example, four-atom-thick layers of partially oxidized Co afforded a current density of up to 10.59 mA cm^{-2} at an η of 0.24 V, significantly outperforming four-atom thick layers of Co, partially oxidized bulk Co, and bulk Co (figure 6.3(f)). Correspondingly, the FE_{HCOO-} at this overpotential for four-atom thick layers of partially oxidized Co shows 90.1% efficiency and incredible stability measurements over 40 h of examination (figure 6.3(g)). Further, ECSA-modified Tafel plots show that prepared layers of partially oxidized Co exhibit different satisfactory kinetics for ECR. Comparison between Co layers and partially oxidized Co verified that higher selectivity values and intrinsic activity were detected for atomic Co_3O_4 instead of atomic Co to reduce CO_2 to formate.

Furthermore, Gao *et al* incorporated oxygen vacancies (OVs) in single unit cell layers (SUCL) of Co_3O_4 (thickness of ~0.84 nm) to investigate the association between ECR performance and the level of OVs [48]. The OV-rich Co_3O_4 SUCL realized a current density 2.7 mA cm^{-2} at −0.87 V versus SCE along with an FE_{HCOO-} of 87.6% (maximum), while OV-poor systems had a current density and FE_{HCOO-} of 1.35 mA cm^{-2} and 67.3% respectively. These outcomes verified the significant role of OVs for advancing ECR experiments. DFT measurements also attested that the occurrence of OVs causes the radical formate anion intermediate to stabilize, as a reduction in the rate-limiting activation barrier was observed from 0.51 to 0.40 eV. Hence, TMO defect engineering at the atomic level can deliver alternative strategies for boosting ECR electrocatalysts.

6.5 Transition metal dichalcogenides

In addition to nanostructured TMOs, the TMDs (i.e. WSe_2 nanoflakes and MoS_2 nanosheets) [49] also present good characteristics for ECR. For example, the remarkable activity of MoS_2 for ECR was reported by Asadi *et al* using an ionic liquid (e.g. EMIM-BF4) [49]. This scheme requires an overpotential (54 mV) for a catalyst to produce CO; it also successfully acquires 98% FE_{CO} at −0.76 V versus RHE. The ECR activity of MoS_2 was verified via experimental and theoretical analysis that showed that the metallic character of the edge Mo atoms, the comparatively small work function, and the greater d-electron density are the factors that advance ECR. Using the same strategy, TMDs for ECR were examined, such as $MoSe_2$, WS_2, and WSe_2 nanoflakes using an EMIM-BF_4 electrolyte. Among these TMDs, W-terminated WSe_2 demonstrated the best ECR activity. Remarkably, at $\eta = 54$ mV, the current density for WSe_2 nanoflakes for ECR was 18.95 mA cm^{-2}, with a TOF of 0.28 s^{-1}, and 24% FE_{CO}. The minor charge transfer resistance of WSe_2, along with its robust structure and extremely active edge W atoms, causes more remarkable ECR characteristics. To explore the additional characteristics TMDs for ECR, a detailed theoretical investigation has been performed on MoS_2 models [50]. In this regard, to bind significant reaction intermediates, Chan *et al* projected a scheme to break scaling interactions by developing different sites (i.e. using atomic doping) [50]. Their studies reveal that S edge Ni-doping of MoS_2 allowed the selective binding of CHO* and COOH* for enhanced CO selectivity [50]. Likewise, DFT measurements also showed the same aspects as reported by Chan *et al* [51]. This standard can offer future insights for developing ECR catalysts that might be further extended to heterogeneous electrocatalysts. Sun *et al* provided effective evidence experimentally using a Mo–Bi bimetallic chalcogenide system as a highly effective ECR catalyst [52]. Here, Mo enables hydrogenation of CO and Bi encourages the CO_2 to CO conversion. Consequently, the as-prepared system yield methanol with 71.2% of FE_{CH_3OH} at −0.7 V versus SHE [53].

6.6 2D-MXenes for CO_2 reduction

Electrochemical manufacturing of value-added compounds utilizing CO_2 as the carbon source has gained interest in recent years [54]. Anthropogenic CO_2 emissions from fossil fuels has upset the natural carbon balance, producing global warming and increasing sea levels. Power is increasingly being produced using renewable energy, producing affordable energy and avoiding unnecessary CO_2 emissions [55, 56]. This energy-powered CO_2RR can convert CO_2 into formic acid, methanol, methane, ethylene, and carbon monoxide (CO) [57, 58]. Due to their large conductivity, various binding sites, and adjustable surface groups, MXenes have been utilized in CO_2 reduction reactions (CO_2RR). MXenes are a family of potential electrocatalyst materials [59, 60]. Li *et al* [59] used theoretic measurement to estimate the activity of TM (group IV, V, VI) carbide M_3C_2. Mo_3C_2 and Cr_3C_2 were the effective catalyst choices for the highly selective translation of CO_2 to CH_4. Because it is currently difficult to synthesize bare MXene experimentally, sound consideration of the dissolution surface (the OH and O functional groups) is required. This will allow one to deduce the stages involved in the CO_2RR reaction. Energy inputs of 1.05 and 1.31 eV are required to convert CO_2 to CH_4 using bare MXene. The energy may be further reduced to 0.35 and 0.54 eV, respectively, when the surface of the MXene molecule is concluded by OH or O. When compared to unmodified MXene, the O-terminated or OH-terminated form of MXene can more effectively assist in the alteration of CO_2 (figure 6.4(a)).

Moreover, Xiao *et al* [61] carried out DFT simulations in conjunction with the CHE model to investigate the hydrogenation process on the surface of M_3C_2 MXenes when CO_2 is present. It has been established that the adsorbed CO_2 becomes activated and can respond to hydrogen on the surface to create bicarbonate species. This leads to the CO_2RR having a higher competitive selectivity than the the HER. While HCHO serves as the primary intermediate of the CO_2RR, the creation of HCO_2 species is the reaction pathway that is energetically the most advantageous for transforming CO_2 to CH_4. All probable mechanisms to convert CO_2 to CH_4 are investigated; this research could influence how CO_2RR catalysts for large-efficacy fuel cells are designed. It is predicted that the CO_2RR catalysts V_3C_2, Mo_3C_2, MO_2TiC_2, and W_3C_2 will reduce CH_4 at 0.74, 0.45, 0.17, and 0.41 V decreased overpotentials, correspondingly. Handoko *et al* [62] used DFT measurements to analyse that on 19 O-terminated M_2XO_2 type MXenes, the electroreduction reactions of CO_2 to CH_4 occurred and concluded that W_2CO_2 and Ti_2CO_2 were highly efficient candidates for M_2XO_2 MXene. Their respective hypothetical overpotentials are 0.52 and 0.69 V, and they exhibit modest selectivity in the development of hydrogen. Because of the readily available *HCOOH pathway, which is dynamically more promising than *CO route, the low overpotential that can be attained on these surfaces is possible since it allows the catalytic substance to be less dependent on C-coordinated intermediates (figures 6.4(b) and (c)).

The O terminating groups found on MXenes are responsible for helping to stabilize reaction intermediates with the help of H coordination. Consequently, the *CO binding strength is no longer the primary CO_2RR performance descriptor for TM catalysts [63]. The most effective catalysts were $Sc_2C(OH)_2$ and $Y_2C(OH)_2$ with potentials of 0.53 and 0.61 V, which were less negative than for Cu. Chen *et al* [64] used the first-principles methodology to discover the CO_2RR characteristics on 17 MXenes with OH termination groups from the kinetics and thermodynamics viewpoint. Simultaneously, the other HER that was competing was disadvantaged because of its robust H binding. MXenes comprising hydrogen in OH-terminated

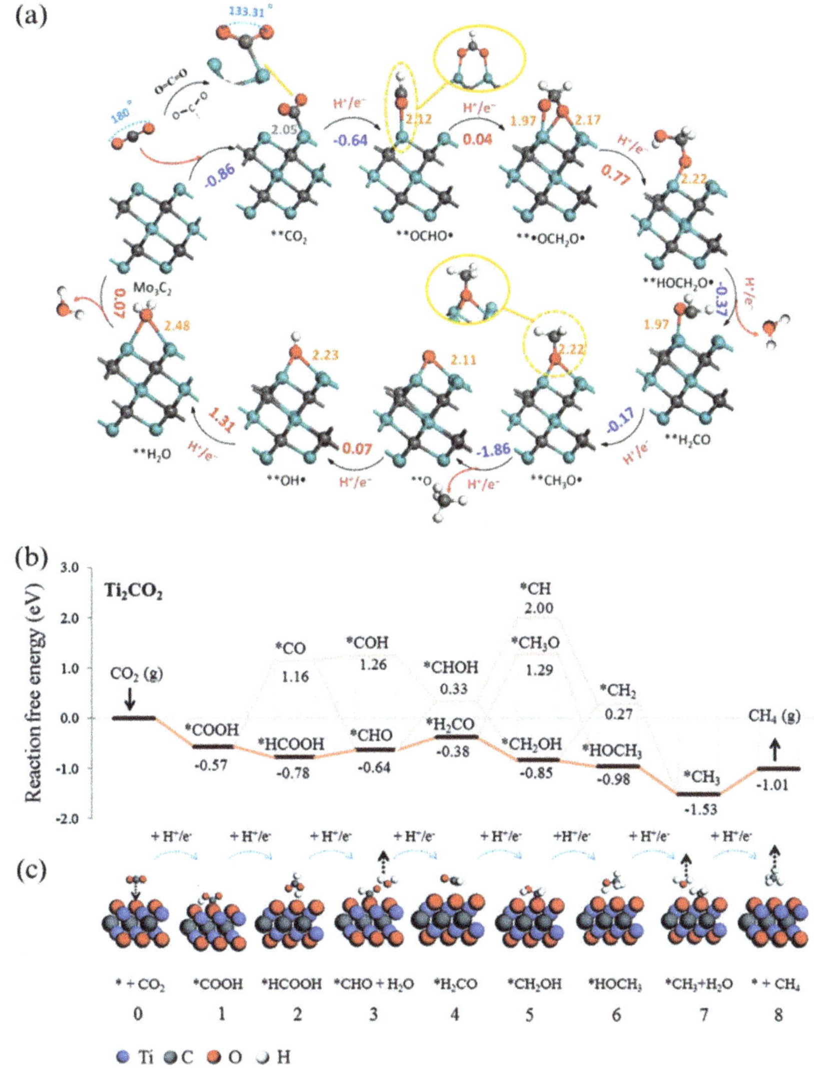

Figure 6.4. (a) Side view of the minimum energy route for the conversion pathways of CO_2 into $**H_2O$ and $*CH_4$ catalyzed by Mo_3C_2 (using PBE/DFT-D_3 measurements). The distribution of various atoms shown by colored spheres is as follows: gray (C), lilac (Mo), red (O), and white (H). (Reproduced with permission from [59]. Copyright 2017 American Chemical Society.) (b) Free energy diagram. (c) Representation of a ball and stick sample for a different probable mechanism to reduce CO_2 into CH_4 over O-terminated Ti_2CO MXene. (Reproduced with permission from [62]. Copyright 2018 The Royal Society of Chemistry.)

form have the potential to be active, which helps the CO_2RR process build structures with more stable intermediates and lowers the overpotential, making it more amenable to the electrocatalytic CO_2RR process. The electrocatalytic CO_2RR reaction mechanism can be intricate when operating under typical circumstances. Following the reaction thermodynamics of each elementary reaction, a wide variety of intermediates and final products can be produced under various potentials. According to the study's findings, as mentioned earlier, the O-terminal or OH-terminal form of MXene can boost the efficiency of the CO_2RR process. Research into MXene-based systems for electrocatalytic CO_2RR is in its infant stages compared to research into other types of 2D materials. Although current research has made significant progress, understanding the nature and catalytic mechanism of MXene's surface active spots is crucial.

6.7 Other reactions

Excluding reactions in carbon and water cycles which can be catalyzed successfully via introducing 2D materials, numerous additional reactions also exist that involve energy conversion. Example of such reactions are the urea oxidation reaction (UOR), formic acid oxidation reaction (FAOR), NRR as well as methanol oxidation reaction (MOR). In comparison with hydrogen and oxygen electrode reactions, electrocatalysts and their character are comparatively less investigated for these reactions [65]. However, many of them hold significant features for highly-efficient energy conversion procedures. For example, liquid fuels (i.e. methanol and formic acid) can be functional in the anode of proton-exchange membrane fuel cells (PEMFCs) in the direction to yield electricity [66]. These liquid fuels can be used as energy sources (i.e. fuel cells) as they can be acquired from available biomass [67]. In addition, instead of hydrogen-based fuel cells, direct-methanol (DMFCs) or direct formic acid (DFAFCs) fuel cells are considerably safer because of their harmless transportation. Hence, the development of higher-activity electrocatalysts for the electrochemical oxidation of these fuels is essential. Alternatively, the nitrogen cycle is also recognized as a modern energy source; for example, NH_3 is a more harmless energy source than hydrogen [68, 69]. Thus, in electrocatalytic studies, UOR and NRR are attractive reactions to be considered. So far, a variety of 2D electrocatalysts exhibit noteworthy progress towards these advanced reactions. This section will summarize 2D materials-based applications for these key electrocatalytic strategies.

6.7.1 Nitrogen reduction reaction

NH_3 is the most vital chemical for energy carriers in the electrocatalytic nitrogen cycle [70]. It has gained significant interest as NH_3 is extensively utilized for producing several chemicals and is a carbon-free energy storage intermediate [71, 72]. Furthermore, the Haber–Bosch route is the leading strategy utilized in industrial processes to produce NH_3; it uses Ru- or Fe-based catalysts for synthesizing NH_3 at high pressures and temperatures via gaseous H_2 and N_2 [73]. This strategy produces 3%–5% of the annual natural gas worldwide, uses ~1% of annual energy worldwide, and discharges an enormous amount of CO_2 into the atmosphere [73]. Alternative strategies (the geochemical route and biological N_2 fixation) for producing NH_3 usually have poor control and demonstrate comparatively slow reaction rates [74]. In addition, proton-supported electrocatalytic NRR provides comparatively higher performance for NH_3 production at ambient conditions, following the equation [68, 75]

$$N_2 + 6H^+ + 6e^- \rightarrow 2NH_3. \qquad (6.3)$$

Notably, the NRR can be combined with renewable wind and solar energy systems as an environmentally aware scheme for producing NH_3 [69, 75]. However, an enormous amount of energy is necessary to break down the solid triple bond between two nitrogen atoms (940.95 kJ mol^{-1}) and nitrogen elements [76]. This favorable scheme usually achieves comparatively little catalytic activity, also, inevitable unwanted and toxic by-products are produced (i.e. N_2H_2 and N_2H_4) that decrease the selectivity toward NH_3 [77, 78]. Similar to the ORR, the NRR pathways can also be classified into two groups, the associative and dissociative pathways, as shown in figure 6.5(a) [79]. Dissociative pathways include the breakage of the triple nitrogen bond before adding any hydrogen that derived into two adsorbed N atoms above the catalyst's surface. Alternatively, in associative pathways, two N atoms are bonded with each other, and the a triple bond is cleaved while each N atom experiences hydrogenation.

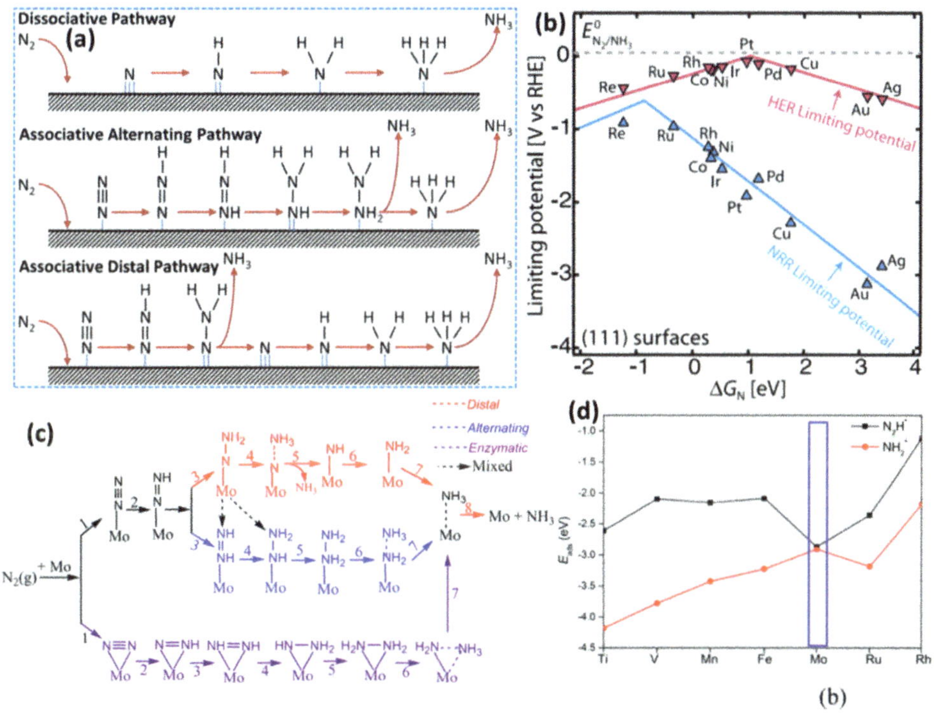

Figure 6.5. (a) Generic pathways for nitrogen reduction to ammonia by heterogeneous catalysts. (Reproduced with permission from [79]. Copyright 2017 Elsevier.) (b) Volcano plot of metals for NRR. (Reproduced with permission from [87]. Copyright 2015 John Wiley and Sons.) (c) Three pathways for N_2 electroreduction to ammonia over different Mo atoms attached on a BN monolayer (defective). (d) Graph for N_2H and NH_2 adsorption energies over different TM atoms reinforced via defective *h*-BN nanosheets. (Reproduced with permission from [82]. Copyright 2017 American Chemical Society.)

Furthermore, the hydrogenation scheme can also arise in associative pathways in two classes, called the associative alternating and distal pathways (figure 6.5(a)). In the first pathway (associative alternating), selective hydrogenation arises over two N atoms above the surface of the catalyst. In contrast, the second NH_3 molecule is discharged after the discharge of the first one. In the second type of pathway (associative distal), hydrogenation arises differently on the N atom farthermost away from the catalyst surface, resulting in the discharging of the distal NH_3 molecule first. Further, the hydrogenation scheme proceeds for the additional bound N to yield a second NH_3. Similar to other electrocatalytic strategies, a volcano plot for the NRR of various transition metal surfaces has been produced using DFT measurements (figure 6.5(b)) [80]. The predicted principle for NRR catalysts is for a stable N_2 adsorption energy with NH* protonation energy to produce NH_2* or the elimination of NH_2* as NH_3 [65]. Competition from the HER needs to be avoided. Catalysts near the top of the volcano plot, metals, for example, Re, Rh, and Ru, bind N_2 neither excessively intensely nor too lightly.

However, the enormous theoretical overpotentials of these metals limit their activity. Furthermore, DFT measurements suggest that transition metal nitrides, for example, ZrN and VN, are very appropriate for the NRR at low onset potentials with higher activity [65, 81]. Generally, the following standards should be considered for the design of NRR electrocatalysts. First, the catalyst should assist in the chemisorption of N_2 molecules for adequate activation of the (inert) triple nitrogen bond N≡N. Second, a catalyst must be able to stabilize the N_2H* selectively, and lastly it should destabilize NH_2* species to decreased the required overpotential [82]. Ensuring these standards, the DFT measurements of distinct transition metal atoms affixed on (defective) *h*-BN monolayers were reported by Zhao *et al* for the NRR (figure 6.5(c)) [82]. The results showed that ΔG values for the N_2 adsorption values affixed with Co, Ag, Sc, Cu, Pd, Zn, Cr, and Ni were positive. The researchers propose that the affixed transition metal atoms are unsuitable for NRR electrocatalysts because of their weak capability to activate N_2. Furthermore, negative adsorption energies were acquired for Mn, Rh, Fe, Ru, Ti, V, and Mo. Hence, these metals are recognized as suitable candidates for additional research into the NRR. As they have a negative N_2 adsorption energy, Rh atoms are omitted from these metals owing to their weaker N_2H* stabilization, as demonstrated in figure 6.5(d).

Likewise, Mn-, V-, Ti-, and Ru atoms are also not appropriate because of their comparatively intense NH_2* interaction (figure 6.5(d)). Therefore, Mo atoms on (defective) *h*-BN nanosheets are recognized as the best electrocatalyst for the NRR that fulfill all the screening standards. Thus, introducing 2D materials is a comparatively novel direction for NRR that has limited examples thus far. Notably, the current research work on NRR primarily depends upon theoretical measurements. Among those which have been explored, the development of a 2D hybrid nanocomposite was proposed by Li *et al* that comprises amorphous Au nanoparticles, rGO and CeO_2 for the NRR [83]. The content of noble metal in the subsequent catalyst was small (1.31 wt% Au); also the presence of CeO_x encouraged the production of low-crystalline/amorphous Au nanoparticles. The Faradaic proficiency and production rate of NH_3 of this catalyst were 10.1% and 8.3 μg

h^{-1} mg^{-1}, respectively, at a potential of −0.2 V versus RHE with no detection of hydrazine. This substantial catalytic activity is similar to the efficiencies and outcomes under higher pressure and temperature [74, 84]. Furthermore, utilizing LDH nanosheets for a photocatalytic NRR was described by Zhao *et al* as photocatalysts for synthetic dinitrogen fixation in ambient conditions. The presence of OVs in the LDH nanosheets causes enhanced activation and adsorption of H_2O and N_2, resulting in outstanding photocatalytic NRR [85]. Moreover, 2D nanosheets of BiOBr also revealed higher activity for the photocatalytic NRR due to the presence of OVs on their surfaces [86]. Motivated by these studies, the mechanism of OVs in 2D electrocatalysts may also be advantageous for boosting the NRR.

6.7.2 Oxidation of carbon fuels

In addition to electrode materials, the extraordinary cost of hydrogen containment produces a limit on the commercialization of the H_2-fed PEMFC strategy; the most probable hazards are present in hydrogen transportation and its small gaseous-phase energy density, despite from years of strong studies [88]. Therefore, greater interest has been focused on innovative fuel cells by spending on liquid fuels, for example, methanol and formic acid [89]. This section will summarize the fabrication of numerous 2D materials for the formic acid (FAOR) and methanol (MOR) reduction reaction routes.

6.7.2.1 Oxidation of methanol

Using the MOR, the DMFCs can transform methanol directly into electricity (adopting a similar primary cell production as PEMFCs) at the anode and the ORR at the cathode [90]. This transformation follows the following reaction:

$$CH_3OH + H_2O - 6e^- \rightarrow 6H^+ + CO_2. \quad (6.4)$$

Formerly, two aspects limited the commercialization of DMFCs; first, the stability of anode operation and, second, the substantial cost linked with precious-metal electrocatalysts (i.e. Pt, figure 6.6(a)) [85, 91]. For example, Pt is recognized as the best candidate for the MOR. Still, it shows low tolerance for certain reaction intermediates (i.e. CO) along with higher cost, with a loss of electrocatalytic activity contained by limited time of the reaction [85]. A predictable approach adopted to solve this difficulty is to introduce Pt alloys and oxophilic metals (figures 6.6(b)–(i)) [85, 91]. In this regard, the Pt–Ru binary system is recognized as the most studied and effective since the Ru supports H_2O dissociative adsorption to produce OH species, which results in increased oxidation of CO molecules on nearby Pt sites; thus, it assists in the redevelopment of active sites. Hence, the following two projected principles have been adopted for MOR electrocatalysts. First, balancing of the co-catalyst via the activity of Pt with OH adsorption energy, and second, regulating the morphological aspects and scattering of noble metal nanostructures for innovative support materials.

Graphene is documented as a potential candidate for MOR owing to its excellent conductivity, higher surface area, and outstanding stability [92]. The growth of

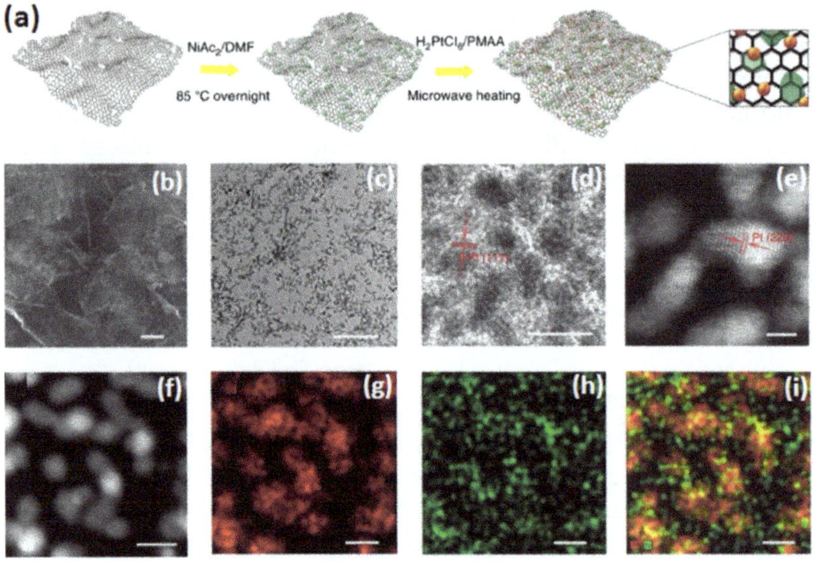

Figure 6.6. The synthesis and microscopic characterizations of Pt/Ni$(OH)_2$/rGO ternary hybrids. (a) Visualization of a two-step solution scheme for the synthesis of ternary hybrid materials. Representative (b) SEM photograph, (c), (d) TEM photograph and (e) ADF-STEM photograph of Pt/Ni$(OH)_2$/rGO-4. Representative (f) STEM photograph and resultant (g) Pt EDS mapping, (h) Ni EDS mapping, and (i) mutual Pt and Ni mapping of Pt/Ni$(OH)_2$/rGO-4. (Reproduced with permission from [85]. Copyright 2015 Springer Nature.)

graphene hybrid nanostructure along with N-doped carbon nanotubes (NCNT-GHN) was reported by Zhang *et al* by way of a carbon platform with a fine dispersion of PtRu nanoparticles (2–4 nm) [93]. Owing to the high nitrogen electron affinity, the strong attachment and uniform scattering of noble metal nanoparticles were realized via activating nitrogen-adjacent carbon atoms in the NCNT-GHN system [94]. Accordingly, the resulting PtRu/NCNT-GHN system exhibited a more substantial oxidation peak current density than commercial Pt/C or PtRu/C catalysts (using similar mass loading). In addition to Ru, the development of a ternary Pt/Ni$(OH)_2$/rGO hybrid system was reported by Huang *et al* due to the excellent interaction between *OH species and Ni$(OH)_2$ [85, 91]. The mechanism of this hybrid system involves the facilitation of Ni$(OH)_2$ for the oxidative elimination of CO via Pt active sites.

Surprisingly, the MOR peak current density for Pt/ Ni$(OH)_2$/rGO revealed at 1236 mA mg^{-1} is superior to commercial Pt/C catalysts. In contrast, rGO offers higher electrical conductivity, which is required for solid electrocatalysis. Furthermore, the as-prepared hybrid system also acquired a more robust capacity for the best interaction of Ni$(OH)_2$ with OH species. Remarkably, the oxidative elimination scheme of CO using a prepared hybrid catalyst follows Langmuir–Hinshelwood adsorption pathways that have quicker kinetics (rather than the Eley–Rideal adsorption pathways that are commonly observed).

6.7.2.2 Oxidation of formic acid

Compared with DMFCs, the DFAFCs that follow FAOR mechanisms as anode reactions are beneficial because of their advanced kinetic activity, less-toxic reactants, and weak crossover effect via the PEM [88]. Thus, the development of active electrocatalysts for FAOR is gaining enormous attraction for researchers. In recent times, specific works confirmed that the introduction of ultrathin 2D materials (particularly 2D metals) causes exceptional activity because of their surface defects and quantum size [95]. Using this concept, the growth of self-supporting Pd nanosheets (2D) was reported by Huang *et al* for the FAOR, providing a higher yield (2.5 times) than Pd black catalysts [96]. Consistent with the novel characteristics of high surface area of 2D electrocatalysts, it yields beneficial aspects for the FAOR. Furthermore, designing 2D metal alloys is also a promising approach for FAOR electrocatalysts to boost stability and activity and decrease the utilization of valuable metals. For example, Yang *et al* developed an ultrathin PdCu alloy nanosheet with an average 2.8 ± 0.3 nm thickness, as shown in figures 6.7(a) and (b) [97]. By adjusting the molar ratio of the precursor, the atomic ratio of Cu and Pd could be adjusted from ~0.14 to ~0.56, and the morphology can be seen in figure 6.7(c). It is worth mentioning that metal alloys possess a different electronic structure to their parent metals because of the small electronegativity value of Cu, as shown in figure 6.7(e) [91]. The Cu/Pd sample ratio (0.37 ± 0.01) possesses advanced mass activity (1628.3 ± 41.4 mA mg^{-1}) as well as higher ECSA (137.6 ± 13.6 m^2 g^{-1}). Thus, higher activity was acquired for the electrocatalytic process of PdCu nanosheets compared to commercial Pd black (figure 6.7(d)). This advanced activity is due to the distinctive electronic structure, ultrathin morphological aspects, and synergistic effect present between Cu and Pd.

6.8 Industrial significance of electrocatalysis for clean energy

The use of naturally available 2D materials as electrocatalysts in energy-related processes is being investigated as part of an effort to obtain energy that is both economical and sustainable. These include the HER, HOR, ORR, and OER processes, which are traditionally catalyzed by costly materials based on Pt or costly metal oxides such as IrO_2 or RuO_2. In the process of electrolyzing water, the cathodic reaction known as the HER is the one in which the half-reaction is

$$2H^+ + 2e \rightarrow H_2. \tag{6.5}$$

The HER process in acidic electrolytes happens in two steps. It starts with proton adsorption, known as the Volmer step (as shown in the equation)

$$H^+ + e^- + M^* \rightarrow M - H. \tag{6.6}$$

In equation (6.6), M* represented an adsorption site on the catalytic material. Desorption of hydrogen gas continues through either the Heyrovsky procedure, which is

$$M - H + H^+ + e^- \rightarrow H_2 + M^*, \tag{6.7}$$

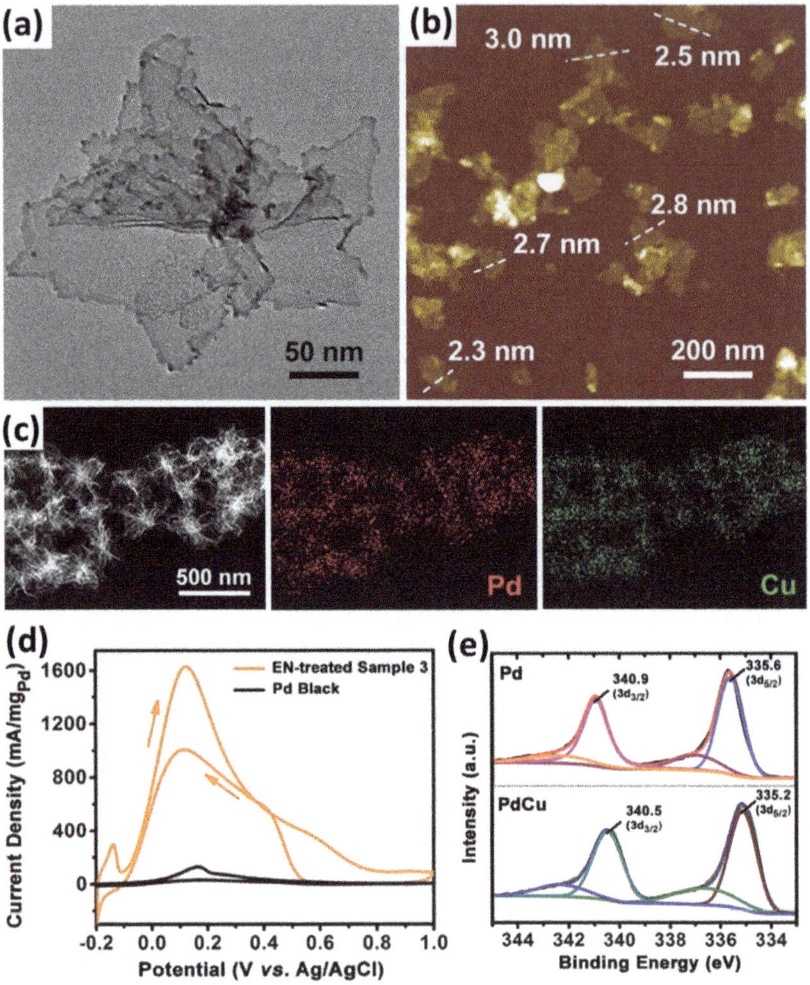

Figure 6.7. Representation of ultrathin PdCu alloy nanosheets. (a) TEM photograph, (b) AFM photograph, (c) DF-STEM-EDS elemental mapping, (d) CV curves of Pd black and EN-treated sample, and (e) XPS analysis. (Reproduced with permission from [97]. Copyright 2017 John Wiley and Sons.)

or the Tafel mechanism, which is

$$2M - H \rightarrow H_2 + M^*. \tag{6.8}$$

The HOR follows the same stages as the HER but in the opposite order. It is essential to emphasize that whereas 2D materials, including TMDs, are highly regarded as HER catalysts, the HOR efficacy of these materials is often relatively low. To this day, the vast majority of HOR catalysts are only available in the form of metals such as Pt and Rh. As a result, investigations into the production of 2D material catalysts for the HOR should be promoted. The OER is the anodic reaction throughout the electrochemical water splitting process. In acidic electrolytes, it converts $2H_2O$ into O_2, four hydrogen ions, and four electrons, whereas in neutral or alkaline conditions, it converts $4OH^-$ into two hydrogen ions, two O_2 atoms, and four electrons. A single molecule of O_2 must undergo four separate proton and electron exchanges during the OER production process, and is generally regarded as the primary contributor to inefficiency in electrolyzer systems [98]. The ORR is a relevant process in the vast majority of energy translation and storing technologies, for example, fuel cells and rechargeable metal–air batteries. One example is the ORR, which can continue via a direct channel containing four electrons, or it can progress via an indirect pathway that contains two electrons and forms an H_2O_2 intermediate. The CO_2RR is another potential method for converting one kind of energy into another. This reaction converts carbon dioxide into valuable chemicals, such as carbon monoxide, formate, or methane. Numerous protons, electrons, and intermediates are involved in CO_2RR as a consequence of the various products that it produces. The electrocatalytic effectiveness of 2D materials is primarily determined by the material's inherent activity, anisotropy, and the effects of mass transfer on the material. There is a strong connection between the surface structure of a material and the effects of mass transfer. The different catalytic sites of the 2D substances determine the electrocatalytic anisotropy factor. A volcano plot connection, a numerical portrayal of Sabatier's principle, is utilized to assess a material's intrinsic efficacy in an electrocatalyst system. This relationship is used to make this evaluation. In a perfect world, an active catalyst would bind to the reaction intermediates in a way that was neither overly strong nor overly weak. The maximization of catalytic activity includes the catalyst supports throughout the design process.

6.8.1 Activity at the edges

The active endings and inert basal phase of 2D materials are excellent examples of the anisotropic components that contribute to the catalytic characteristics of these materials. Atoms located on the ends of 2D constituents are subjected to different chemical conditions than those located in other areas of the materials. This environment has a more significant tendency toward unsaturated coordination than the basal phase, which usually has saturated coordination. Because the edge sites are the ones that account for the catalytic activity of 2D substances, it is necessary to improve the edge surface to improve their overall efficacy. This section summarizes some essential investigations conducted on edges as active centers of 2D substances in various catalytic reactions. Experiments have demonstrated that the ORR efficacy of graphene edges is much greater than that of the basal plane [99]. Oxidation or the initial graphite material can cause graphene edges to have O_2 groups at their terminals. Graphene edges also frequently have O_2 groups. Because these O_2 groups have the potential to affect ORR activity, graphene with low O_2 content was obtained using the liquid-assisted mechanical exfoliation of graphite. This allowed the effect of O_2 groups on ORR [100] to be reduced to a significant extent. Using nano-sized graphene they were able to demonstrate, via the use of this approach, that the quantity of edge sites is the fundamental cause for better ORR efficacy. Given favorable thermodynamics, the active sites are the zigzag edges of the structure. The significant adsorption of the –OH species, which blocks the active sites for O_2 binding [101], led researchers to conclude that the edges of armchairs are inactive and should not be used. Ar plasma etching [102] generated an edge-rich and pristine graphene that demonstrated effective ORR activity. This was done in order to reduce the influence that dopants have. In ORR electrocatalysis, the edge-rich graphene performed better than the pure graphene by having a lower onset potential and a greater current density.

An ideal model is provided for determining the function of edges as ORR active sites on graphene by the pristine graphene generated through plasma etching. The edges of the *h*-BN nanosheets, much like graphene's edges, are expected to be the active sites for ORR. This study was carried out on Au electrodes and a lower ORR overpotential was detected for the *h*-BN nanosheets compared to BN nanotubes [103]. In contrast to the extensive presence of edge structures in BN nanosheets, BN nanotubes exhibited a limited number of B- and N-edge structures. According to experimental evidence obtained from 2H-MoS_2, the catalytic sites in HER electrocatalysis are located at the margins of TMDs. HER activity was shown to have a linear association with the number of end sites on the MoS_2 catalyst [104] when a monolayer of MoS_2 nanoparticles placed on an Au(111) surface was investigated. The MoS_2 nanoparticles were of varying sizes. A notable example of reaching a significant fraction of uncovered edge sites is creating a mesoporous MoS_2 structure with a double-gyroid shape (figures 6.8(a)–(c)) [105]. The binary-gyroid MoS_2 catalyst has a curvature that provided a great density of edge sites, leading to increased HER efficacy. An additional factor to consider is the orientation of the TMDs on the substrate. For example, MoS_2, WSe_2, and $MoSe_2$ films were positioned vertically on substrates to achieve the greatest possible edge termination, increasing the catalytic HER efficiency [106].

However, studies have revealed that the addition of edges at the basal surface of the metallic 1T-phase of TMDs also promotes the HER [107, 108]. Because there are more active sites and the 1T-phase of TMDs has higher conductivity [109], exfoliated 1T-phase group VIB TMDs nanosheets displayed better catalytic HER performance than the 2H-phase. This can be seen in the low Tafel slope of 1T-MoS_2, which was 43 mV dec^{-1}. MXenes furnished with O* or OH* termination are likewise exciting candidates for use as HER electrocatalysts. However, bare MXenes are only sometimes employed as catalysts because of their low activity. It is worth noting that the basal phase O* of the vast majority of MXenes is energetic for HER, which is an intriguing fact. For example, delamination of Mo_2CT_x improved HER activity even though it had a large proportion of visible basal surfaces [110]. Surprisingly, the edges of TMDs are active for CO_2RR when ILs are present as co-catalysts. This is the case despite boundaries being the active sites for the HER, a reaction competing with CO_2RR when the circumstances are acidic. As a result of their metallic nature and high d-electron density [49], CO_2 may be reduced to CO at the Mo-terminated ends of MoS_2. Through the production of complexes, the ionic liquid called 1-ethyl-3-methylimidazolium tetrafluoroborate (EMIM-BF_4) can both stabilize CO_2 and inhibit HER. Enhanced CO_2RR catalysis [111] may be seen in a

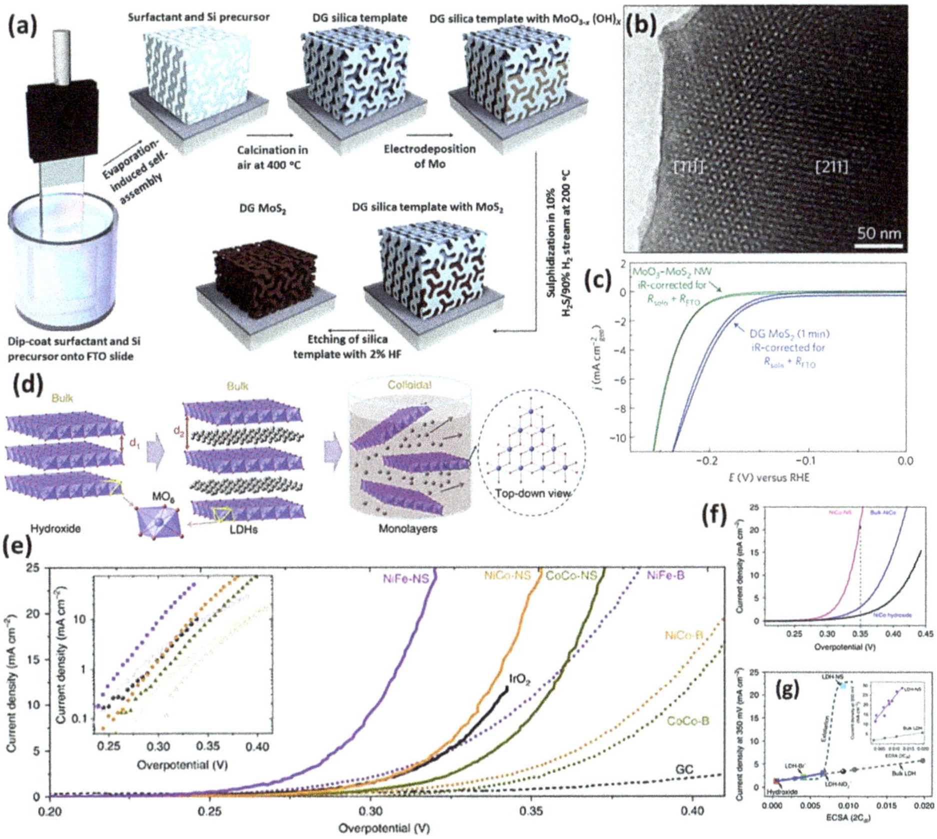

Figure 6.8. Visualization of enhancing active sites towards the electrocatalysis of energy-associated reactions. (a) Synthesis mechanism of double-gyroid (DG) mesoporous MoS_2. (b) Corresponding TEM photograph. (c) Evaluation of the HER activity of DG MoS_2 and MoO_3–MoS_2 nanowires. (Reproduced with permission from [105]. Copyright 2012 Springer Nature.) (d) Representation of the exfoliation approach for bulk LDHs. (e) Evaluations of OER activity for bulk counterpart and exfoliated LDHs. (f) A plot of polarization curves for NiCo LDH–NO_3, NiCo LDH nanosheets, and NiCo hydroxide (0.07 $mg\,cm^{-2}$ loading, 5 $mV\,s^{-1}$ scan rate; the arrow points to current density at $\eta = 350$ mV). (g) Charging current density variances ($\Delta j = j_a - j_c$) plotted against scan rates. (Reproduced with permission from [112]. Copyright 2014 Springer Nature.)

group of TMD nanoflakes consisting of WS_2 and WSe_2 with a greater edge density than bulk equivalents.

This is seen most clearly in the case of WSe_2 nanoflakes terminating in W atoms. Because of their exceptional OER efficiency, layered metal oxides and LDHs are the types of 2D materials most commonly used as OER catalysts. This is because the edges of these materials are responsible for their OER activity. Exfoliating bulk LDHs (NiFe, NiCo, and CoCo) into single-layers via liquid-phase exfoliation produced up to 4.5 times greater OER catalysis than previously, equivalent to that of IrO_2 catalysts (figures 6.8(d)–(g)) [112]. In addition, by breaking mono-layered NiFe LDHs into ultrafine nanosheets of <3 nm thickness with maximum visible edges [113], OER activity may be increased even further. Moreover, OER-active sites are created by liquid exfoliation followed by centrifugation, which thins BP into nanosheets. It has been suggested that these newly produced sites are the edges [114]. The OER performance of the nanosheets is significantly enhanced compared to that of bulk phosphorus, with an onset at 1.45 V and a Tafel slope of 88 mV dec^{-1}, respectively [114].

6.8.2 Mass transport

Highly active catalysts need beneficial mass transport because of the quick reduction of interfacial reactant species (H^+ or OH^-) and the production of gaseous materials both slow down reaction speeds. Therefore, maintaining a constant supply of reactants and a quick release of gas is necessary to maintain high reaction efficiency. For mass transfer in liquid and gaseous phases [115, 116], the interstitial spaces between adjacent sheets were used as 2D channels in 2D catalysts. By introducing spacers into MoS_2 nanosheets, open, robust, and connected channels were produced. This allowed for the achievement of accessible surface area as well as increased ion diffusion, which resulted in an overall improvement in HER catalytic activity (figures 6.9(a) and (b)) [116].

6.8.3 Introducing impurities and defects

Although 2D materials have an inherent electrocatalytic performance arising from their active sites, electrocatalysis in their natural conditions is limited by the quantity of active sites. As a result, introducing impurities such as dopants or even additional functional groups to 2D materials for catalysis might enhance the intrinsic activity of the materials. Doping or incorporating functional groups to the ends of 2D materials might potentially increase changes to their catalytic activity. This is because the boundaries of 2D materials are the catalytically active portions. Modifying the intrinsic activity of dormant basal planes by either adding or functionalizing the basal sites is possible. The enhancement of the inherent activity of active sites can be accomplished using a technique known as defect engineering. Structural defects, such as edges with low coordination numbers, are characterized by dangling bonds and the presence of atom vacancies. Doping materials such as graphene and *h*-BN is a common practice in 2D materials, and it is used to improve these materials' ORR catalytic capabilities. Incorporating graphene sheets with heteroatoms results in increased electrocatalytic activity for the ORR. The introduction of heteroatoms may modulate graphene's chemical reactivity and electrical characteristics. N incorporated graphene is the type of heteroatom-doped graphene that has received the most research attention. In alkaline conditions, N-doped graphene, synthesized using CVD, was shown to be an effective metal-free ORR electrocatalyst, according to research carried out by Dai and colleagues [117]. This particular ORR electrocatalyst follows a direct four-electron route.

The ORR may be traced back to its origin in graphitic and pyridinic N moieties. Graphitic N shifts its electron density to the graphene lattice, increasing the carbon atoms' nucleophilicity and making O_2 adsorption easier. The pyridinic N synergistic

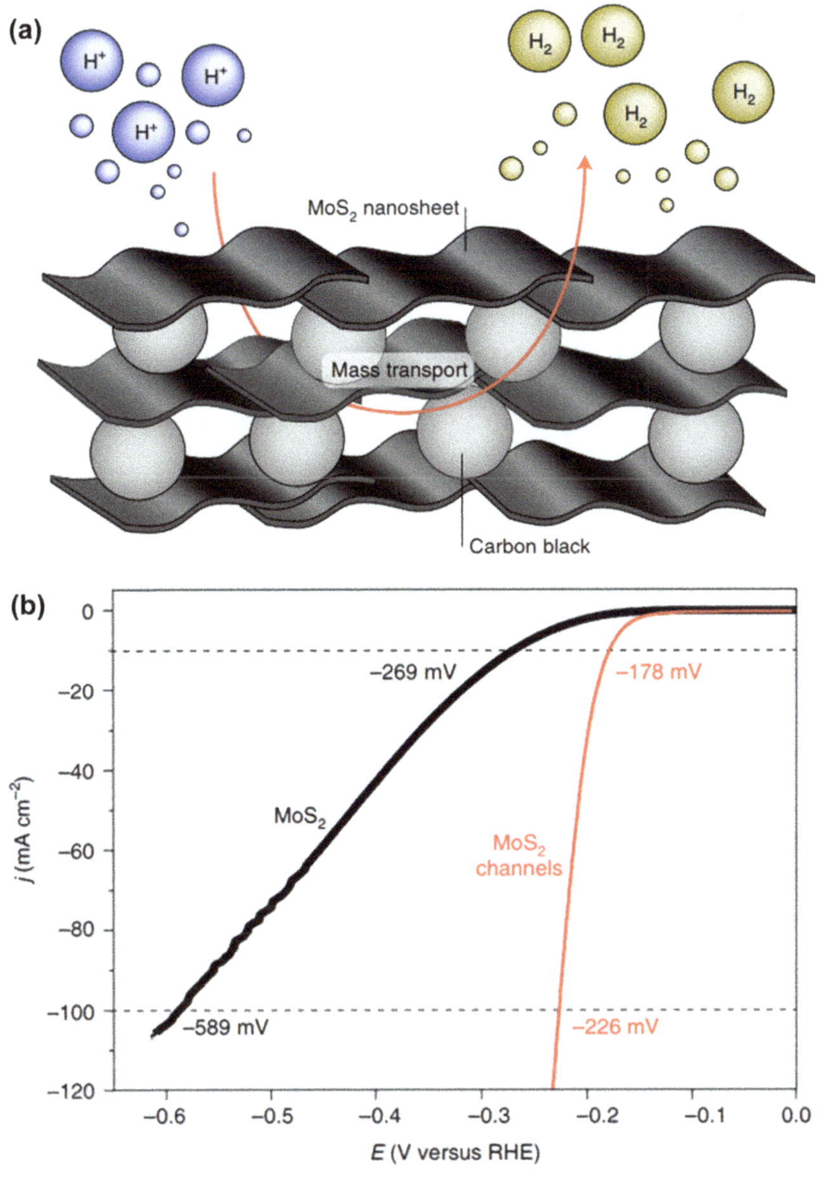

Figure 6.9. The mass transport disturbs the electrocatalytic activity of 2D substances. (a) Illustration of MoS_2 showing the engineered channels. (b) Results of an HER experiment for MoS_2 nanosheets and engineered channel MoS_2. (Reproduced with permission from [116]. Copyright 2018 American Chemical Society.)

effects provide an ORR active site brought about by introducing dual-doped graphene by substituting several heteroatoms in graphene [118]. For example, the co-doping of graphene with boron and N produced an ORR that was approximately identical to that of Pt, which was superior to the ORR produced by graphene doped with just boron or N. N is an electron-drawing group that activates B to become the active site inside a B–C–N hetero-ring in graphene. C and N are the two other ring components. This synergy is reduced to the extent that the distance between the pyridinic N atoms and the B atoms increases. The presence of metal dopants can also improve graphene's electrocatalytic activity. ORR activity can be attributed to the minute amounts of manganese (0.0018% by wt.%) present in metal-free graphene [119]. The doping of *h*-BN nanosheets with carbon results in large spin and charge density, which decreases the energy gap. This decrease in the energy gap promotes O_2 adsorption and facilitates an ORR route that requires a lower barrier than a Pt-based catalyst [120]. Because of the stability of the O atom adsorbed by the Au_8 cluster reinforcing the four-electron pathway [121], the ORR overpotential was decreased by 0.5 V after Au nanoparticles were decorated on the surface of *h*-BN nanosheets. This was in contrast to the situation before the decoration. Porous BCN nanosheets have been produced using a polymer sol–gel method [122]. These nanosheets incorporate the benefits of graphene and *h*-BN.

In addition to the large concentration of pyridinic N, which speeds up the process of O_2 adsorption, the B–N–C groups located around the edges make OH adsorption and protonation of O_2 possible. Therefore, absorbent BCN nanosheets display unusual catalytic ORR behavior in acidic and alkaline environments, rivaling Pt/C. The inherent HER catalytic activity of TMDs may be effectively improved by the incorporation of metals as well as nonmetallic atoms. One of the essential criteria for estimating the intrinsic performance of the HER is associated with the G of adsorbed H (ΔG_H), where it has been determined that a value up to zero is optimum for an HER catalyst. Consequently, by doping the S edge of MoS_2 and WS_2 with cobalt, the ΔG_H of the S edge in both TMDs became more promising for the HER. This was proved both theoretically and experimentally [123]. P-doping also activated the inert S edges and basal surface of 2H-MoS_2, which was demonstrated by a decrease in the valence charge of molybdenum as well as a drop in the group gap energy (ΔG_H) of the surrounding phosphorous atoms, approaching zero [124]. P-doped 2H-MoS_2 had an onset potential of 130 mV and a low Tafel slope of 49 mV dec^{-1}; these characteristics indicate it improved the HER electrocatalytic activity compared to undoped 2H-MoS_2. Doping graphene with heteroatoms causes the electrical characteristics to be adjusted to improve the HER. This is the case, even though the basal plane of graphene has a significant positive ΔG_H, which hinders HER. Qiao and colleagues used experimental and DFT investigations in their study [125] to assess a variety of single and dual heteroatom-doped graphenes. It was determined that the decrease of ΔG_H on the basal planes of B-, N-, S-, and P-doped graphene over a range of 0.6 to 1 eV occurred for the C atom close to the dopant element, which is considered to be the HER-active site on graphene. This site is located on the basal plane. The adsorption step was shown to be RDS in these singly doped graphene materials, as indicated by the Tafel slope of $\sim$120 mV dec^{-1}. Compared to the singly doped reference sample, the overpotential of the dual-doped variants of graphene (N, S-doped graphene and N, P-doped graphene) is lower. This indicates that dual doping can potentially boost the inherent HER activity of graphene.

In addition, it is essential to recognize that the HER performance of these doped graphene materials was subpar in contrast to that of MoS_2. It is possible that the HER activity of MXenes can be changed by introducing functional groups in the basal plane [111, 126]. Recent research used experimental and theoretical techniques to investigate how functionalization of T_x influences the HER of MXenes based on Ti and Mo. A decreased HER activity level was observed despite the higher fluorine coverage on the basal plane of $Ti_3C_2T_x$. In a similar vein, it was discovered that Mo_2CT_x, with a poor fluorine coverage on the basal plane, functions as an active and stable HER catalyst, exhibiting an overpotential of 189 mV at –10 mA cm^{-2} in the reaction. Numerous investigations have also been conducted on using heteroatom-doped graphene catalysts for CO_2RR. Both formate and carbon monoxide may be efficiently produced by graphene doped with either boron or N (figure 6.10) [22, 28]. Because of B-graphene's asymmetric charge and spin density, active chemisorption sites were found in the B and C atoms. On the other hand, pyridinic N was found to be the active adsorption site for intermediate COOH*, which led to the formation of CO. Therefore, the kind of dopant may allow a potential point of control to induce selective CO_2RR for desired products. In OER catalysis, doped 2D materials are the primary material.

Researchers have developed N-introduced, O-functionalized, and ends-rich graphene nanosheets on carbon cloth [127]. This work is considered to be the

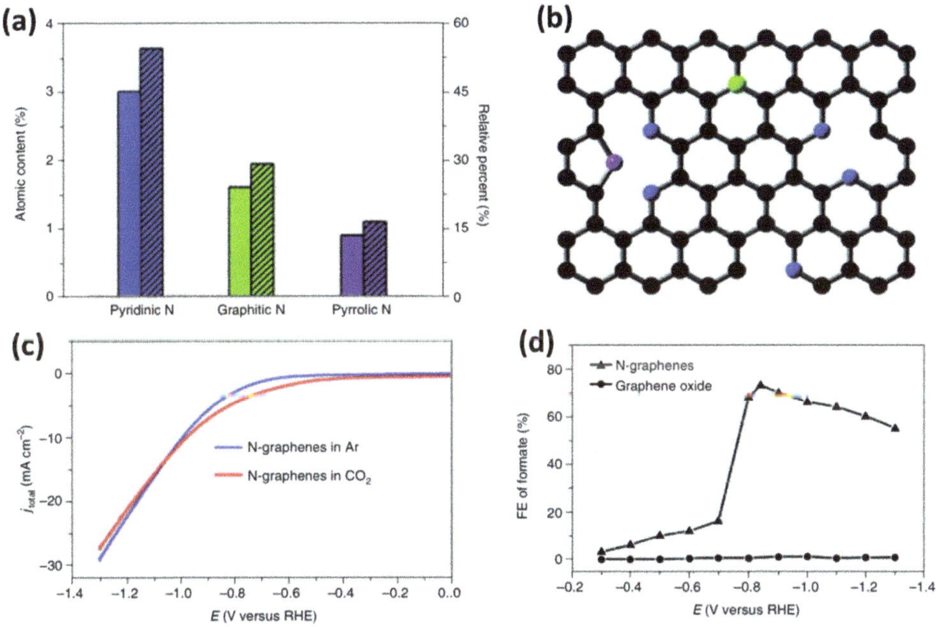

Figure 6.10. Aggregated intrinsic experiment for improved electrocatalysis in energy-based reactions. (a) Statistical analysis of N atomic content versus comparative percentage. (b) A variety of N arrangements in N incorporated graphene. (c) LSV curves of N incorporated graphene catalyst in CO_2- and Ar-saturated $KHCO_3$. (d) Assessment of faradaic efficiency (FE) for N-graphene and GO across applied potentials. (Reproduced with permission from [28]. Copyright 2016 The Royal Society of Chemistry.)

most important among the graphene-based OER catalysts. This material displayed exceptionally high OER activity due to a significant amount of functional groups and active sites on the interface of the graphene nanosheets. This catalyst design outperformed pristine graphene with a 351 mV overpotential and a Tafel slope of 38 mV dec^{-1}, outperforming even the most advanced OER catalysts such as RuO_2. Doping BP, a comparatively novel addition to the class of 2D materials, has also been investigated to enhance the material's potential to act as an OER catalyst. The OER onset potential of Te-doped BP nanosheets was much lower than that of the undoped nanosheets [128]. These nanosheets were obtained via chemical vapor deposition and monitored using liquid exfoliation. Interestingly, g-C_3N_4, when combined with TMs, exhibits bi-functional capabilities as an ORR and OER catalyst. These properties were proven in some metal-C_3N_4 complexes [129]. The molecule Co–C_3N_4, in which the Co–N_2 moiety is incorporated in the C_3N_4 matrix, is one example of this. The adsorption energy of the ORR and OER intermediates became more favorable when the d-band position shifted. Although it is common knowledge that incorporating structural defects into 2D materials can boost the action of various electrocatalytic reactions, the most common application of this strategy is on LDHs, which is used to change the electronic structure and obtain higher OER efficacy (figures 6.11(a)–(c)). Because of how they distribute electrons

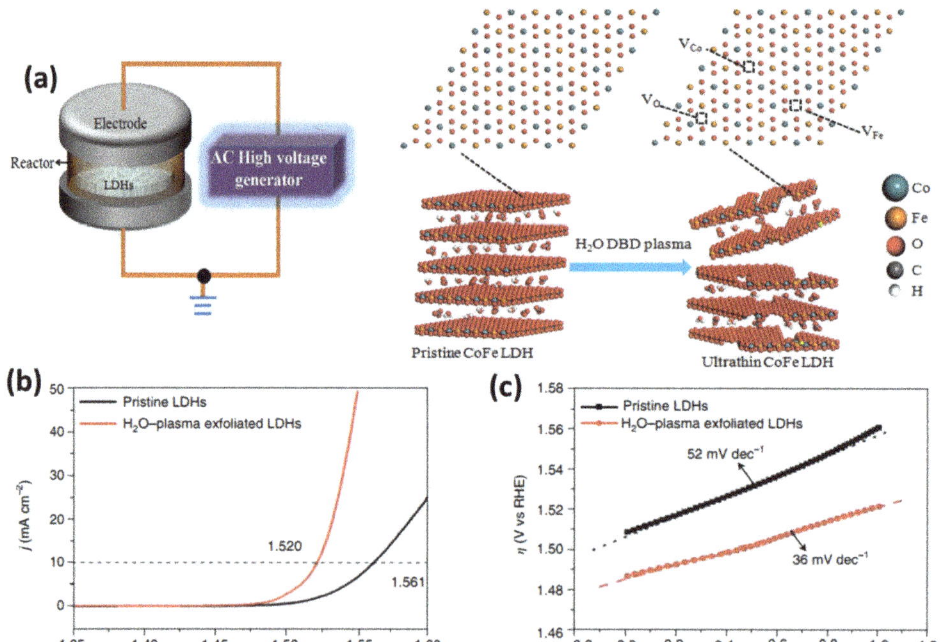

Figure 6.11. (a) Diagram of exfoliation for CoFe-LDH (water plasma enabled) with manifold vacancies produced. (b) LSV curves and (c) Tafel plots referring to OER activity between bare and exfoliated LDHs (water plasma enabled). (Reproduced with permission from [130]. Copyright 2017 John Wiley and Sons.)

and orbitals, surface defects in the form of O_2 and metal vacancies may be used to customize the surface's electrical characteristics. Metal vacancies raise the valency of other metal centers that are nearby, which in turn improves OER activity. Plasma methods were used to exfoliate CoFe-LDHs, which resulted in numerous vacancies of cobalt, iron, and O_2 [130, 131]. It is hypothesized that these vacancies will reduce the energy required for water adsorption and thus make OER easier. The water plasma exfoliated CoFe-LDH nanosheets displayed outstanding OER catalytic activity, with a 232 mV overpotential and a Tafel slope of 36 mV dec^{-1}.

6.9 Conclusion, challenges, and perspectives

In conclusion, to actualize the usage of 2D materials in real-life CO_2 conversion systems on an industrial scale, it will be necessary to rapidly manufacture 2D materials that have superior selectivity, higher activity, and higher stability. However, manufacturing 2D materials is still challenging because of the aggregation and restacking electrostatic and vdW interactions. In addition, productivity is still not high enough for economical manufacturing and the introduction of 2D materials for the conversion of CO_2. When designing the next-generation of 2D nanocatalysts, it is crucial to consider whether they can be manufactured in large quantities.

1. In recent decades, 2D materials have been the subject of much research and are being exploited in the electrochemical reduction of CO_2 as an electrocatalyst to produce value-added products. Consequently, through a variety of novel 2D catalysts, extraordinary progress has been made using various methods toward the improvement of selectivity and overall performance. These advancements have provided promising paradigms for developing the next generation of CO_2 conversion systems. Despite this, the fabrication of 2D catalysts aimed at CO_2 reduction via ECR still faces many hurdles, opening up some exciting research potential in this sector. The vast majority of the currently ongoing research concentrates exclusively on the materials features of catalysts (i.e. morphological aspects, defects, different materials and compositions). On the other hand, catalytic systems consist of many constituents that must be examined while investigating 2D catalysts. The overall performance of CO_2 via ECR may be determined by a wide variety of characteristics (i.e. the electrolytes used, the design of the reactor, and the local reaction environment).
2. In CO_2 ECR, a confined reaction atmosphere may be formed via solid–gas–liquid or solid–liquid interfaces. These types of interfaces are possible as various phenomena, including desorption and adsorption, occur at the interfaces. Moreover, interactions taking place between different reactants and solutions can have a substantial influence on CO_2 activity. For example, the different sizes and types of cations and anions found in electrolytes have been identified to have governing effects on the selectivity of CO_2 ECR. In addition, the use of ILs and the electrical fields that are generated in the Helmholtz layer have been shown to have a variety of consequences. Organic surfactants such as cetyltrimethylammonium bromide, which inhibit the

competing HER at the catalyst's surface, have also been shown to improve CO_2 ECR. The performance of CO_2 ECR could be determined using various aspects, for example, the interactions between solid catalysts and liquid electrolytes and the liquid–solid interfaces that are produced. 2D nanocatalysts, with their one-of-a-kind morphologies in both dimensions and their many dopant and vacancy sites, may elicit diverse reactions in a highly localized reaction atmosphere. To further improve the performance of 2D CO_2 catalysts, investigations of confined reaction environments are necessary.

3. The studies reported by researchers on 2D materials also examine further unique enhancement tactics (i.e. strain engineering and catalyst modifiers), among other possibilities. Polymers have recently been proven to offer outstanding optimizing capabilities. These capabilities include stabilizing metal nanocatalysts and modifying heterogeneous catalytic surfaces. Further, hydrophilic polymeric modifiers have the potential to improve the selectivity for formic acid, whereas cationic modifiers have the potential to encourage the production of CO on copper catalysts. Furthermore, polymeric N-heterocyclic carbenes (NHC; i.e. monodentate and polydentate) show stabilizing properties for Pd and Au nanocatalysts. The reason behind this is the development of metal–carbene interactions that produce limitations for nanoclusters of nanoparticles. These outcomes are mainly related to the utilization of 2D catalysts. It is possible to perform organic changes at the atomic scale using 2D nanocatalysts by applying the processes and methodologies presented in innovative optimization paradigms. Lattice stresses have shown the capacity to change the 2D band structures and disrupt linear scaling association for CO_2 ECR. As a result, the selectivity and activity of the material can be increased. However, it is still difficult to investigate the processes at play due to the challenges associated with 2D materials regarding decoupling strain from additional electrical properties with a mismatched lattice. When investigating optimization behaviors, it is necessary to combine simulation and experimentation in many ways. In addition, applying strain on 2D catalysts in the correct distribution and magnitude is still challenging.
4. Gas-phase electrolyzers have improved the stability and current density of CO_2 ECR due to the use of polymer ion-exchange membranes and gas diffusion layers within MEA electrolyzers. Remarkably, certain recently developed all-solid-state reactors, which make use of polymer electrolytes, can greatly simplify the process of liquid product separation in continuous operation. However, to date, the vast majority of research on 2D catalysts has been carried out using electrolyzers that operate in the liquid phase. To make progress while investigating 2D materials for CO_2 ECR, it is necessary to make the most of the potential presented by modern reactors. Solid-state reactors include dynamics between solid electrolytes and solid catalysts. The mass transport, desorption, and adsorption of reaction species and the confined reaction atmosphere are different at solid–solid interfaces compared to the typical liquid–solid interfaces. Characterizing and optimizing the

performance of 2D catalysts using these sophisticated reaction atmospheres and reactors is necessary to obtain more remarkable CO_2 ECR activity and selectivity of CO_2 ECR in the future that will be used on an industrial scale.

5. To produce 2D nanocatalysts that can be employed in industrial-scale electrolyzers, *in situ* and operando experiments require specifically built equipment in extremely high quantities. The intricacy of the catalytic processes causes many scientific problems in terms of time and length scales. As a result, operando multimodal and *in situ* characterization approaches are strongly recommended to obtain evidence for complete and informative characterization platforms. Particularly when combined with up-to-date DFT simulations, these aforementioned characterizations of 2D nanocatalysts can contribute to the development of our basic understanding of CO_2 ECR processes. The validated evidence may show the fundamental reaction active sites, local environment, channels, intermediates, and morphological variations for 2D catalysts. Even if such characterizations frequently need sophisticated equipment such as electron microscopy, synchrotron x-ray sources, and free electron lasers, the facts that are corroborated can reveal all of these things.

Bibliography

[1] Schlögl R 2015 Heterogeneous catalysis *Angew. Chem. Int. Ed.* **54** 3465–520

[2] Lim R J, Xie M, Sk M A, Lee J-M, Fisher A, Wang X and Lim K H 2014 A review on the electrochemical reduction of CO_2 in fuel cells, metal electrodes and molecular catalysts *Catal. Today* **233** 169–80

[3] Loiudice A, Lobaccaro P, Kamali E A, Thao T, Huang B H, Ager J W and Buonsanti R 2016 Tailoring copper nanocrystals towards C_2 products in electrochemical CO_2 reduction *Angew. Chem. Int. Ed.* **55** 5789–92

[4] Tang Q, Lee Y, Li D-Y, Choi W, Liu C W, Lee D and Jiang D-E 2017 Lattice-hydride mechanism in electrocatalytic CO_2 reduction by structurally precise copper-hydride nanoclusters *J. Am. Chem. Soc.* **139** 9728–36

[5] Zhang L, Zhao Z-J and Gong J 2017 Nanostructured materials for heterogeneous electrocatalytic CO_2 reduction and their related reaction mechanisms *Angew. Chem. Int. Ed.* **56** 11326–53

[6] Sun Z, Ma T, Tao H, Fan Q and Han B 2017 Fundamentals and challenges of electrochemical CO_2 reduction using two-dimensional materials *Chem.* **3** 560–87

[7] Alvarez-Guerra M, Albo J, Alvarez-Guerra E and Irabien A 2015 Ionic liquids in the electrochemical valorisation of CO_2 *Energy Environ. Sci.* **8** 2574–99

[8] Kauffman D R, Thakkar J, Siva R, Matranga C, Ohodnicki P R, Zeng C and Jin R 2015 Efficient electrochemical CO_2 conversion powered by renewable energy *ACS Appl. Mater. Interfaces* **7** 15626–32

[9] Appel A M *et al* 2013 Frontiers, opportunities, and challenges in biochemical and chemical catalysis of CO_2 fixation *Chem. Rev.* **113** 6621–58

[10] Shen J, Kolb M J, Göttle A J and Koper M T M 2016 DFT study on the mechanism of the electrochemical reduction of CO_2 catalyzed by cobalt porphyrins *J. Phys. Chem.* C **120** 15714–21

[11] Lee C W, Yang K D, Nam D-H, Jang J H, Cho N H, Im S W and Nam K T 2018 Defining a materials database for the design of copper binary alloy catalysts for electrochemical CO_2 conversion *Adv. Mater.* **30** 1704717

[12] Jia M *et al* 2018 Carbon-supported Ni nanoparticles for efficient CO_2 electroreduction *Chem. Sci.* **9** 8775–80

[13] Larrazábal G O, Martín A J and Pérez-Ramírez J 2017 Building blocks for high performance in electrocatalytic CO_2 reduction: materials, optimization strategies, and device engineering *J. Phys. Chem. Lett.* **8** 3933–44

[14] Bandi A 1990 Electrochemical reduction of carbon dioxide on conductive metallic oxides *J. Electrochem. Soc.* **137** 2157–60

[15] Ma T, Fan Q, Tao H, Han Z, Jia M, Gao Y, Ma W and Sun Z 2017 Heterogeneous electrochemical CO_2 reduction using nonmetallic carbon-based catalysts: current status and future challenges *Nanotechnology* **28** 472001

[16] Vasileff A, Zheng Y and Qiao S Z 2017 Carbon solving carbon's problems: recent progress of nanostructured carbon-based catalysts for the electrochemical reduction of CO_2 *Adv. Energy Mater.* **7** 1700759

[17] He H and Jagvaral Y 2017 Electrochemical reduction of CO_2 on graphene supported transition metals—towards single atom catalysts *Phys. Chem. Chem. Phys.* **19** 11436–46

[18] Li Q, Zhu W, Fu J, Zhang H, Wu G and Sun S 2016 Controlled assembly of Cu nanoparticles on pyridinic-N rich graphene for electrochemical reduction of CO_2 to ethylene *Nano Energy* **24** 1–9

[19] Tao H, Gao Y, Talreja N, Guo F, Texter J, Yan C and Sun Z 2017 Two-dimensional nanosheets for electrocatalysis in energy generation and conversion *J. Mater. Chem.* A **5** 7257–84

[20] Hao G-P, Sahraie N R, Zhang Q, Krause S, Oschatz M, Bachmatiuk A, Strasser P and Kaskel S 2015 Hydrophilic non-precious metal nitrogen-doped carbon electrocatalysts for enhanced efficiency in oxygen reduction reaction *Chem. Commun.* **51** 17285–8

[21] Duan X, Xu J, Wei Z, Ma J, Guo S, Wang S, Liu H and Dou S 2017 Metal-free carbon materials for CO_2 electrochemical reduction *Adv. Mater.* **29** 1701784

[22] Sreekanth N, Nazrulla M A, Vineesh T V, Sailaja K and Phani K L 2015 Metal-free boron-doped graphene for selective electroreduction of carbon dioxide to formic acid/formate *Chem. Commun.* **51** 16061–4

[23] Tao H, Yan C, Robertson A W, Gao Y, Ding J, Zhang Y, Ma T and Sun Z 2017 N-doping of graphene oxide at low temperature for the oxygen reduction reaction *Chem. Commun.* **53** 873–6

[24] Ma T, Fan Q, Li X, Qiu J, Wu T and Sun Z 2019 Graphene-based materials for electrochemical CO_2 reduction *J. CO_2 Util.* **30** 168–82

[25] Huang S-F, Terakura K, Ozaki T, Ikeda T, Boero M, Oshima M, Ozaki J-I and Miyata S 2009 First-principles calculation of the electronic properties of graphene clusters doped with nitrogen and boron: analysis of catalytic activity for the oxygen reduction reaction *Phys. Rev.* B **80** 235410

[26] Kiuchi H, Shibuya R, Kondo T, Nakamura J, Niwa H, Miyawaki J, Kawai M, Oshima M and Harada Y 2016 Lewis basicity of nitrogen-doped graphite observed by CO_2 chemisorption *Nanoscale Res. Lett.* **11** 127

[27] Wu J *et al* 2016 Incorporation of nitrogen defects for efficient reduction of CO_2 via two-electron pathway on three-dimensional graphene foam *Nano Lett.* **16** 466–70

[28] Wang H, Chen Y, Hou X, Ma C and Tan T 2016 Nitrogen-doped graphenes as efficient electrocatalysts for the selective reduction of carbon dioxide to formate in aqueous solution *Green Chem.* **18** 3250–6

[29] Li W, Seredych M, Rodríguez-Castellón E and Bandosz T J 2016 Metal-free nanoporous carbon as a catalyst for electrochemical reduction of CO_2 to CO and CH_4 *ChemSusChem* **9** 606–16

[30] Wang Y, Hou P, Wang Z and Kang P 2017 Zinc imidazolate metal–organic frameworks (ZIF-8) for electrochemical reduction of CO_2 to CO *ChemPhysChem* **18** 3142–7

[31] Zou X, Liu M, Wu J, Ajayan P M, Li J, Liu B and Yakobson B I 2017 How nitrogen-doped graphene quantum dots catalyze electroreduction of CO_2 to hydrocarbons and oxygenates *ACS Catal.* **7** 6245–50

[32] Liu Y, Zhao J and Cai Q 2016 Pyrrolic-nitrogen doped graphene: a metal-free electrocatalyst with high efficiency and selectivity for the reduction of carbon dioxide to formic acid: a computational study *Phys. Chem. Chem. Phys.* **18** 5491–8

[33] Chai G-L and Guo Z-X 2016 Highly effective sites and selectivity of nitrogen-doped graphene/CNT catalysts for CO_2 electrochemical reduction *Chem. Sci.* **7** 1268–75

[34] Xu J *et al* 2016 Revealing the origin of activity in nitrogen-doped nanocarbons towards electrocatalytic reduction of carbon dioxide *ChemSusChem* **9** 1085–9

[35] Yang H, Wu Y, Lin Q, Fan L, Chai X, Zhang Q, Liu J, He C and Lin Z 2018 Composition tailoring via N and S co-doping and structure tuning by constructing hierarchical pores: metal-free catalysts for high-performance electrochemical reduction of CO_2 *Angew. Chem. Int. Ed.* **57** 15476–80

[36] Xie J, Zhao X, Wu M, Li Q, Wang Y and Yao J 2018 Metal-free fluorine-doped carbon electrocatalyst for CO_2 reduction outcompeting hydrogen evolution *Angew. Chem. Int. Ed.* **57** 9640–4

[37] Navalon S, Dhakshinamoorthy A, Alvaro M and Garcia H 2016 Metal nanoparticles supported on two-dimensional graphenes as heterogeneous catalysts *Coord. Chem. Rev.* **312** 99–148

[38] Song Y *et al* 2016 High-selectivity electrochemical conversion of CO_2 to ethanol using a copper nanoparticle/n-doped graphene electrode *ChemistrySelect* **1** 6055–61

[39] Lv K, Fan Y, Zhu Y, Yuan Y, Wang J, Zhu Y and Zhang Q 2018 Elastic Ag-anchored N-doped graphene/carbon foam for the selective electrochemical reduction of carbon dioxide to ethanol *J. Mater. Chem.* A **6** 5025–31

[40] Li Y, Su H, Chan S H and Sun Q 2015 CO_2 electroreduction performance of transition metal dimers supported on graphene: a theoretical study *ACS Catal.* **5** 6658–64

[41] Kim D, Resasco J, Yu Y, Asiri A M and Yang P 2014 Synergistic geometric and electronic effects for electrochemical reduction of carbon dioxide using gold–copper bimetallic nanoparticles *Nat. Commun.* **5** 4948

[42] Kim S, Dong W J, Gim S, Sohn W, Park J Y, Yoo C J, Jang H W and Lee J-L 2017 Shape-controlled bismuth nanoflakes as highly selective catalysts for electrochemical carbon dioxide reduction to formate *Nano Energy* **39** 44–52

[43] Dai L *et al* 2017 Ultrastable atomic copper nanosheets for selective electrochemical reduction of carbon dioxide *Sci. Adv.* **3** e1701069

[44] Dai L, Xue Y, Qu L, Choi H-J and Baek J-B 2015 Metal-free catalysts for oxygen reduction reaction *Chem. Rev.* **115** 4823–92

[45] Li F, Chen L, Knowles G P, MacFarlane D R and Zhang J 2017 Hierarchical mesoporous SnO_2 nanosheets on carbon cloth: a robust and flexible electrocatalyst for CO_2 reduction with high efficiency and selectivity *Angew. Chem. Int. Ed.* **56** 505–9

[46] Gao S *et al* 2016 Ultrathin Co_3O_4 layers realizing optimized CO_2 electroreduction to formate *Angew. Chem. Int. Ed.* **55** 698–702

[47] Gao S, Lin Y, Jiao X, Sun Y, Luo Q, Zhang W, Li D, Yang J and Xie Y 2016 Partially oxidized atomic cobalt layers for carbon dioxide electroreduction to liquid fuel *Nature* **529** 68–71

[48] Gao S *et al* 2017 Atomic layer confined vacancies for atomic-level insights into carbon dioxide electroreduction *Nat. Commun.* **8** 14503

[49] Asadi M *et al* 2014 Robust carbon dioxide reduction on molybdenum disulphide edges *Nat. Commun.* **5** 4470

[50] Chan K, Tsai C, Hansen H A and Nørskov J K 2014 Molybdenum sulfides and selenides as possible electrocatalysts for CO_2 reduction *ChemCatChem.* **6** 1899–905

[51] Hong X, Chan K, Tsai C and Nørskov J K 2016 How doped MoS_2 breaks transition-metal scaling relations for CO_2 electrochemical reduction *ACS Catal.* **6** 4428–37

[52] Sun X, Zhu Q, Kang X, Liu H, Qian Q, Zhang Z and Han B 2016 Molybdenum–bismuth bimetallic chalcogenide nanosheets for highly efficient electrocatalytic reduction of carbon dioxide to methanol *Angew. Chem. Int. Ed.* **55** 6771–5

[53] Long X, Li G, Wang Z, Zhu H, Zhang T, Xiao S, Guo W and Yang S 2015 Metallic iron–nickel sulfide ultrathin nanosheets as a highly active electrocatalyst for hydrogen evolution reaction in acidic media *J. Am. Chem. Soc.* **137** 11900–3

[54] Li Z and Wu Y 2019 2D early transition metal carbides (MXenes) for catalysis *Small* **15** 1804736

[55] Luc W, Ko B H, Kattel S, Li S, Su D, Chen J G and Jiao F 2019 SO_2-induced selectivity change in CO_2 electroreduction *J. Am. Chem. Soc.* **141** 9902–9

[56] Haegel N M *et al* 2017 Terawatt-scale photovoltaics: trajectories and challenges *Science* **356** 141–3

[57] Zhang S, Fan Q, Xia R and Meyer T J 2020 CO_2 reduction: from homogeneous to heterogeneous electrocatalysis *Acc. Chem. Res.* **53** 255–64

[58] Zhang S, Kang P, Ubnoske S, Brennaman M K, Song N, House R L, Glass J T and Meyer T J 2014 Polyethylenimine-enhanced electrocatalytic reduction of CO_2 to formate at nitrogen-doped carbon nanomaterials *J. Am. Chem. Soc.* **136** 7845–8

[59] Li N, Chen X, Ong W-J, MacFarlane D R, Zhao X, Cheetham A K and Sun C 2017 Understanding of electrochemical mechanisms for CO_2 capture and conversion into hydrocarbon fuels in transition-metal carbides (MXenes) *ACS Nano* **11** 10825–33

[60] Li P, Zhu J, Handoko A D, Zhang R, Wang H, Legut D, Wen X, Fu Z, Seh Z W and Zhang Q 2018 High-throughput theoretical optimization of the hydrogen evolution reaction on MXenes by transition metal modification *J. Mater. Chem.* A **6** 4271–8

[61] Xiao Y and Zhang W 2020 High throughput screening of M_3C_2 MXenes for efficient CO_2 reduction conversion into hydrocarbon fuels *Nanoscale* **12** 7660–73

[62] Handoko A D, Khoo K H, Tan T L, Jin H and Seh Z W 2018 Establishing new scaling relations on two-dimensional MXenes for CO_2 electroreduction *J. Mater. Chem.* A **6** 21885–90

[63] Huang Y, Handoko A D, Hirunsit P and Yeo B S 2017 Electrochemical reduction of CO_2 using copper single-crystal surfaces: effects of CO* coverage on the selective formation of ethylene *ACS Catal.* **7** 1749–56

[64] Chen H, Handoko A D, Xiao J, Feng X, Fan Y, Wang T, Legut D, Seh Z W and Zhang Q 2019 Catalytic effect on CO_2 electroreduction by hydroxyl-terminated two-dimensional MXenes *ACS Appl. Mater. Interfaces* **11** 36571–9

[65] Seh Z W, Kibsgaard J, Dickens C F, Chorkendorff I, Nørskov J K and Jaramillo T F 2017 Combining theory and experiment in electrocatalysis: insights into materials design *Science* **355** eaad4998

[66] Chen Y, Bellini M, Bevilacqua M, Fornasiero P, Lavacchi A, Miller H A, Wang L and Vizza F 2015 Direct alcohol fuel cells: toward the power densities of hydrogen-fed proton exchange membrane fuel cells *ChemSusChem* **8** 524–33

[67] Ning S, Ding L, Lin Z, Lin Q, Zhang H, Lin H, Long J and Wang X 2016 One-pot fabrication of Bi_3O_4Cl/BiOCl plate-on-plate heterojunction with enhanced visible-light photocatalytic activity *Appl. Catal.* B **185** 203–12

[68] Chen G-F, Cao X, Wu S, Zeng X, Ding L-X, Zhu M and Wang H 2017 Ammonia electrosynthesis with high selectivity under ambient conditions via a Li^+ incorporation strategy *J. Am. Chem. Soc.* **139** 9771–4

[69] Rittle J and Peters J C 2016 An Fe-N_2 complex that generates hydrazine and ammonia via Fe=NNH_2: demonstrating a hybrid distal-to-alternating pathway for N_2 reduction *J. Am. Chem. Soc.* **138** 4243–8

[70] Jia H-P and Quadrelli E A 2014 Mechanistic aspects of dinitrogen cleavage and hydrogenation to produce ammonia in catalysis and organometallic chemistry: relevance of metal hydride bonds and dihydrogen *Chem. Soc. Rev.* **43** 547–64

[71] Pickett C J and Talarmin J 1985 Electrosynthesis of ammonia *Nature* **317** 652–3

[72] Brown K A *et al* 2016 Light-driven dinitrogen reduction catalyzed by a CdS:nitrogenase MoFe protein biohybrid *Science* **352** 448–50

[73] van der Ham C J M, Koper M T M and Hetterscheid D G H 2014 Challenges in reduction of dinitrogen by proton and electron transfer *Chem. Soc. Rev.* **43** 5183–91

[74] Marnellos G and Stoukides M 1998 Ammonia synthesis at atmospheric pressure *Science* **282** 98–100

[75] Bao D, Zhang Q, Meng F-L, Zhong H-X, Shi M-M, Zhang Y, Yan J-M, Jiang Q and Zhang X-B 2017 Electrochemical reduction of N_2 under ambient conditions for artificial N_2 fixation and renewable energy storage using N_2/NH_3 cycle *Adv. Mater.* **29** 1604799

[76] Cheng M-J, Kwon Y, Head-Gordon M and Bell A T 2015 Tailoring metal-porphyrin-like active sites on graphene to improve the efficiency and selectivity of electrochemical CO_2 reduction *J. Phys. Chem.* C **119** 21345–52

[77] Bauer N 1960 Theoretical pathways for the reduction of N_2 molecules in aqueous media: thermodynamics of $N_2H_n{}^1$ *J. Phys. Chem.* **64** 833–7

[78] Shilov A E 2003 Catalytic reduction of molecular nitrogen in solutions *Russ. Chem. Bull.* **52** 2555–62

[79] Shipman M A and Symes M D 2017 Recent progress towards the electrosynthesis of ammonia from sustainable resources *Catal. Today* **286** 57–68

[80] Skúlason E, Bligaard T, Gudmundsdóttir S, Studt F, Rossmeisl J, Abild-Pedersen F, Vegge T, Jónsson H and Nørskov J K 2012 A theoretical evaluation of possible transition metal electro-catalysts for N_2 reduction *Phys. Chem. Chem. Phys.* **14** 1235–45

[81] Abghoui Y, Garden A L, Hlynsson V F, Björgvinsdóttir S, Ólafsdóttir H and Skúlason E 2015 Enabling electrochemical reduction of nitrogen to ammonia at ambient conditions through rational catalyst design *Phys. Chem. Chem. Phys.* **17** 4909–18

[82] Zhao J and Chen Z 2017 Single Mo atom supported on defective boron nitride monolayer as an efficient electrocatalyst for nitrogen fixation: a computational study *J. Am. Chem. Soc.* **139** 12480–7
[83] Li S-J, Bao D, Shi M-M, Wulan B-R, Yan J-M and Jiang Q 2017 Amorphizing of Au nanoparticles by CeO_x–RGO hybrid support towards highly efficient electrocatalyst for N_2 reduction under ambient conditions *Adv. Mater.* **29** 1700001
[84] Zheng X *et al* 2016 Robust ultra-low-friction state of graphene via moiré superlattice confinement *Nat. Commun.* **7** 13204
[85] Huang W *et al* 2015 Highly active and durable methanol oxidation electrocatalyst based on the synergy of platinum–nickel hydroxide–graphene *Nat. Commun.* **6** 10035
[86] Li H, Shang J, Ai Z and Zhang L 2015 Efficient visible light nitrogen fixation with BiOBr nanosheets of oxygen vacancies on the exposed {001} facets *J. Am. Chem. Soc.* **137** 6393–9
[87] Montoya J H, Tsai C, Vojvodic A and Nørskov J K 2015 The challenge of electrochemical ammonia synthesis: a new perspective on the role of nitrogen scaling relations *ChemSusChem* **8** 2180–6
[88] Yu X and Pickup P G 2008 Recent advances in direct formic acid fuel cells (DFAFC) *J. Power Sources* **182** 124–32
[89] Feng L, Chang J, Jiang K, Xue H, Liu C, Cai W-B, Xing W and Zhang J 2016 Nanostructured palladium catalyst poisoning depressed by cobalt phosphide in the electro-oxidation of formic acid for fuel cells *Nano Energy* **30** 355–61
[90] Winter M and Brodd R J 2004 What are batteries, fuel cells, and supercapacitors? *Chem. Rev.* **104** 4245–70
[91] Liu J, Huang Z, Cai K, Zhang H, Lu Z, Li T, Zuo Y and Han H 2015 Clean synthesis of an economical 3D nanochain network of PdCu alloy with enhanced electrocatalytic performance towards ethanol oxidation *Chem.—Eur. J.* **21** 17779–85
[92] Huang H and Wang X 2014 Recent progress on carbon-based support materials for electrocatalysts of direct methanol fuel cells *J. Mater. Chem.* A **2** 6266–91
[93] Lv R *et al* 2011 Open-ended, N-doped carbon nanotube–graphene hybrid nanostructures as high-performance catalyst support *Adv. Funct. Mater.* **21** 999–1006
[94] Li Y-H, Hung T-H and Chen C-W 2009 A first-principles study of nitrogen- and boron-assisted platinum adsorption on carbon nanotubes *Carbon* **47** 850–5
[95] Jiang Y, Yan Y, Chen W, Khan Y, Wu J, Zhang H and Yang D 2016 Single-crystalline Pd square nanoplates enclosed by {100} facets on reduced graphene oxide for formic acid electro-oxidation *Chem. Commun.* **52** 14204–7
[96] Huang X, Tang S, Mu X, Dai Y, Chen G, Zhou Z, Ruan F, Yang Z and Zheng N 2011 Freestanding palladium nanosheets with plasmonic and catalytic properties *Nat. Nanotechnol.* **6** 28–32
[97] Yang N *et al* 2017 Synthesis of ultrathin PdCu alloy nanosheets used as a highly efficient electrocatalyst for formic acid oxidation *Adv. Mater.* **29** 1700769
[98] Chia X and Pumera M 2018 Characteristics and performance of two-dimensional materials for electrocatalysis *Nat. Catal.* **1** 909–21
[99] Yuan W, Zhou Y, Li Y, Li C, Peng H, Zhang J, Liu Z, Dai L and Shi G 2013 The edge- and basal-plane-specific electrochemistry of a single-layer graphene sheet *Sci. Rep.* **3** 2248
[100] Benson J, Xu Q, Wang P, Shen Y, Sun L, Wang T, Li M and Papakonstantinou P 2014 Tuning the catalytic activity of graphene nanosheets for oxygen reduction reaction via size and thickness reduction *ACS Appl. Mater. Interfaces* **6** 19726–36

[101] Deng D, Yu L, Pan X, Wang S, Chen X, Hu P, Sun L and Bao X 2011 Size effect of graphene on electrocatalytic activation of oxygen *Chem. Commun.* **47** 10016–8
[102] Tao L, Wang Q, Dou S, Ma Z, Huo J, Wang S and Dai L 2016 Edge-rich and dopant-free graphene as a highly efficient metal-free electrocatalyst for the oxygen reduction reaction *Chem. Commun.* **52** 2764–7
[103] Uosaki K, Elumalai G, Noguchi H, Masuda T, Lyalin A, Nakayama A and Taketsugu T 2014 Boron nitride nanosheet on gold as an electrocatalyst for oxygen reduction reaction: theoretical suggestion and experimental proof *J. Am. Chem. Soc.* **136** 6542–5
[104] Jaramillo T F, Jørgensen K P, Bonde J, Nielsen J H, Horch S and Chorkendorff I 2007 Identification of active edge sites for electrochemical H_2 evolution from MoS_2 nanocatalysts *Science* **317** 100–2
[105] Kibsgaard J, Chen Z, Reinecke B N and Jaramillo T F 2012 Engineering the surface structure of MoS_2 to preferentially expose active edge sites for electrocatalysis *Nat. Mater.* **11** 963–9
[106] Wang H, Kong D, Johanes P, Cha J J, Zheng G, Yan K, Liu N and Cui Y 2013 $MoSe_2$ and WSe_2 nanofilms with vertically aligned molecular layers on curved and rough surfaces *Nano Lett.* **13** 3426–33
[107] Tsai C, Chan K, Nørskov J K and Abild-Pedersen F 2015 Theoretical insights into the hydrogen evolution activity of layered transition metal dichalcogenides *Surf. Sci.* **640** 133–40
[108] Voiry D, Salehi M, Silva R, Fujita T, Chen M, Asefa T, Shenoy V B, Eda G and Chhowalla M 2013 Conducting MoS_2 nanosheets as catalysts for hydrogen evolution reaction *Nano Lett.* **13** 6222–7
[109] Lukowski M A, Daniel A S, English C R, Meng F, Forticaux A, Hamers R J and Jin S 2014 Highly active hydrogen evolution catalysis from metallic WS_2 nanosheets *Energy Environ. Sci.* **7** 2608–13
[110] Seh Z W, Fredrickson K D, Anasori B, Kibsgaard J, Strickler A L, Lukatskaya M R, Gogotsi Y, Jaramillo T F and Vojvodic A 2016 Two-dimensional molybdenum carbide (MXene) as an efficient electrocatalyst for hydrogen evolution *ACS Energy Lett.* **1** 589–94
[111] Asadi M *et al* 2016 Nanostructured transition metal dichalcogenide electrocatalysts for CO_2 reduction in ionic liquid *Science* **353** 467–70
[112] Song F and Hu X 2014 Exfoliation of layered double hydroxides for enhanced oxygen evolution catalysis *Nat. Commun.* **5** 4477
[113] Zhao Y *et al* 2018 Sub-3 nm ultrafine monolayer layered double hydroxide nanosheets for electrochemical water oxidation *Adv. Energy Mater.* **8** 1703585
[114] Ren X *et al* 2017 Few-layer black phosphorus nanosheets as electrocatalysts for highly efficient oxygen evolution reaction *Adv. Energy Mater.* **7** 1700396
[115] Menzel N, Ortel E, Kraehnert R and Strasser P 2012 Electrocatalysis using porous nanostructured materials *ChemPhysChem* **13** 1385–94
[116] Wang G, Tao J, Zhang Y, Wang S, Yan X, Liu C, Hu F, He Z, Zuo Z and Yang X 2018 Engineering two-dimensional mass-transport channels of the MoS_2 nanocatalyst toward improved hydrogen evolution performance *ACS Appl. Mater. Interfaces* **10** 25409–14
[117] Qu L, Liu Y, Baek J-B and Dai L 2010 Nitrogen-doped graphene as efficient metal-free electrocatalyst for oxygen reduction in fuel cells *ACS Nano* **4** 1321–6
[118] Guo D, Shibuya R, Akiba C, Saji S, Kondo T and Nakamura J 2016 Active sites of nitrogen-doped carbon materials for oxygen reduction reaction clarified using model catalysts *Science* **351** 361–5

[119] Wang L, Ambrosi A and Pumera M 2013 'Metal-free' catalytic oxygen reduction reaction on heteroatom-doped graphene is caused by trace metal impurities *Angew. Chem. Int. Ed.* **52** 13818–21

[120] Zhao J and Chen Z 2015 Carbon-doped boron nitride nanosheet: an efficient metal-free electrocatalyst for the oxygen reduction reaction *J. Phys. Chem.* C **119** 26348–54

[121] Elumalai G, Noguchi H, Lyalin A, Taketsugu T and Uosaki K 2016 Gold nanoparticle decoration of insulating boron nitride nanosheet on inert gold electrode toward an efficient electrocatalyst for the reduction of oxygen to water *Electrochem. Commun.* **66** 53–7

[122] Wang J, Hao J, Liu D, Qin S, Portehault D, Li Y, Chen Y and Lei W 2017 Porous boron carbon nitride nanosheets as efficient metal-free catalysts for the oxygen reduction reaction in both alkaline and acidic solutions *ACS Energy Lett.* **2** 306–12

[123] Bonde J, Moses P G, Jaramillo T F, Nørskov J K and Chorkendorff I 2009 Hydrogen evolution on nano-particulate transition metal sulfides *Faraday Discuss.* **140** 219–31

[124] Huang X, Leng M, Xiao W, Li M, Ding J, Tan T L, Lee W S V and Xue J 2017 Activating basal planes and S-terminated edges of MoS_2 toward more efficient hydrogen evolution *Adv. Funct. Mater.* **27** 1604943

[125] Jiao Y, Zheng Y, Davey K and Qiao S-Z 2016 Activity origin and catalyst design principles for electrocatalytic hydrogen evolution on heteroatom-doped graphene *Nat. Energy* **1** 16130

[126] Handoko A D, Fredrickson K D, Anasori B, Convey K W, Johnson L R, Gogotsi Y, Vojvodic A and Seh Z W 2018 Tuning the basal plane functionalization of two-dimensional metal carbides (MXenes) to control hydrogen evolution activity *ACS Appl. Energy Mater.* **1** 173–80

[127] Li D, Ren B, Jin Q, Cui H and Wang C 2018 Nitrogen-doped, oxygen-functionalized, edge- and defect-rich vertically aligned graphene for highly enhanced oxygen evolution reaction *J. Mater. Chem.* A **6** 2176–83

[128] Zhang Z, Khurram M, Sun Z and Yan Q 2018 Uniform tellurium doping in black phosphorus single crystals by chemical vapor transport *Inorg. Chem.* **57** 4098–103

[129] Zheng Y, Jiao Y, Zhu Y, Cai Q, Vasileff A, Li L H, Han Y, Chen Y and Qiao S-Z 2017 Molecule-level g-C_3N_4 coordinated transition metals as a new class of electrocatalysts for oxygen electrode reactions *J. Am. Chem. Soc.* **139** 3336–9

[130] Liu R, Wang Y, Liu D, Zou Y and Wang S 2017 Water-plasma-enabled exfoliation of ultrathin layered double hydroxide nanosheets with multivacancies for water oxidation *Adv. Mater.* **29** 1701546

[131] Wang Y, Zhang Y, Liu Z, Xie C, Feng S, Liu D, Shao M and Wang S 2017 Layered double hydroxide nanosheets with multiple vacancies obtained by dry exfoliation as highly efficient oxygen evolution electrocatalysts *Angew. Chem. Int. Ed.* **56** 5867–71

IOP Publishing

Muhammad Ikram, Ali Raza, Jahan Zeb Hassan and Salamat Ali

Chapter 7

Strategies to improve electrocatalytic activity

Parts of this chapter have been reprinted with permission from [17]. Copyright 2023 Elsevier.

Some unique ways to improve the electrocatalytic activity of geometrical 2D materials have been presented. These include increasing the number of active sites, heteroatom doping, phase engineering, heterostructure formation, and synergistic modulation to improve the electrode material's electrical conductivity, exposure to active catalyst sites, and potential reaction barriers. The final section provides insight and valuable suggestions for developing efficient 2D TMDs electrocatalysts. This chapter offers various strategies for the improvement of electrocatalytic activity for the HER, ORR, and ECR as well. Carrier dissociation and migration kinetics are critical challenges in improving electrocatalysis and photocatalysis, and can be significantly associated with the structure–activity relationship of catalysts. So far, 2D layered materials, for example, graphene and graphite-like materials, continue to face numerous challenges, which limit their practicability and functionality in various applications such as sensors, semiconductors, and catalysis. Thus, many diverse routes should be discovered and investigated to yield 2D layered material-centered catalysts that result in boosted photocatalysis and, most significantly, electrocatalysis activity. Evidently, 2D catalysts/nanocatalysts present numerous advantages over 3D bulk catalysts, which are described. Finally, remarks regarding various strategies for the improvement of electrocatalytic materials are also provided.

7.1 The hydrogen evolution reaction

7.1.1 Engineering protocols

The 2D electrocatalysts, such as MoS_2 [1], $MoSe_2$, WS_2 [2], and NbS_2 [3], etc, have shown themselves to be attractive electrocatalysis candidates over the last few years. This is in response to the drawbacks associated with noble metals and similar

doi:10.1088/978-0-7503-5291-8ch7

compounds. Some innovative methods have been recommended to boost the performance of the HER further to maximize the usage of the one-of-a-kind features possessed by TMDs.

7.1.1.1 Defect engineering

7.1.1.1.1 Size control and edge enrichment

Reducing the lateral size of monolayers can enhance catalytic activity by increasing the density of edge sites. Lin *et al* [4] developed a method for producing monolayered 1H-MoS_2 nanoclusters with lateral dimensions of roughly 12.5 nm (figure 7.1) [4]. The Tafel slope and onset overpotential of the 1H-MoS_2 nanoclusters (on a glassy

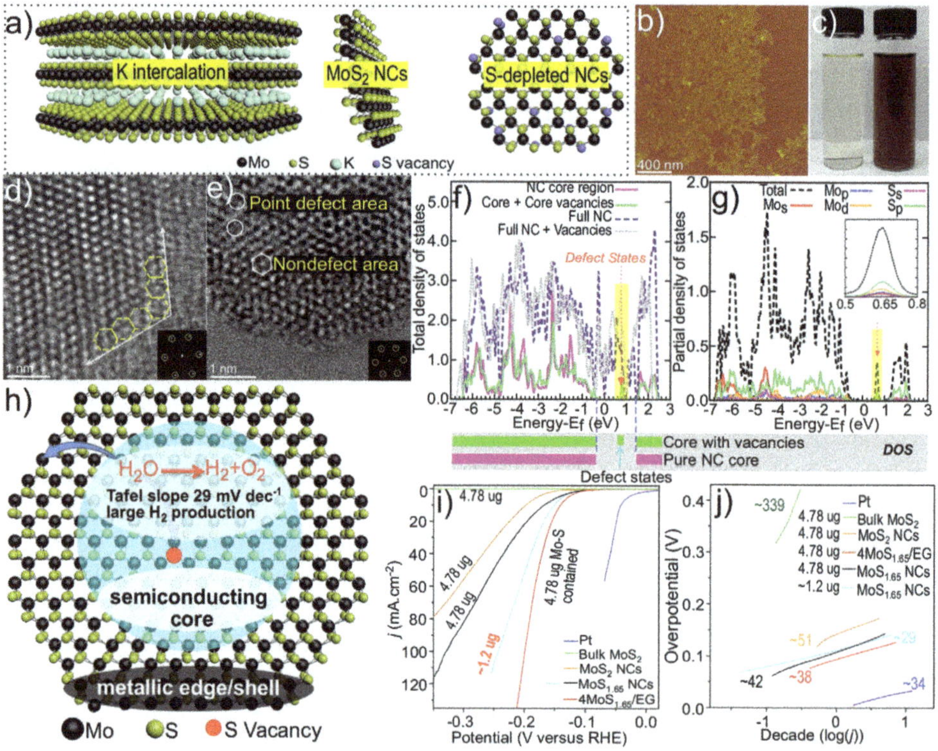

Figure 7.1. (a) The synthesis of monolayer MoS_2 nanoclusters. (Reproduced with permission from [7]. Copyright 2017 Elsevier.) (b) AFM photograph of the prepared nanoclusters. (c) The optical appearance of the prepared product (left) and after S depletion (right). (d) TEM photograph of the prepared product and (e) after S depletion (the insets correspond to FFT patterns). (f) Evaluated DOS for MoS_2 in the edge and/or core region and the entire NC. (g) Decomposition of the entire DOS for MoS_2 in the core with S depletion (S vacancies) and edge regions into limited DOS of the Mo and S orbitals. (h) Demonstration of near-edge regions, the metallic edge, and the semiconducting core of MoS_2 nanoclusters. (i) Polarization curves of different catalysts in 0.5 M H_2SO_4 and (j) Tafel plots. (Reproduced with permission from [4]. Copyright 2016 American Chemical Society.)

carbon electrode) were 51 mV dec^{-1} and 120–140 mV, respectively (figures 7.1(i) and (j)), which are superior to monolayered 1H-MoS_2 sheets (on a Au foil electrode) [5] and comparable to monolayered 1T-WS_2 nanosheets (on a glassy carbon electrode) [6]. Theoretical calculations (figures 7.1(f) and (g)) revealed that the 1H-MoS_2 NC has a metallic surrounding zone that wraps around the semiconducting core, which enhances charge transport at the catalytically active regions [7]. The formation of pores inside larger monolayers can further enrich the catalytically active edge sites. Ajayan and colleagues demonstrated this hypothesis using 1H-MoS_2 monolayer triangles generated by a CVD technique [8]. These massively crystalline 1H monolayers (up to 100 μm in length) were largely entirely inactive for the HER (a Tafel slope of 342 mV dec^{-1} and an onset overpotential of roughly 500 mV) [8]. The samples were activated for the HER by developing O_2 plasma pores, which produced a Tafel slope of 162–171 mV dec^{-1} and an initial overpotential of 400 mV. At the freshly created inside edges of the rims, S- and Mo-terminated structures were found in equal quantities. Another method for creating pores and edges inside the monolayer is to anneal MoS_2 monolayers in the presence of hydrogen gas. These edge-rich monolayers' Tafel slope and onset overpotential were measured to be 117 mV dec^{-1} and 300 mV. The HER performance may be improved even further by starting with lower crystal sizes (100–200 nm), with a Tafel slope of 50 mV dec^{-1} and an initial overpotential of 120 mV [9].

To create pores, ultrathin sheets of TaS_2 were also treated with O_2 plasma [10]. Chemical exfoliation was used to create the first extremely crystalline TaS_2 sheets. The lateral dimension of these TaS_2 sheets was roughly 15 μm, and the thickness ranged between one and three layers [10]. After the formation of pores, the Tafel slope and onset overpotential were decreased, going from values of 215 mV dec^{-1} and 310 mV, respectively, to values of 125–142 mV dec^{-1} and 225 mV. This change occurred after the Tafel slope had been decreased [10]. In addition, oxidation with H_2O_2 could be consolidated with the typical liquid exfoliation process for MoSe2 to generate chemically induced pores [11]. The thin sheets of exfoliated $MoSe_2$ were oxidized and etched throughout this technique to create a porous structure with high porosity. The onset overpotential of the porous $MoSe_2$ was decreased from 220 to 75 mV by this pore generation process, while the Tafel slope remained intact at 80 mV dec^{-1} [12]. Edge enrichment and the formation of monolayered nanoclusters with tiny lateral sizes (e.g. 5–10 nm) are additional appealing strategies. Small lateral size systems, even for multilayered TMDs nanoclusters with partial 1T phase, exhibit excellent HER activity [13]. The synthesis of monolayered TMDs nanoclusters is an inspiring strategy for achieving optimal HER catalytic efficiency. This is especially true when recent synthesis technique advancements are considered [14, 15]. Even though the stability of the 1T/1T′ TMDs still needs to be suitably determined, the edge richness and size strategic planning can at the very least be supplemented with the phase engineering described above. After this phase engineering is complete, the following sections will discuss how various strategies, in addition to controlling the HER's edge and size, may be implemented in order to enhance its efficiency.

7.1.1.1.2 Defects and strains

Although phase engineering is a powerful technique for harnessing the catalytic potential of TMDs, there is much benefit in researching and attempting to optimize stable 1H/2H TMDs thermodynamically for the HER. As previously stated, one of the most crucial variables to consider is that the basal plane of 1H/2H TMDs is not highly active for the HER, which adds a considerable amount to the overall surface area. Wang and colleagues calculated ΔG_{H*} during the Volmer reaction and suggested that different types of sulfur vacancies on the basal plane of MoS_2 could boost HER activity; these new S-depleted sites being suited for both Tafel and Heyrovsky reactions [16, 17]. Even though proper management of these S vacancies remains challenging, many published studies have shown that MoS_2 monolayers may be capable of creating some advantageous S vacancies [18]. In 2016, studies on the HER catalysis of monolayered MoS_2 nanoclusters (see figure 7.1(a)) were published [4]. Cation exchange resin was used on these newly formed 1H-MoS_2 nanoclusters, which had lateral sizes of approximately 12.5 nm, to generate S-depleted $MoS_{1.65}$ nanoclusters. When compared to the catalytic properties of pure MoS_2 nanoclusters, the HER activity saw a substantial increase (Tafel slope of 51 mV dec^{-1}, onset overpotential of 120–140 mV of pure nanoclusters). The Tafel slope of these S-depleted nanoclusters was 29 mV dec^{-1}, the onset overpotential was 60–75 mV (as determined by the Tafel plot and a low Tafel slope of approximately 30 mV dec^{-1}) which suggested a Volmer–Tafel mechanism reaction of the Mo–S nanoclusters, which is incredibly exciting and one of the most potent Mo–S alkaline catalysts ever established. DFT calculations revealed the additional consequences of defect generation, such as increased lattice strain and distinctive electronic states found underneath the conduction band, which increased the effectiveness of electron transport while also facilitating HER activity [17]. It is first necessary to create controlled S depletion and the related HER catalytic potential to enhance activity. Zheng and colleagues [19] performed an in-depth investigation to identify the link between the MoS_2 point S vacancies and the HER catalytic performance on the defective sites (figure 7.2(a)). If there is no S depletion on the substratum, the expected value of ΔG_{H*} on the basal plane is approximately 2eV (catalytic insert for the HER). With a surface S depletion of 3.12% (figure 7.2(b)), this value drops to ±0.18 eV and approaches 0.08 eV with 9.38%–18.75% S vacancies. More sophisticated handling of the S vacancies can result in thermo neutral ΔG_{H*}, which is an improvement over the best-configured edge sites. The ΔG_{H*} on the basal plane can also be adjusted during the Volmer methodology to be negligible and close to 0 eV, which results in more efficient hydrogen atom adsorbent (figure 7.2(b)).

A small amount of applied strain can enhance stability while decreasing the number of S depletions for the greatest achievable HER, given that strain in the basal plane improves HER catalytic efficiency due to greater hydrogen binding [20], Further research was conducted on the ΔG_{H*} of MoS_2 monolayers subjected to uniaxial strain energy. As seen in figure 7.2(c), the ΔG_{H*} decreased with increasing strain in all the S vacancy concentrations tested. This finding is useful since the stability of MoS_2 monolayers decreases as the number of S vacancies grows. The modeled electronic constructions demonstrated that S depletion resulted in the formation of a defect level that was situated between the gap and beneath the

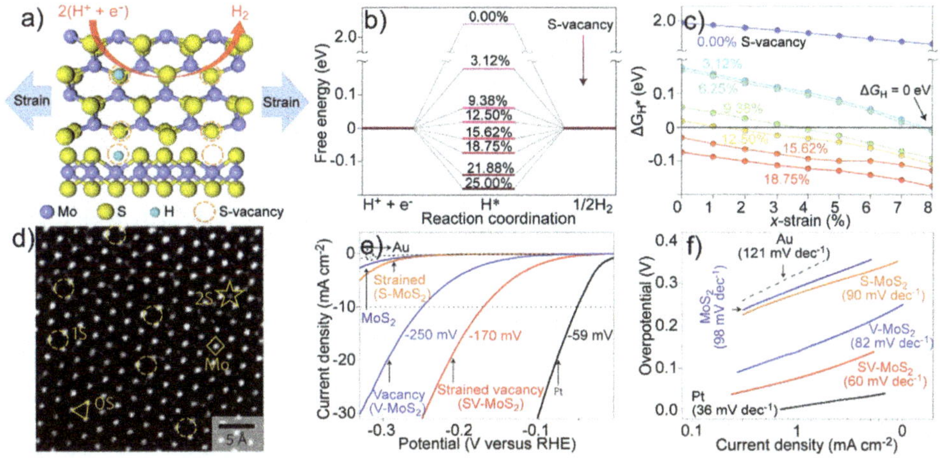

Figure 7.2. (a) Top and bottom views of MoS_2 with strained S depletion (S vacancies) on the basal plane. Theoretical assessment for ΔG_{H^*} versus (b) reaction coordination of HER by introducing S vacancies and (c) x-strain with S vacancies. (d) The aberration-corrected TEM photograph of MoS_2 monolayer (with 4×4 nm dimensions and 43 S vacancies (~11.3%)). (e), (f) Polarization curves and the corresponding Tafel plots of various species (S-MoS_2, V-MoS_2, and SV-MoS_2 are the MoS_2 monolayers with 0% S vacancy, 12.5 2.5% S vacant positions, and both 1.35 0.15% strain and 12.5 2.5% S vacancies, in both). (Reproduced with permission from [19]. Copyright 2016 Springer Nature.)

bottom of the conduction band minimum [7, 19]. A single-layer of 1H-MoS_2 is an example of an n-type semiconductor, and its Fermi level should be close to the conduction band [7]. These newly produced gap states cause an increase in the amount of hydrogen adsorption that occurs on the S-depleted Mo sites of the basal plane (localized around the S vacancy) [17]. The progressively enhanced H binding may be logically explained by an increase in the number of S vacancies and a greater closeness to the Fermi level (as well as an increase in the number of gap states) (figures 7.2(b) and (c)). Tensile strain also affects the gap states, which aids in hydrogen bonding [6]. Zheng and his colleagues created a massive 1H-MoS_2 monolayer sheet with the lowest feasible edge ratio. Their approach was founded on the previously mentioned theoretical understandings. Both strained (1.35% ± 0.15%) and unstrained (1.35% ± 0.15%) MoS_2 single-layers have been produced with varying percentages of S vacancies (for example, 21.88%, 18.75%, 12.50%, 8.00%, and 6.25%, relative to the total number of S atoms).

Additionally, theoretical simulation results show that the overstretched MoS_2 monolayer with 12.5% S vacancies has the best-matched ΔG_{H^*} to 0 eV. The strain in this monolayer is 1.35% ± 0.15% [19]. The strained pure MoS_2 sheets had poor HER catalytic activity, as demonstrated by an initial overpotential larger than 350 mV (as measured by the current density at 10 mV cm^{-2}), and the Tafel slope was 90 mV dec^{-1}. The TEM of a MoS_2 monolayer with 43 S vacancies/S depletion (approximately 11.3%) shows that Mo atoms, two S atoms (2S, one S atom above the other underneath the Mo plane), one S atom (1S, only one S below the Mo plane), and zero S atoms (0S, both the S atoms below and above the Mo plane have been depleted, indicating that the S atoms were missing, see figure 7.2(d)). This HER performance outperformed that of unstrained pure MoS_2 (e.g. the slope of Tafel 98 mV dec^{-1}). The onset increases in strength and the Tafel slope of the MoS_2 monolayer initially decreased to 250 mV and 82 mV dec^{-1}, respectively, after S depletion (12.5% S vacancies) [17]. Consequently, they have been reduced even lower to 170 mV and 60 mV dec^{-1} with a cumulative 1.35% ± 0.15% strain (figures 7.2(e) and (f)). The best HER catalytic potential illustrated is less than the phase-engineered and edge-enriched TMDs discussed in the previous section [6, 21]. However, the performance enhancement brought about by S depletion and strain demonstrates a fascinating, potentially fruitful route for HER enhancement. The basal plane ΔG_{H*} values of the 1H and 1T 2D TMDs are all larger than 0 eV. Surface chalcogen vacancies have the potential to increase H binding while decreasing ΔG_{H*}, resulting in a more thermodynamically neutral state. This improvement has already been illustrated on the 2D $MoSe_2$ and $MoSSe_2$ nano-clusters with Se vacant positions [13]. TMDs' stability is decreased due to surface vacancies and strain, preventing further applications of the technique. To address this issue, additional approaches, such as surface doping, can be used to stabilize the structure while obtaining the optimal ΔG_{H*}.

7.1.1.2 Phase engineering

The TMDs are one-of-a-kind layered materials, as previously stated. TMDs can exist in three states: 1H, 1T, or 1T′. These phases correlate to the TMDs, which are 2D monolayered building units. The electrical properties and catalytic activity of these 2D TMDs could not be more different. 1H represents the most thermodynamically stable form of molybdenum disulfide, tungsten disulfide, molybdenum disulfide-selenide, and tungsten disulfide-selenide (or 2H) [17]. These materials are semiconducting; hence 1H (or 2H) has a very low electrical conductivity. By manipulating these materials, one can produce 1T metallic TMDs, each of which has a distinct catalytic active site and significantly improved electron transport effectiveness. In addition, theoretical simulations demonstrate that the basal plane of a number of 1T TMDs (for instance, MoS_2, NbS_2, VS_2, TaS_2, WS_2, and TiS_2) is responsible for the catalytic HER, whereas the basal plane of their 1H/2H phases is commonly static because of high ΔG_{H*} [22]. The key challenges concerning 1T TMD monolayers that need to be studied are stability and low-cost manufacturing techniques. Chhowalla and colleagues [6] reported the HER of chemically exfoliated WS_2 single-layers with both 1T and 1H structural configuration. In the study of thin films of 1H-WS_2, the onset overpotentials were found to be between 150 and 200 mV, and the Tafel slope was at least 110 mV dec^{-1}. The HER performance improved steadily when the concentration of 1T-WS_2 increased and the ratio of 1H-WS_2 decreased (figures 7.3(a)–(d)) [6]. Among the numerous WS_2 samples, 1T-WS_2 monolayers had excellent potential for HER. They possessed a Tafel slope of 55 mV dec^{-1} (60 mV dec^{-1} without IR correction), an initial overpotential of 80–100 mV, and a J_o of 2×10^{-5} A cm^{-2}. Because the catalyst was placed on a glassy carbon electrode with a thickness greater than five monolayers, the HER performance of

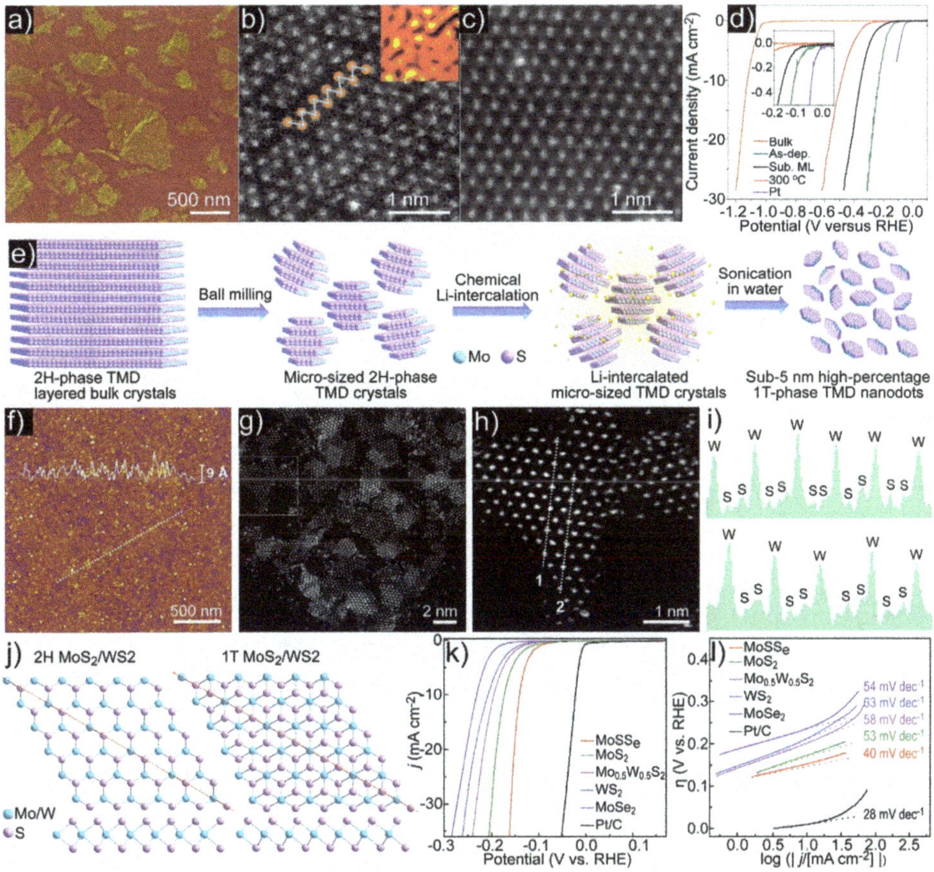

Figure 7.3. (a) AFM photograph of WS_2 monolayers. HAADF-STEM photograph of WS_2 monolayer with distorted (b) 1T and (c) 1H structure. (d) Polarization curves for bulk and WS_2 monolayers (including 1H and 1T phases), sub-monolayer WS_2, and Pt nanoparticles. (Reproduced with permission from [29]. Copyright 2013 Springer Nature.) (e) Illustration of the synthesis scheme of monolayer 1T TMDs acquired from 2H bulk counterparts. (f) AFM photograph of as-prepared monolayer WS_2 nanocrystals and (g) HAADF-STEM photograph. (h) Equivalent L2D-WF-ABSF filtered photograph (from g). (i) Brightness patterns with the dotted lines (from h) (line 1: top; line 2: bottom). (j) Illustration of 1H/2H and 1T TMDs structures. (k), (l) Polarization curves and resultant Tafel plots for certain nanocrystals. (Reproduced with permission from [30]. Copyright 2018 John Wiley and Sons.)

WS_2 monolayers (particularly 1H monolayers) may have been overstated. In the 1H semi-conductive monolayers, the diameter of the catalyst on the electrode must ideally be the same as the monolayer. It will ensure that the transfer of electrons between the electrode and the highest ranked surface of the catalyst is as quick as possible. The HER performance will suffer if the monolayers have been restacked in their previous positions after being eliminated. As a consequence of this, the connection between the level of electrocatalytic activity and the quantity of monolayered Mo–S nanocrystals loaded onto the glassy carbon electrode is of the utmost importance.

The double-layer capacitance, commonly known as C_{dl} (which reflects the effective total active sites of the surface area), decreased progressively as the catalyst loading increased [4]. Including more catalysts reduced the C_{dl} when the bulk density was greater than 100 μg cm^{-2}. Because of all of these features, agglomeration or re-stacking of monolayered nanocrystals resulted in a considerable loss in catalytic efficiency, in particular, the slope of the Tafel plot was exacerbated [4]. The performance of the HER, 1H-WS_2 monolayers of various sizes (edge lengths ranging from 400 to 800 nm) has also been studied [23]. Monolayers on gold foil were created using CVD process, and these monolayers were then used directly for the HER. These monolayers showed Tafel slopes ranging from 102 to 104 mV dec^{-1} and J_o values ranging from 6.31 to 17.78 μA cm^{-2}. Because of the contact between the Au foil and the WS_2, the current exchange density is equivalent to 1T-WS_2 nanosheets operating at 2×10^{-5} A cm^{-2} [6, 23]. For example, 1H-MoS_2 monolayers initially grown on Au foils have a Tafel slope of 61 mV dec^{-1} and a J_o of 38.1 μA cm^{-2}. This represents a considerable advancement in comparison to WS_2 of a smaller size [5].

The lateral dimension of the 1T-MoS_2 nanosheets generated by $LiBH_4$ exfoliation can be as tiny as 1 μm, with a 1T content of up to 80% [24]. When the 1T nanosheets were annealed, the Tafel slope was enhanced from 40 to 75–85 mV dec^{-1}, causing the onset overpotential to increase. The low ΔG_{H*} value at the basal plane of 1T-MoS_2 sheets (roughly 0.12 eV), which was comparable to the ΔG_{H*} value at the edge of 1H/2H-MoS_2 sheets (0.08 eV), was primarily responsible for the favorable HER achievement of 1T-MoS_2 sheets [22, 25]. By integrating phase engineering and edge enrichment, researchers created 1T-MoS_2 monolayers with smaller diameters for the HER. Li *et al* performed an exfoliation of MoS_2 with the assistance of butyllithium intercalant (*n*-butyllithium) in 2017 [26]. The very few MoS_2 sheets were exfoliated, fragmented, and changed into 1T-MoS_2 nanosheets (>70% phase content, 100–200 nm lateral dimension) with the assistance of an ultra-sonication treatment. The Tafel slope and onset corrosion potential of the synthesized 1T-MoS_2 nanosheets were 42.7 mV dec^{-1} and 156 mV, respectively. Because the 1T:1H phase ratio is lower and multilayered sheets have been found in the sample, the performance of these 1T-MoS_2 sheets is inferior to that of the previously mentioned 1T-MoS_2 sheets [17, 26]. Despite this, the demonstrated performance is greatly superior to several 1H-MoS_2 monolayers [4, 27]. He *et al* established a solvothermal process to produce vertically aligned 1T-WS_2 sheets with lateral dimensions of several micrometers to expose the greatest number of active surface sites. These sheets have a phase composition of roughly 70% [28]. The Tafel slope of the 1T-WS_2 sheets parallel to the ground was 43 mV dec^{-1}, and the onset overpotential was 118 mv (at a *j* of 10 mA cm^{-2}).

The flat 1T-WS_2 sheets placed on the glassy carbon electrode display a Tafel slope equal to 52 mV dec^{-1} and a benchmark corrosion potential equal to 230 mV. Because of the increased visibility of the 1T basal plane, such an improvement could be achieved. Regarding 1T TMDs, basal plane exposure is essential for a successful

HER. In contrast, most 1H/2H TMDs only have catalytically active edges. Zhang and colleagues [13] recently announced monolayered 1T TMDs nanocrystals. Ball-milling was used to first reduce bulk TMDs (tens of micrometers) to micro-sized particles and intercalated by *n*-butyllithium to produce different monolayered TMDs nanodots ($MoSe_2$, MoS_2, WS_2, MoSSe, $Mo_{0.5}$, and $W_{0.5}S_2$) with a high proportion of 1T phase (67%–80%, see figures 7.3(e)–(l)). According to figures 7.3(k) and (l), these nanodots were smaller than 5 nm, and their Tafel slope ranged from 40 to 63 mV dec^{-1}. The lowest Tafel slope was created by the MoSSe dots, which might be attributed to the development of Se-depleted sites during the initial stage [13]. Compared to those of 1H monolayered nanocrystals and 1T monolayered sheets, the Tafel slope of the 1T-MoS_2 nanocrystals was significantly less steep due to undetermined preparations and purifications of the 1T nanocrystals that may need to be improved further [4]. The distinction between the 1T phase and the 1T′ phase, which have dramatically different electrical properties and, as a result, HER pathways, is critical to recall when undertaking phase engineering on TMDs. Sokolikova and colleagues [22] recently reported the solution phase synthesis of kinetically stable 1T′-WSe_2 on arbitrary substrates (figure 7.4). The Tafel slope of the 1T′-WSe_2 was only 150 mV dec^{-1} at ultralow mass loadings of 40 $\mu g\,cm^{-2}$; however, after thermal conversion to 1H/2H-WSe_2, it significantly increased to 232 mV dec^{-1} (figure 7.4(f)). This method illustrates the significance of phase engineering to increase catalytic activity when done in an intelligent manner, even though the Tafel slope and *j* at 10 mA cm^{-2} were significantly better than for other optimized TMDs (figures 7.4(f) and (g)). Additional methods have also been suggested, which will be discussed below, to enhance the catalytic process's performance even further.

7.1.1.3 Interface engineering

The construction of high-performance electrocatalysts also requires the use of another crucial protocol, which is interface engineering. When two different components interact at an interface, both of those materials' electronic states and chemical characteristics will undergo concurrent transformations due to those interactions. This has a significant effect on the electrocatalytic performances of the materials and has been demonstrated by several experiments [32, 33]. The field of interface engineering may be broken down into two distinct subfields: heterostructure engineering and synergistic interaction. These two subfields are inextricably linked to one another and constantly interact. While heterojunction engineering frequently involves the formation of complex chemical bonds between two dissimilar materials, synergic communication refers to the physical contact of two substances with limited electron transfer [34]. The synthesis of high activity electrocatalysts with one-of-a-kind physical and chemical characteristics may be accomplished effectively through interface engineering. For instance, Zheng and colleagues hybridized g-C_3N_4 with N-graphene *in situ* to produce an ultrathin nanosheet (figure 7.5(a)) [34]. The XAS spectra provided evidence that the layers of g-C_3N_4 and N-graphene had a robust interfacial interaction, manifesting as the formation of out-of-plane orientated C–N bonds (figure 7.5(b)). These interlayer bonds can affect the electronic

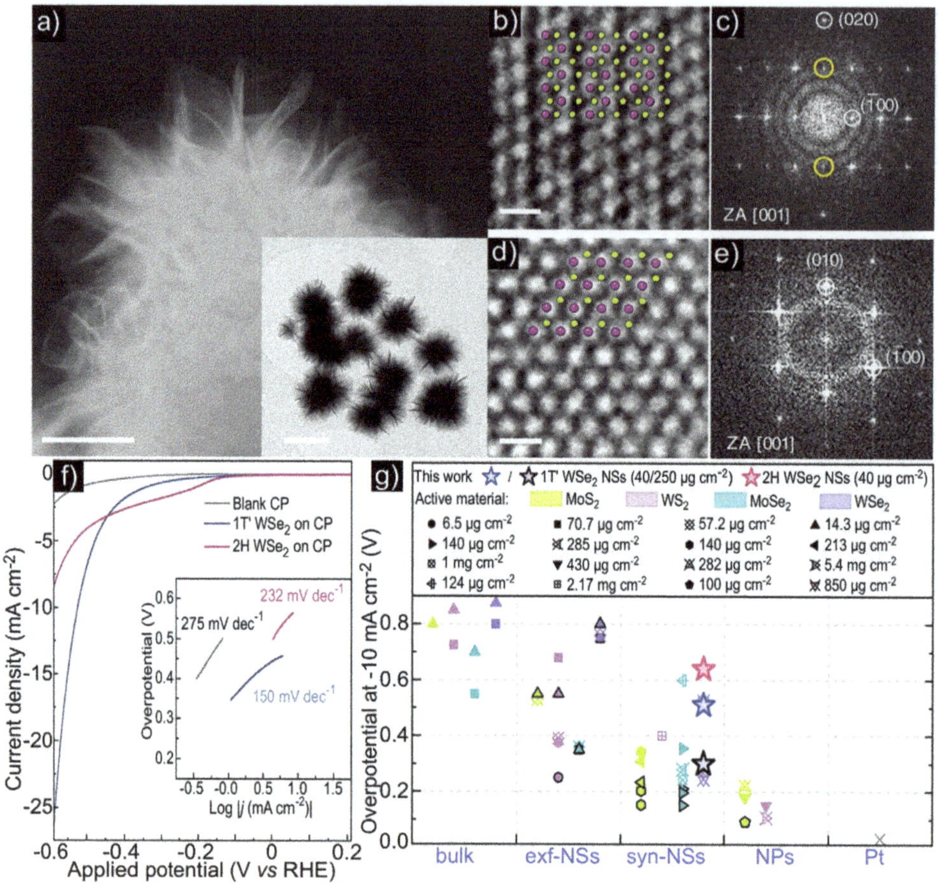

Figure 7.4. (a) Annular dark-field scanning TEM photograph of WSe_2 branched nanoflowers (100 nm scale bar), the inset reveals an overview photograph representing a collection of WSe_2 nanoflowers (200 nm scale bar). (b), (d) Zoomed photographs (0.5 nm scale bar) with an overlapped crystal model of (b) 1T′ and (d) 2H. (c), (e) FFT patterns. (f) Polarization curves of 2H- and 1T′-WSe_2 nanosheets developed over carbon paper; the inset corresponds to parallel Tafel plots. (g) Evaluation plots summarize the described TMD group VI for the HER using electrocatalysts (syn-NSs: synthesized nanosheets; exf-nanosheets: exfoliated nanosheets). (Reproduced with permission from [31]. Copyright 2019 Springer Nature.)

state of a composite and make it easier for electrons to go from the N-graphene layer to the g-C_3N_4 layer (figure 7.5(c)). Consequently, enhanced HER performance was accomplished on the hybridized C_3N_4/NG sample compared to the physically mixed sample (figure 7.5(d)). This increase occurs because of the synergistic effect of chemical and electronic couplings, which, as the results of certain DFT simulations have shown, leads to a more advantageous proton adsorbent kinetic model on the C_3N_4@NG surface (figure 7.5(e)) [34]. Additionally, this activity pattern proved the predictive power of the DFT model utilized in areas other than metals.

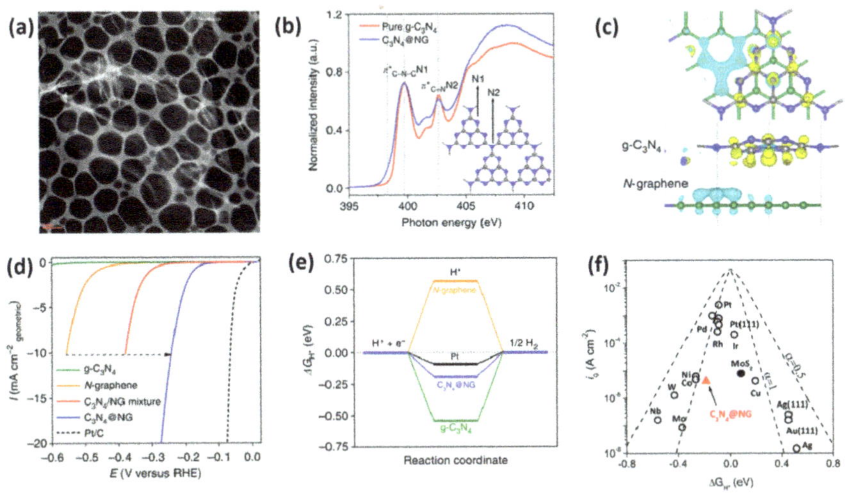

Figure 7.5. (a) HAADF-STEM photographs of a C_3N_4@NG nanosheet. (b) Nitrogen K-edge NEXAFS spectra for different catalysts (the inset reveals two kinds of nitrogen species in the g-C_3N_4 network). (c) Interfacial electron transfer in C_3N_4@NG. Yellow shows electron accumulation and the cyan iso-surface shows electron depletion. (d) HER polarization curves for different electrocatalysts using acidic media. (e) Free energy illustration at equilibrium potential for the HER of three metal-free catalysts. (f) Volcano plots of i_0 as a function of the ΔG_{H*} for freshly grown C_3N_4@NG and common metal-based catalysts. (Reproduced with permission from [34]. Copyright 2014 Springer Nature.)

The most significant discovery made by this research is that comparable to the precious metals, metal-free materials that were rationally developed also offer significant potential for a highly efficient electrocatalytic HER (figure 7.5(f)). Similarly, Duan *et al* prepared a 3D free-standing hybrid film of g-C_3N_4 and N-graphene. This film demonstrated a performance equivalent to that of the 2D heterostructured C_3N_4@NG hybrid [34, 35]. This technique for creating 3D porous films may also be used in the production of other 2D materials [36, 37]. Film electrodes have been effective for two reasons and through their combination: (i) the increased porosity in three dimensions may enhance the exposure of active catalytic sites; and (ii) the formation of hierarchical intralayer and interlayer pores in the film may make it possible to increase both the surface area of the film and the rate at which mass is transferred during the electrocatalytic processes. In contrast to pure MoS_2 and $CoSe_2$ units, the catalytic performance of a hybrid structure consisting of MoS_2/$CoSe_2$ with an interface resembling a heterojunction was significantly improved [38].

It is interesting to note that the long-term stability of the MoS_2/$CoSe_2$ hybrid structure is superior to pure MoS_2, demonstrating the excellent integration of the stability of MoS_2 with the strong catalytic capabilities of $CoSe_2$. A similar phenomenon has also been seen with amorphous Co–Mo–S_x chalcogels [39]. It was discovered that materials containing CoS_x are less stable, although these compounds have a higher activity level than those containing MoS_x. Enhanced HER activity was obtained by combining the more active CoS_x building blocks with the more stable MoS_x units into a compact framework, allowing optimal use of the available space. The researchers postulate that the link between the activity and the stability of TMD materials is driven by a synergy between electrical and structural properties. This synergy between the two types of effects is what they believe is responsible for the link between activity and stability (the number of defects) [17, 39]. Other hybrid systems that can include rGO/WS_2 [40], molybdenum disulfide (MoS_2)/Au [41], and rGO/MoS_2, have been created utilizing a mechanism that is quite similar [42]. The design principles for creating heterostructures in 2D electrocatalysts are guided by several key parameters: (1) all 2D nanomaterials involved should have similar crystal structures to enable the formation of stable heterostructures; (2) at least one of the nanomaterials should demonstrate potential activity for the desired electrocatalytic process; (3) the hybrid structures should exhibit good conductivity to ensure efficient electron transport to reactive intermediates; (4) the properties of one material should compensate for the deficiencies of the other, such as poor intermediate absorption energy, low conductivity, or limited stability.

7.1.2 Heteroatom doping

Incorporating heteroatoms is a beneficial method frequently utilized to improve the HER efficacy of TMD electrocatalysts [17, 43]. Doped heteroatoms efficiently affect electronic structure, notably the d-band of products comprising TMD, and reduce the value of ΔG_{H*} of the electrocatalysts, making the HER method simpler to carry out. It is feasible to partially replace either the metal sites or the silicon sites in TMDs [44] or the nonmetal sites [45, 46] to increase the HER performance. This opens up further opportunities to adjust the fundamental features of the host materials.

7.1.2.1 Metal doping

Other metal elements, including vanadium, cobalt, iron, nickel, copper, and zinc, have been employed effectively as incorporated elements for TMDs containing electrocatalysts. Doping can, in general, speed up the HER process by enhancing the number of catalytic sites [47], boosting the conductivity [48], and/or improving the electronic structure [44]. However, while choosing the doping components, attention should be paid since various dopants may influence the HER activity of TMD electrocatalysts. Ni and Co doping is frequently used to enhance the HER performance of MoS_2 [47]. These two components can potentially lower the value of ΔG_{H*} and increase the density of active sites, ultimately improving HER functionality. Xiong *et al* established a one-step hydrothermal technique to manufacture Co-added MoS_2 [49], which displayed an improvement of the HER compared to pristine MoS_2 (figure 7.6(a)). In alkaline conditions with a potential difference of 90 mV, the sample with the highest HER performance was Co–MoS_2-0.5. This was the case when the quantity of Co source doping was 0.5 mmol. This improvement can be attributed to the reduction in ΔG_{H*} (shown in figure 7.6(b)) and the controlled electronic structure generated by adding cobalt. Xie *et al* discovered that adding a V

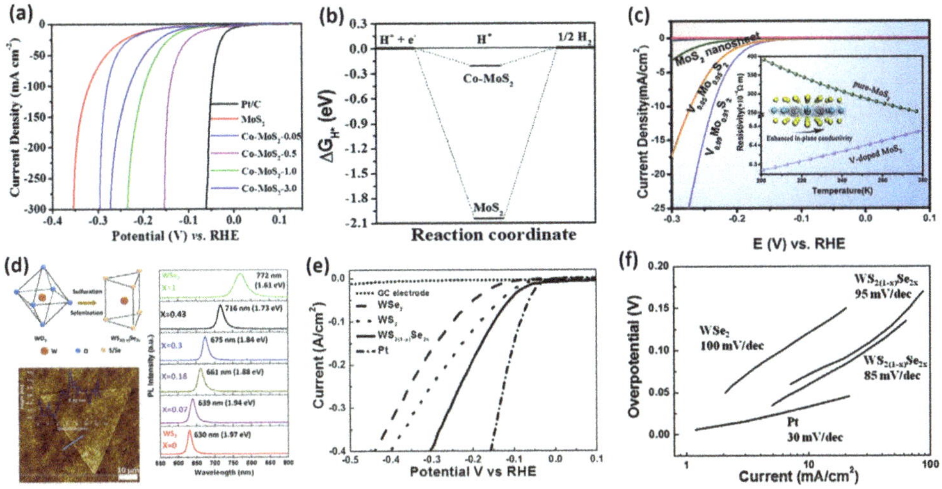

Figure 7.6. (a) LSV curves for the HER and (b) evaluated ΔG_{H*} for pristine and Co-doped MoS_2. (Reproduced with permission from [49]. Copyright 2018 The Royal Society of Chemistry.) (c) LSV curves for pristine and V-doped MoS_2 nanosheets. (Reproduced with permission from [48]. Copyright 2014 The Royal Society of Chemistry.) (d) Growth of h-$WS_{2(1-x)}Se_{2x}$ using a synchronized selenization and sulfurization reaction of monoclinic WO_3. AFM photograph and corresponding PL spectra (at room temperature) for the grown $WS_{2(1-x)}Se_{2x}$ NS. (e) Assessment of HER experimental results for monolayer WS_2, WSe_2, Pt, glassy carbon, and finally $WS_{2(1-x)}Se_{2x}$ ($x = 0.43$). (f) Corresponding slope values. (Reproduced with permission from [46]. Copyright 2015 John Wiley and Sons.)

atom increased the conductivity of MoS_2 nanosheets but did not enhance the solidity of active sites. This is in contrast to the Co and Ni doping, which increased the density of active sites (figure 7.6(c)) [48].

7.1.2.2 Nonmetal doping

There are several varieties of nonmetal-doped TMDs, each of which demonstrates distinctive properties and outstanding HER performance. In contrast to the metal doping technique, the nonmetal doping strategy optimizes ΔG_{H*} and produces crystal deformation or an amorphous structure with an abundance of active sites [50, 51]. Hydrothermal synthesis was used by Xie *et al* to produce O_2-doped MoS_2 nanosheets [52]. Because of the low temperature at which the MoS_2 nanosheets were synthesized, they naturally only had a minimal quantity of Mo–O bonds. In addition, the results of DFT showed that the MoS_2 sheet that had O_2 inserted into it displayed a smaller bandgap, measuring 1.30 eV, whereas the value of the 2H-MoS_2 sheet was 1.75 eV. The absorption of O_2 into MoS_2 nanosheets caused the bandgap to shrink, resulting in a larger carrier density and enhanced conductivity in the material. Consequently, the optimized O_2-incorporated MoS_2 catalyst displayed a relatively high level of HER, as evidenced by a value of $\eta = 120$ mV (at 1 mA cm^{-2}) for the Tafel slope and a value of 55 mV dec^{-1} for the HER activity. In this study, Fu *et al* evaluated the effect that Se doping had on $WS_{2(1-x)}Se_{2x}$ single-layer nanosheets that they had synthesized and analysed [46]. Crystal distortion in WS_2

caused by the giant Se atom led to the generation of a polarized electric field that sped up the bond-breaking process in H_2O molecules. The process began when S atoms were replaced with a larger Se atom. Therefore, single-layer $WS_{2(1-x)}Se_{2x}$ NS demonstrated a substantially lower of $\eta = 80$ mV to produce 10 mA cm^{-2}, whereas WSe_2 and WS_2 nanosheets required 150 and 100 mV, respectively, to achieve the same value (figures 7.6(d)–(f)). Doping the MoS_2 and $MoSe_2$ catalysts with chlorine (Cl) was discovered to effectively boost their HER efficacy and adjust the electronic arrangement of amorphous $MoSe_2$ and MoS_2, as found by Jin's group. This was the case for both MoS_2 and $MoSe_2$ catalysts (figures 7.7(a) and (b)) [53]. MoS_xCl_y and $MoSe_xCl_y$'s HER performance was ultimately improved due to the ideal electrical structure combined with multiple active sites (figures 7.7(c) and (d)) [53]. In addition to this, amorphous MoQ_xCl_y (Q = S and Se) was developed on the n^+pp^+Si substrate with micro pyramids (MPs), which together constitute a highly effective PEC-HER photocathode (figure 7.7(e)).

These PEC-HER photocathodes illustrated HER efficacies of 43 and 38.8 mA cm^{-2} at 0 V against RHE for MoS_xCl_y/Si and MoS_xCl_y/Si, respectively. This was determined by comparing it to a reversible hydrogen electrode. These values were even superior to those obtained from Pt/Si photocathodes (figure 7.7(f)). In general, incorporating heteroatoms could improve the HER efficiency of TMD catalytic activity in two ways. To begin, adding an element might result in deformation of the crystal structure and an increase in the number of catalytic sites. Second, because

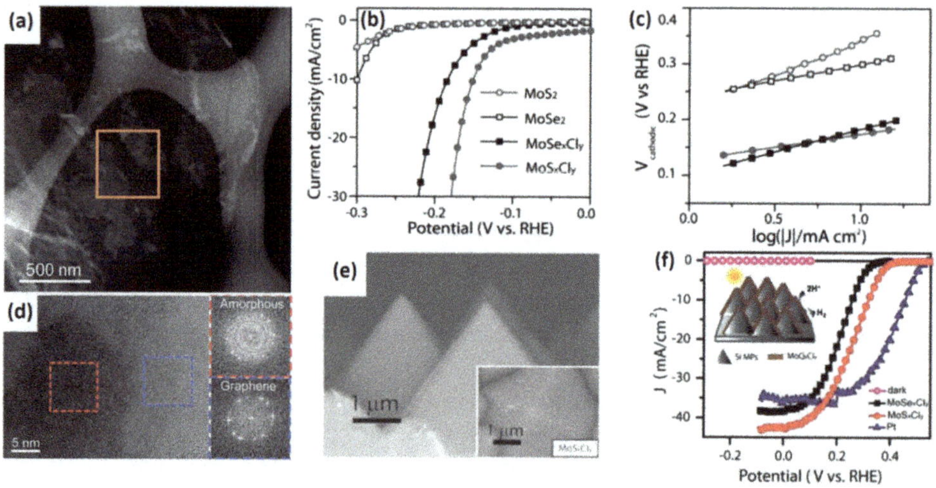

Figure 7.7. (a) EDS and element mapping of partially shielded single-layer graphene NS by MoS_xCl_y. (b) HR-TEM photographs and corresponding local FFT images of MoS_xCl_y vertical graphene (VG) sheet. (c) Tafel analysis, (d) HR-TEM and local FFT image of a MoS_xCl_y-VG sheet. (e) Top-down SEM images of MoQ_xCl_y/Si. (f) PEC-HER performance of $MoSe_xCl_y$/Si MP (squares), MoS_xCl_y/Si MP (circles), and Pt/Si MP (triangles) photocathodes. 0.5 M H_2SO_4 accompanied the experiments under 1 Sun irradiation. ((a) and (d) Reproduced with permission from [48]. Copyright 2015 The Royal Society of Chemistry. (b), (c) and (e) Reproduced with permission from [54]. Copyright 2015 John Wiley and Sons. (f) Reproduced with permission from [53]. Copyright 2015 John Wiley and Sons.)

various dopants have varied configurations of electrons, the electronic arrangement of the catalysts may be greatly modulated by the dopants. Depending on the scenario, this might optimize ΔG_{H*} or the corresponding degree of energy level. However, it still requires careful selection because different atoms will affect the catalytic activity differently, and unsuitable elements may negatively affect the overall performance [43, 47]. In our opinion, this may result from various preparation procedures and amounts of doping, which leads to the dopants anchoring on various locations of the TMDs nanosheets. On the other hand, the inner mechanism requires more hypothetical and experimental inquiry.

7.1.3 Formation of heterostructures

7.1.3.1 Increase in electronic conductivity

TMD electrocatalysts can be coupled with conducting species, including graphene, to provide more efficient reactions [40, 55]. Carbon paper [56] and metal substrates [57, 58] are a strategy that is often employed to boost the HER activity of TMDs. Including TMDs containing electrocatalysts with conductive substrates produces a conductive framework with various interior electron shifting channels and increases the effective ECSA and active sites for the HER process. For instance, Dai's team successfully developed a practical solvothermal strategy in 2011 for synthesizing a composite electrocatalyst of rGO and MoS_2 nanosheets [42]. The HER efficiency of the composite MoS_2/rGO material was significantly higher than that of the basic MoS_2 particles, which displayed $\eta = 100$ mV. In this particular instance, the improved HER performance was determined to be attributable to active sites near the margins of the MoS_2 that were exposed correctly. Additionally, the interaction with the graphene network should enhance the efficiency with which carriers migrate from the MoS_2 to the electrodes. The 3D highly conductive substrates, which can include Cu, Ni, Fe, and Ti foam, are able to provide a somewhat larger surface area than that of the combination of TMDs and other carbon-based materials (for example, rGO, CNTs, and carbon nanofibers). This is advantageous for significantly increasing the packing quantity of catalysts because it allows for more surface area to be utilized [59, 60]. Specifically, coupling with conducting scaffolds is the sole way to boost catalytic effects since it increases the catalytic surface area and improves the transfer of electrons between the targets and the catalysts. Despite this, the catalyst's intrinsic activity remains typically constant once the optimization procedure is completed [61, 62].

7.1.3.2 Optimization of kinetics

From the point of view of semiconductor physics, the significant function of heterostructures is in facilitating the quickening of the electrons' motion in a certain direction through appropriately corresponding band structures in the various collected materials. This is one of the critical roles that heterostructure plays [63]. The HER is a surface electrochemical method that relies heavily on the catalyst's ability to effectively interact with hydrogen present in the electrolyte [64, 65]. As discussed in section 2.2.1, ΔG_{H*} is an essential descriptor when determining the extent to which an HER catalyst may be utilized. The effective optimization of the kinetic reaction process is achieved directly from the construction of heterostructure catalysts. A straightforward hydrothermal process used by Yu *et al* in the production of the $MoS_2/CoSe_2$ heterostructure (figure 7.8(a)) reveals the heterostructure of MoS_2 and $CoSe_2$ [38]. In its as-prepared state, the hybrid electrocatalyst required a threshold voltage of 75 mV (to achieve 10 mA cm^{-2}) with a Tafel slope of 36 mV dec^{-1}. These were both less than for the MoS_2 and the $CoSe_2$, and comparable to the standard Pt/C electrocatalyst (figure 7.8(b)). According to this research, the heterostructure composed of $MoS_2/CoSe_2$ serves not one but two crucial functions in increasing the amount of catalytic activity. The creation of MoS_2 and $CoSe_2$ interfaces would result in an abundance of interfaces containing rich catalytic sites.

The heterostructure, on the other hand, is capable of optimizing ΔG_{H*} (figure 7.8 (c)) and bringing its value near Pt (111) [66]. The mild ΔG_{H*} reaction has the potential to effectively lower the reaction barrier, allowing for the easier formation of H_2 molecules from H atoms. Furthermore, the adsorption energy of the H_2 molecules was as low as 0.5 eV, which would accelerate the evacuation of H_2 from the catalytically active site and increase the HER mechanism's speed. Hydrogen production using chloralkali electrolysis is extensively used in industrial settings [67]. However, compared to the HER process that occurs in an acidic medium, the HER process that occurs in alkaline media is often much slower. This is primarily

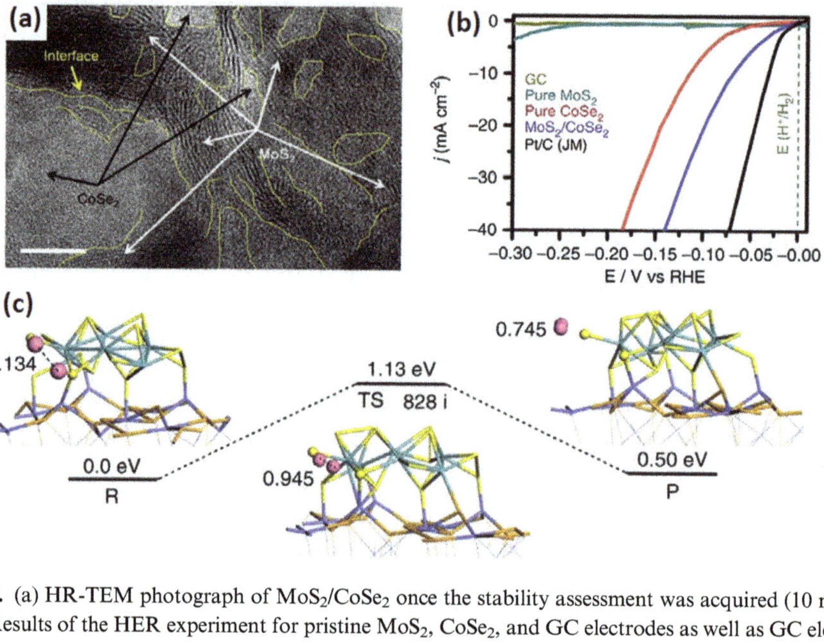

Figure 7.8. (a) HR-TEM photograph of $MoS_2/CoSe_2$ once the stability assessment was acquired (10 nm scale bar). (b) Results of the HER experiment for pristine MoS_2, $CoSe_2$, and GC electrodes as well as GC electrodes improved with $MoS_2/CoSe_2$ hybrid, and finally Pt/C catalyst. (c) Estimated energy barrier and corresponding mechanisms for the HER approach over the $MoS_2/CoSe_2$ hybrid surface referring to Volmer–Tafel pathways. In (c), the orange, yellow, blue, azure, and pink spheres show the Se, S, Co, Mo, and H atoms, respectively. (Reproduced with permission from [38]. Copyright 2015 Springer Nature)

attributable to the sluggish kinetics of water-splitting [68]. To cut down on the amount of energy that is used in the chloralkali sector, it is necessary to research high-performance alkaline HER catalysts. Because an alkaline electrolyte does not contain H^+, the HER process must begin with the dissociation of water; however, this requires an additional energy source because it produces protons [17]. The production of heterostructure catalysts containing numerous functional sites is an effective strategy for accelerating the process.

7.2 Oxygen reduction reactions

7.2.1 Defect engineering

The ORR involves four proton–electron transfers, which can occur through associative or dissociative pathways depending on the oxygen dissociation barriers on the catalyst surface (*) to convert O_2 to H_2O [62]:

$$\text{Dissociative:} \quad O_2 + 2^{\cdot} \rightarrow 2O^{\cdot} \tag{7.1}$$

$$2O^{\cdot} + 2H^+ + 2e^- \rightarrow 2OH^{\cdot} \tag{7.2}$$

$$2OH^{\cdot} + 2H^+ + 2e^- \rightarrow 2H_2O + 2^{\cdot} \tag{7.3}$$

$$\text{Associative:} \ O_2 + * \ \rightarrow O_2^{\cdot} \tag{7.4}$$

$$O_2^{\cdot} + H^+ + e^- \rightarrow OOH^{\cdot} \tag{7.5}$$

$$OOH^{\cdot} + H^+ + e^- \rightarrow O^{\cdot} + H_2O \tag{7.6}$$

$$O^{\cdot} + H^+ + e^- \rightarrow OH^{\cdot} \tag{7.7}$$

$$OH^{\cdot} + H^+ + e^- \rightarrow H_2O + *. \tag{7.8}$$

The free energy of O_2 adsorption, also known as ΔG_o, is connected to the total activity directed toward the ORR. In a manner analogous to the HER, if a catalyst binds O_2 excessively, the activity will be restricted due to the transfer of proton electrons to either O* or OH*. For catalysts that bind O_2 too weakly, performance is restricted by proton–electron shifting to O_2* (associative process) or breaking of O_2 comprising the O–O bond (dissociative process) [62]. The onset potential, the slope of the Tafel plot, and half-wave potential are examples of the parameters frequently used to assess the catalytic efficacy toward the ORR [69, 70]. These parameters include the potential corresponding to half of the current restricted by diffusion. Carbon materials have been the focus of a significant amount of study as electrocatalysts for organic reductive decomposition, because of their low manufacturing costs, good corrosion resistance, and enormous conductance [71]. By incorporating heteroatoms, graphene-containing substances have been utilized as active electrocatalysts for the ORR [72]. It is possible to produce intense ORR activity by loading particles of noble-metal-based catalysts onto graphene substrates with higher conductivity and surface area. N-incorporated graphene (Pt NTs/NG-PM) exhibits extremely high activity and endurance for the ORR in acidic conditions when paired with PtNTs [73]. Unlike the half-wave potential of the industrial standard Pt/C catalyst, the Pt NT/NG catalyst has a 0.903 V versus RHE value that is 28 mV more positive (0.875 V versus RHE). Xie *et al* [74] successfully dispersed monodisperse Pt nanocrystals of a tiny size (2.8 nm) into a 3D mesopore-rich N-incorporated graphene aerogel (NGA), as illustrated in figures 7.9(a) and (b). The Pt nanocrystal@NGA has a 1750 m^2 g^{-1} surface area, dense mesopores, and a hefty 3.93% N content, allowing for strong ORR electrocatalytic efficacy. Each of the ORR catalysts mentioned in the preceding section still requires the addition of noble metals (i.e. Pt).

To prepare for significant applications in future energy and storing systems, it is necessary to design ORR catalysts that do not include Pt. Many researchers are concentrating their efforts on graphene-based materials and the possibility of using them as electrocatalysts that do not require the utilization of noble metals [69]. Enhancing the ORR catalytic efficacy of carbon-based products through defect engineering techniques, including heteroatom doping, can be a significant contributor [75]. Wang *et al* [76] created a FeN_x/graphene flake that displayed high electrocatalytic stability and efficacy for ORR in essential solutions. This can be seen in figures 7.9(c) and (d); catalytic activity increased 100 times for every metal and ~30 times per gram compared to Pt/C, as represented in figures 7.9(e) and (f). Wu *et al* [77] successfully produced N, S incorporated 3D N–S–GAs that exhibited high ORR performance. The remarkable performance can be described by the

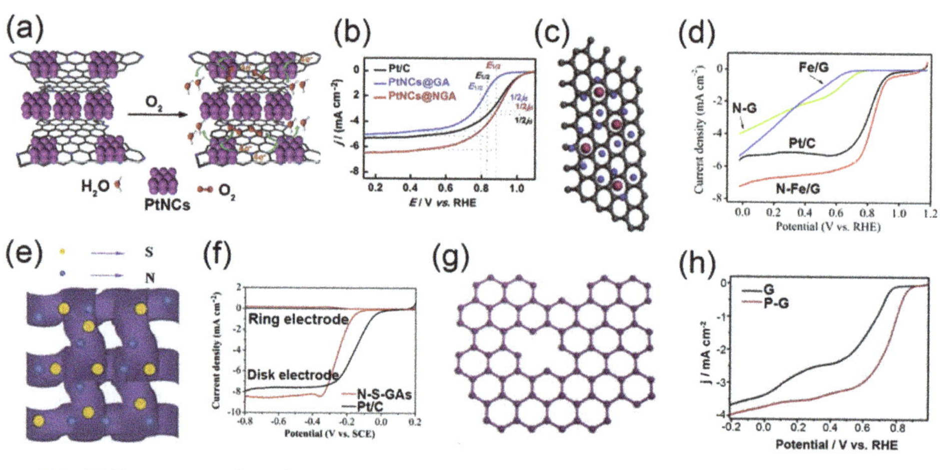

Figure 7.9. (a) Pt nanocrystals and NGA. (b) LSV curves for the electrocatalytic ORRs for different samples. (Reproduced with permission from [74]. Copyright 2017 The Royal Society of Chemistry.) (c) Structural illustration of N–Fe/G. (d) LSV curves of various catalysts including Fe/G (Fe atom loaded graphene), N–G (metal-free N-doped graphene), N–Fe/G, and Pt/C catalysts (in 0.1 M KOH). (Reproduced with permission from [79]. Copyright 2018 Elsevier.) (e) Synthesis illustration and (f) LSV curves for co-doped (N and S) 3D graphene aerogels (in 0.1 M KOH). (Reproduced with permission from [77]. Copyright 2016 The Royal Society of Chemistry.) (g) Illustration of the plasma structure that etched the graphene (P–G). (h) LSV curves for bare graphene and P–G (in 0.1 M KOH). (Reproduced with permission from [80]. Copyright 2016 The Royal Society of Chemistry.)

increased exposure of electrocatalytically active sites brought about by the linked porosity network and the synergistic impact brought about by the incorporation of doped N and sulfur into graphene sheets. For many reasons, 2D material edges are highly significant for electrocatalysts [78] and they are more highly active than the basal surface. This makes edges very important for electrocatalysts. As shown in figures 7.9(g) and (h), plasma etching produces graphene with a high number of edges and no dopants, which results in high ORR activity. With a 0.737 V half-wave potential versus RHE, which is more positive, compared to 0.572 V, the edge-excessive graphene electrode exhibits exceptional ORR electrocatalytic performance. Pure graphene RHE, an *n* (number of the electron) value of 3.85 for fuel selectivity with a shifted electron quantity per O_2, and improved stability with a slight cost (12%) to the density of current after 20 000 test series were achieved. The durability is superior to the Pt/C electrodes currently available on the market.

7.2.2 Nonmetal doping

Literature analysis reveals that the introduction of non-metallic elements as a dopant in the assembly of MoS_2 nanosheets causes ORR experiments to improve because of transformed electronic structures caused by the doped atoms. Further, incorporating heteroatoms in the MoS_2 assembly, for instance, P, O, N, etc, resulted in diminishing bandgap energy, modified electronic structure, advanced conductivity, and finally offered catalytically active sites. The growth of MoS_2 nanosheets and enhanced edge defects was acquired by Huang *et al* via g-C_3N_4 as a self-sacrificial species; nanosheets were acquired using the hydrothermal strategy as well as by applying heat treatment; lastly, oxygen was incorporated into MoS_2 employing H_2O_2 as displayed in figure 7.10(a). The as-prepared sample that exhibited advanced catalytic activity had a half-wave electric potential of 0.80 V and a primary potential of 0.94 V in 0.1 M KOH (figure 7.10(b)). Interestingly, the ORR selectivity of O–MoS_2 changed from $2e^-$ route to $4e^-$. This change can be attributed to the electronegative O-atom polarizing the nearby unsaturated Mo atom (figure 7.10(c)). Furthermore, at the edge side, the Mo atom produced a supplementary positive charge that was favorably adsorbed through O-molecules, accelerating the ORR [81].

According to DFT, Xie *et al* compared bare 2H-MoS_2 nanosheets (1.75 eV) with O-bound MoS_2 nanosheets (with a narrow bandgap energy of 1.30 eV), as displayed in figure 7.10(d). The outcomes reveal that a combination of MoS_2 nanosheets and oxygen may produce additional carriers and enhance the intrinsic conductivity of the material, thus producing higher ORR and HER activity [52]. Additionally, introducing P and N (low-electronegative atoms) into MoS_2 has been suggested to improve ORR experiments, and certain justifications exist for the advanced activity. The development of P-doped MoS_2 nanosheets was done by Huang *et al* which yields enhanced ORR activity and also provides the reason behind improved ORR activity via boundary molecular orbital theory, as shown in figures 7.11(a) and (b). Further, the energy levels of HOMO and LUMO for P-doped MoS_2 nanosheets are advanced and linked with bare MoS_2 nanosheets [82]. In general, for semiconductors, the increased orbital energy of the frontier is advantageous for electron

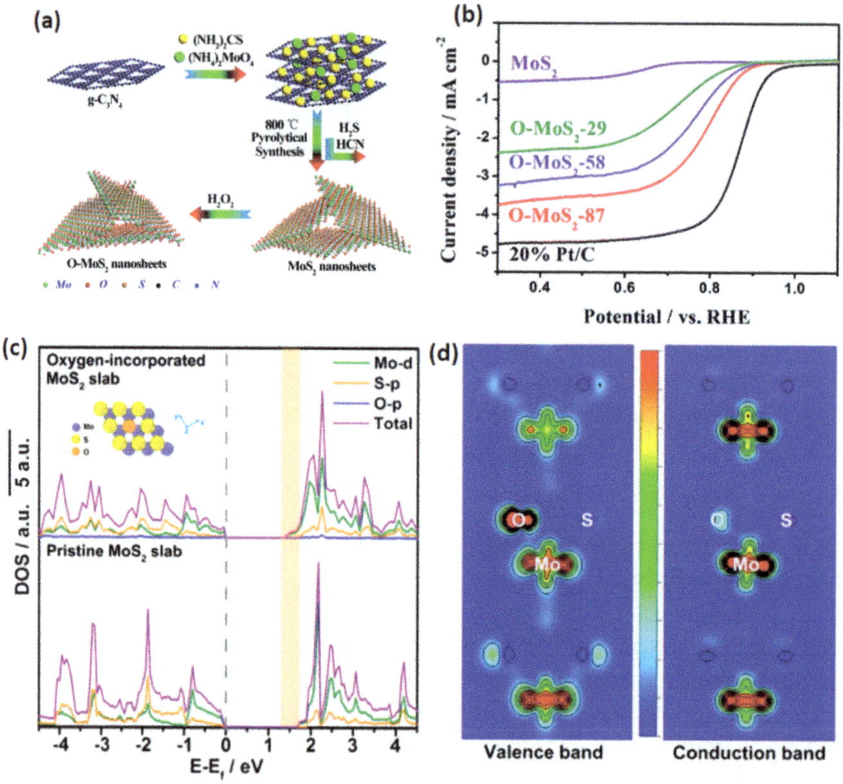

Figure 7.10. (a) A suggested synthetic procedure for integrating oxygen into ultrathin sheets of MoS_2 nanomaterials. (b) ORR efficiency test performed on O–MoS_2 and pure MoS_2 nanosheets, in addition to 20% Pt/C, in O_2 saturated 0.1 M KOH. (Reproduced with permission from [81]. Copyright 2015 The Royal Society of Chemistry.) (c) DOS was estimated for an oxygen-incorporated MoS_2 sheet at the top and a pure 2H-MoS_2 sheet at bottom. The decrease in bandgap that occurred due to oxygen incorporation is represented in orange. (d) In oxygen-incorporated MoS_2 ultrathin nanosheets, the charge density distribution of the valence band (left) and the conduction band (right) near the oxygen atom. (Reproduced with permission from [52]. Copyright 2013 The American Chemical Society.)

contribution [83], which produces oxygen adsorption and the growth of intermediate product (OH^-), thus speeding up the rate of reaction. Moreover, the measurements of overpotential in ORR reactions for P- or N-doped MoS_2 (monolayer) were reported by Zhang *et al*, which shows that the replacement of S atoms (from monolayer MoS_2) with N or P atoms can produce elevated values of spin density for the MoS_2 basal plane, thus refining its capability to trigger O_2 and more improvements regarding the activation of the $4e^-$ mechanism [84]. It is worth noting that the ORR activity of P-doped MoS_2 is poorer than for N-doped MoS_2 because of the exceptionally higher adsorption sites for the intermediary product (for P-doped MoS_2). However, these results do not correspond with the fact that the P-doped product has no projected applications.

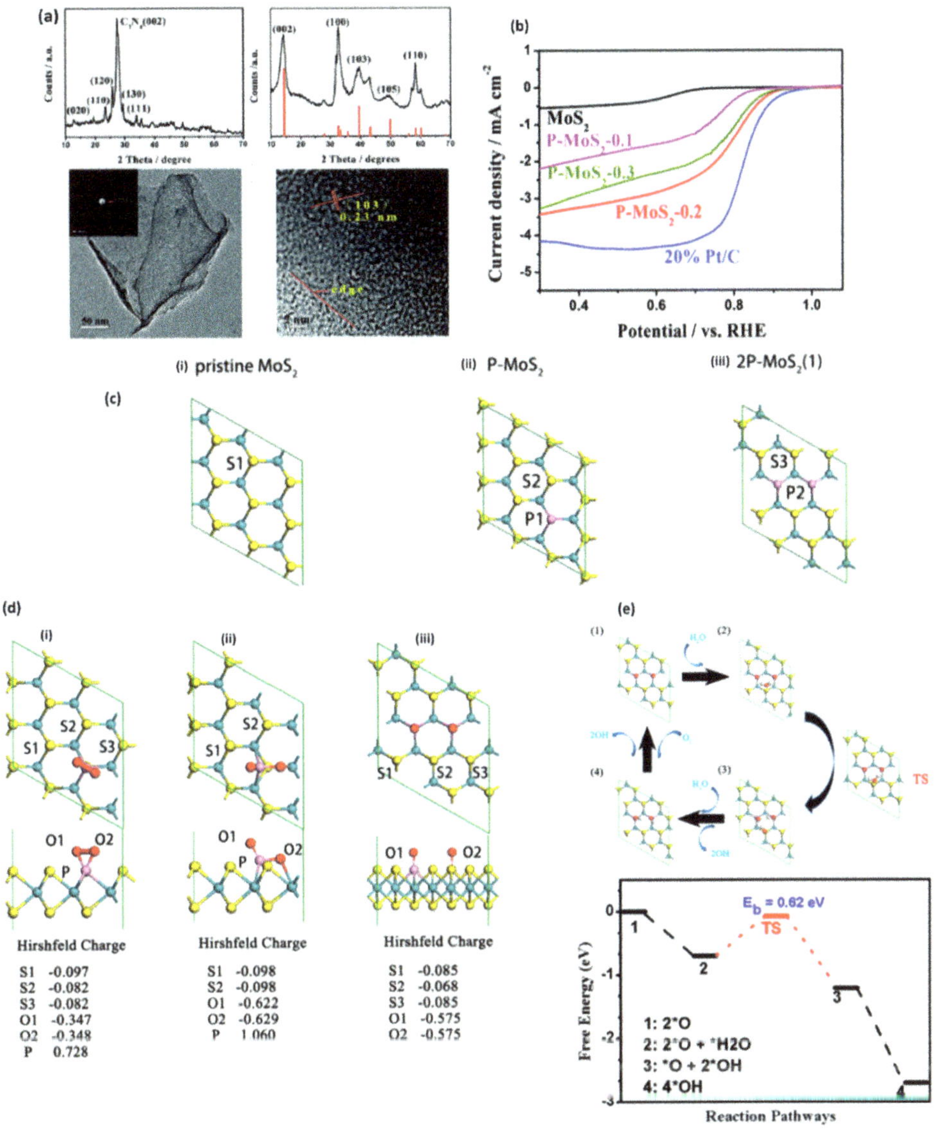

Figure 7.11. (a) XRD analysis of the MoO_3/g-C_3N_4 intermediary and ultrathin P-MoS_2 nanosheets and TEM and HR-TEM images of ultrathin P-MoS_2 nanosheets. (b) ORR electrocatalysis for P-MoS_2-0.1, P-MoS_2-0.2, P-MoS_2-0.3, and MoS_2 catalysts, as well as 20% Pt/C. (Reproduced with permission from [82]. Copyright 2015 The Royal Society of Chemistry.) (c) Top view of the optimized structure of (i) pristine MoS_2 nanosheets, (ii) single P-doped MoS_2 nanosheets, and (iii) double P-doped MoS_2 nanosheets. (d) Top and side views of the optimized adsorption configurations of (i) O_2-adsorbed P-MoS_2, (ii) 2O-adsorbed P-MoS_2, and (iii) 2O-adsorbed 2P-MoS_2. Additionally, the Hirshfeld charge strengths of the atoms that correlate is provided here. (e) On a 2P-MoS_2 sheet in a highly alkaline environment, an optimized structural arrangement of a reagent, intermediary, and product (1–4) and the related reaction mechanisms of the ORR. (Reproduced with permission from [85]. Copyright 2018 Springer Nature.)

The investigations of ORR pathways along with significant active sites of P-doped MoS_2 (monolayer) were done by Liu *et al* using DFT measurements, wherein the MoS_2 with certain structures, as well as P-dopant, lies in an alkaline medium, as shown in figure 7.11(c). The results clearly show that better activity was gained by doubly P-doped MoS_2 (having higher P content) than by single P-doped. The catalytic active for doubly P-doped product is the S_2 atom that is neighbored by two P atoms, thus reducing the charge of the S_2 site and offering a durable hydrogen bond which in turn yields improved adsorption of H_2O and OH^- groups. Concurrently, compared with single P-doped content, doubly doped content has suitable intermediary product adsorption energy, as demonstrated in figure 7.11(d). The best P-doping content of the catalyst as calculated via DFT was 5.5% for the ORR experiment, closer to the observed experimental value (4.7%) [85]. Therefore, the study of Liu *et al* provides a rational explanation for why P-elements should be supported in catalytic tests, which had until then left researchers baffled.

7.2.3 Metal doping

In addition to non-metallic elements, metallic doping can also be introduced into the assembly of MoS_2 to support the ORR experiment. It is worth mentioning that Co, Cu, and additional transition metals dopants when introduced in the S vacancy of MoS_2 produce an enriched ORR [17]. The theoretical measurements for the ORR experiment were performed by Xiao *et al* for (monolayer) MoS_2 doped with Co/Ni. Incorporating metal dopants into the basal plane of (monolayer) MoS_2 can regulate the energy required to adsorb the chemical intermediates onto the catalyst's surface, and thus aggregate the catalytic performance. In Co/Ni-doped MoS_2, Co/MoS_2 (monolayer) was comparable to the FeN_4 catalytically active site in the M–N–C, and Ni/MoS_2 was comparable to the CoN_4 catalytically active site. Thus, Co/Ni-doped (monolayer) MoS_2 had improved ORR activity compared to bare MoS_2 [86]. Furthermore, Urbanova and He *et al* introduced metal compounds (i.e. Mn, V, Ti, and Fe) into MoS_2 and established a noteworthy development in ORR activity (figures 7.12(a) and (b)) [87, 88]. Although numerous studies have been completed towards incorporating metal as a dopant in the assembly of MoS_2, selecting the kinds of metals is a simple trial-and-error process. In this regard, the exploration of ORR activity for a variety of transition metal doped MoS_2 monolayers was performed by Wang *et al* using DFT measurements. The outcomes reveal that doping transition metal atoms causes substitution of the S vacancy of MoS_2, leading to considerable electronic structure alteration. As an ORR intermediate, the Cu-doped MoS_2 reveals the greatest theoretical binding strength; thus, it corresponds to the best ORR activity, as represented in figures 7.12(c) and (d) [89]. As the S-bonds in metal are considered reactive active sites in metal-doped TMDs, growing the metal atom density over the MoS_2 surface up to a maximum extent and dropping the metal accumulation during heat treatment can be recognized as a significant parameter to improve the catalyst's reactivity [17]. However, these parameters are challenging to accomplish. The main difficulty is controlling the metal doping over

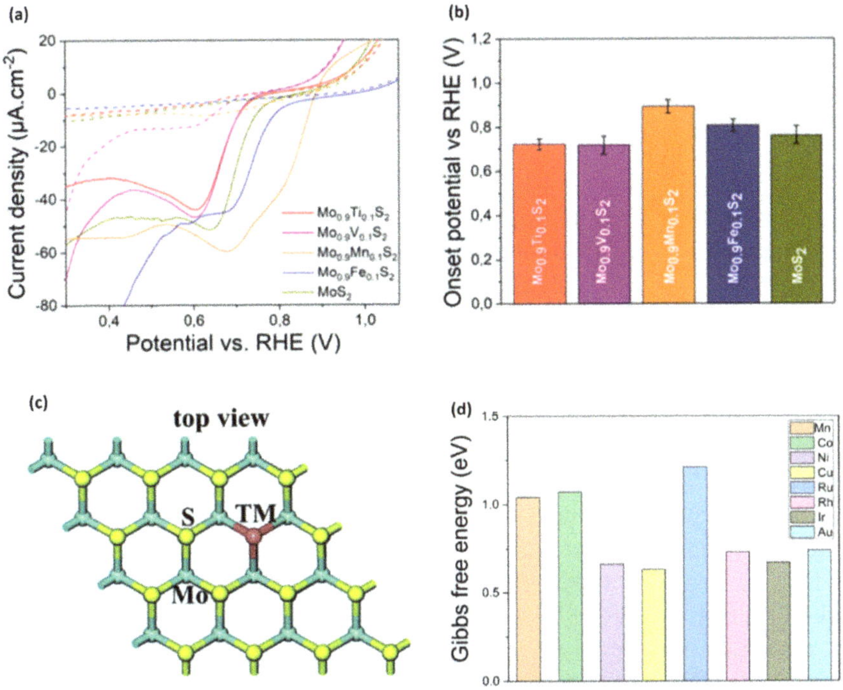

Figure 7.12. (a) Normal curves of LSV point toward ORRs and (b) corresponding values of onset potentials. (Reproduced with permission from [87]. Copyright 2020 American Chemical Society.) (c) A diagrammatic representation of a transitional atom that integrates into the empty S slot in the MoS_2 monolayer. (Reproduced with permission from [89]. Copyright 2017 The Royal Society of Chemistry.) (d) ORR on transition metals embedded in MoS_2 monolayers using the OOH association process in an acidic medium, depicted as a schematic of the Gibbs free energy diagrams. According to findings from previous research in the data from [89].

the MoS_2 surface, which is quite low. Hence, decreasing these problems is critical to increasing the ORR activity of metal-doped MoS_2.

7.3 CO_2 reduction reactions

7.3.1 Surface modification

The surfaces of 2D nanosheets with a high degree of uncovered low-coordination atoms are straightforward candidates for modification with metallic and nonmetallic species. Multiple metals can be combined to increase the chemical intermediates' binding energies, such as CO. This can also eliminate scale relationships between intermediates that have effectively absorbed to lower overpotential [90]. For example, *CO firmly attaches to metals only through the C atom, but including an oxophilic atom on the material's surface will selectively stabilize *COOH and *CHO over *CO. Adding adatoms like cobalt, manganese, bismuth, and others can significantly alter the electrical and geometric properties of nanosheets. It is possible to modulate the activity of CO_2 conversion while maintaining its selectivity by doping, which allows the tuning of the binding strengths of various intermediates and the reaction energies. Doping Pd with Te enables extraordinarily efficient electrocatalytic conversion of aqueous CO_2 to CO. DFT studies show Te atoms attaching preferentially to Pd's terrace sites, suppressing hydrogen growth. On the other hand, the formation of CO is caused by adsorption and activating on high-index sites of Pd. Chemical doping of 2D materials, particularly those with a configuration and doping amount that can be controlled, continues to be a problem. Surface functional groups, such as amines, are essential to the activity and selectivity of chemical reactions. Adding an amine to a catalyst can concentrate CO_2 on the electrode surface, raising the concentration of CO_2 in the immediate area [17]. In addition, amine can prevent $CO_2^{\cdot -}$ from becoming $CO_2^{\cdot -}$ via an interaction involving an H-bond. This lowers the onset potential for converting CO_2 to $CO_2^{\cdot -}$ by producing more stable surroundings. Similarly, amine can assist in bonding *CO, which in turn can facilitate the significant reduction of CO to methanol and the C–C coupling necessary for creating C_2 [91]. Pristine graphene-type carbon materials have limited CO_2 activation catalytic activity, but doping them with heteroatoms such as B, N, and/or Ni can substantially reduce CO_2 barrier systems for adsorption and make the electrocatalytic CO_2 reduction more feasible [17]. At a relatively lower doping concentration level of 0.6 atom percent, quaternary N doping may significantly increase the carrier concentration up to 2.6×10^{13} cm^{-2}, which is four times greater than for unmodified graphene [92]. This point should be emphasized because it is crucial to note that further research is required. A deeper investigation of the ECR characteristics of single- and co-doped nanomaterials, including a variety of p-block and d-block components in addition to B, N, and Ni, is still required.

7.3.2 Surface-structure tuning

The primary means by which structural tuning of 2D nanosheets may be accomplished is by adjusting four crucial elements: surface imperfections, surface permeability, accessible crystal facets, and surface phase [17]. Creating micro-, meso-, and macroporous structures on 2D nanosheets enhances CO_2 adsorption and conversion efficiency [17]. It is possible to increase the quantity of CO_2 collected by metal oxides or hydroxides through surface adjusting with an oxygen vacancy. Because one oxygen atom of CO_2 is positioned by two oxygen atoms' vacancy defects, the energy threshold for CO_2 activation is decreased. This is because CO_2 molecules tend to attach at oxygen vacancies [93]. Similarly, chalcogen vacancies on chalcogenide nanosheets have the potential to function as adsorption sites for CO_2, hence aiding in the selective reduction of CO_2 to CO [94]. The surface vacancy density can be altered by lowering the surface temperature or doping the material with a lower-valent foreign element. It is easier to transfer and diffuse reactants and products over electrodes when macropores ($>$ 50 nm) are present. On the other hand, mesopores (2–50 nm) and micropores (2 nm) give a much greater surface for active sites with significant dispersion. In addition, the molecular sizes of the reactants or products can be used to tune the widths of pores or channels, which can then alter the selectivity accordingly. The facets of the crystal are an important factor in influencing the activity and selectivity of the reaction. CO_2 adsorption and activation are affected by the variable Lewis acidity levels and polarizing power displayed by the various surface facets. In electrocatalytic processes, it is recommended to employ facets with intense adsorption energies yet low activation barriers to achieve the best results [17]. The catalytic activity and selectivity of TMDs, as well as a large number of metals, are extremely sensitive to the surface arrangement and composition. TMD's octahedral (1T) phase has an intrinsically low charge-transfer resistance because of its structure. When contrasted with the prismatic (2H) trigonal phase, the outcome of strained lattice distortion is an increase in active-site density. The electrical characteristics and activity of catalysts can be tuned by manipulating the adsorbed ions near vacancies on the surface. Specifically, incorporating the surface of metal electrode materials that have been partially oxidized, notably Sn, Bi, and Co, gives greater activity than pristine metals because of the increased surface area [95].

7.3.3 Optimization of the electrolyte and electrolyzer

There is a significant relationship between the CO_2 electrolysis product and the type, content, and amount of the electrolytes [96]. Alkaline electrolytes, including potassium hydroxide (KOH), are advantageous for reducing CO_2 because they decrease undesired HER and reduce the activation energy barriers for CO_2 reduction and CO–CO coupling [97]. It was shown that employing IL electrolytes in conjunction with various electrodes can decrease CO_2 [98, 99]. It was proposed that ion pairing between the IL cation and the CO_2 radical anion could lower the activation overpotential [100]. The current density that can be accomplished in ILs is approximately ten times greater than that in aqueous electrolytes [98]. Interestingly, most ILs have a significantly higher CO_2 solubility than water, which favors a high concentration of CO_2 at the interface in both the electrode and the electrolyte.

Additionally, there is evidence that the adsorption of ILs at the interface can effectively inhibit the development of hydrogen. Consequently, aqueous ILs are diluted and include ion-pairing and HER-inhibiting capabilities of potential for CO_2 reduction in commercial processes [101]. On the other hand, the challenge lies in the possibility that ILs will become unstable when subjected to the high current densities required for commercial systems. Because carbon dioxide is not very soluble in aqueous electrolytes, it is possible to construct a flow-cell electrolyzer with a gas diffusion electrode (GDE) as the cathode to circumvent the consequences of mass transport constraint [102]. This electrolyzer supplies gaseous carbon dioxide to the electrode–electrolyte interface to produce a triple-phase boundary. Because of this, it is possible to transform carbon dioxide at a rapid pace. Interface engineering should be used to reduce flooding and salt accumulation to facilitate the operation of stable cells and the commercialization of CO_2 electrolyzers. In the meantime, it is essential to have a sturdy electrode–electrolyte interface to make it possible for gaseous CO_2 and products to quickly move in and out of the GDE without causing the ionic and electrical permeability to deteriorate at high reaction rates [17].

7.4 Conclusion, challenges and perspectives

In this chapter descriptions of various approaches for the improvement of electrocatalytic activity in the direction of the HER, ORR, and ECR are provided. Generally, carrier dissociation along with migration kinetics are common protocols for advancing the activity of electrocatalysis and photocatalysis. Beyond these strategies and the characteristics that are described in previous chapters, this chapter reveals that engineering protocols such as defect engineering, phase engineering, interface engineering, and various kinds of doping techniques such as heteroatom doping, etc, can be used for the improvement of electrocatalytic activity. The results acquired from experimental and theoretical data clearly reveal that these routes cause superior activity electrocatalysts for electrolysis. These novel strategies surely suggest that electrocatalysts can reach the industrial and practical level by adopting these techniques to obtain economical, environmentally friendly and highly efficient catalysts.

Bibliography

[1] Lin J, Wang P, Wang H, Li C, Si X, Qi J, Cao J, Zhong Z, Fei W and Feng J 2019 Defect-rich heterogeneous MoS_2/NiS_2 nanosheets electrocatalysts for efficient overall water splitting *Adv. Sci.* **6** 1900246

[2] Yang Y, Fei H, Ruan G, Li Y and Tour J M 2015 Vertically aligned WS_2 nanosheets for water splitting *Adv. Funct. Mater.* **25** 6199–204

[3] Yang J *et al* 2019 Ultrahigh-current-density niobium disulfide catalysts for hydrogen evolution *Nat. Mater.* **18** 1309–14

[4] Lin L, Miao N, Wen Y, Zhang S, Ghosez P, Sun Z and Allwood D A 2016 Sulfur-depleted monolayered molybdenum disulfide nanocrystals for superelectrochemical hydrogen evolution reaction *ACS Nano* **10** 8929–37

[5] Shi J, Ma D, Han G-F, Zhang Y, Ji Q, Gao T, Sun J, Song X, Li C and Zhang Y 2014 Controllable growth and transfer of monolayer MoS_2 on Au foils and its potential application in hydrogen evolution reaction *ACS Nano* **8** 10196–204

[6] Voiry D, Yamaguchi H, Li J, Silva R, Alves D C, Fujita T, Chen M, Asefa T, Shenoy V B and Eda G 2013 Enhanced catalytic activity in strained chemically exfoliated WS_2 nanosheets for hydrogen evolution *Nat. Mater.* **12** 850–5

[7] Lin L, Miao N, Huang J, Zhang S, Zhu Y, Horsell D D, Ghosez P, Sun Z and Allwood D A 2017 A photocatalyst of sulphur depleted monolayered molybdenum sulfide nanocrystals for dye degradation and hydrogen evolution reaction *Nano Energy* **38** 544–52

[8] Ye G, Gong Y, Lin J, Li B, He Y, Pantelides S T, Zhou W, Vajtai R and Ajayan P M 2016 Defects engineered monolayer MoS_2 for improved hydrogen evolution reaction *Nano Lett.* **16** 1097–103

[9] Tao L, Duan X, Wang C, Duan X and Wang S 2015 Plasma-engineered MoS_2 thin-film as an efficient electrocatalyst for hydrogen evolution reaction *Chem. Commun.* **51** 7470–3

[10] Li H, Tan Y, Liu P, Guo C, Luo M, Han J, Lin T, Huang F and Chen M 2016 Atomic-sized pores enhanced electrocatalysis of TaS_2 nanosheets for hydrogen evolution *Adv. Mater.* **28** 8945–9

[11] Coleman J N, Lotya M, O'Neill A, Bergin S D, King P J, Khan U, Young K, Gaucher A, De S and Smith R J 2011 Two-dimensional nanosheets produced by liquid exfoliation of layered materials *Science* **331** 568–71

[12] Lei Z, Xu S and Wu P 2016 Ultra-thin and porous $MoSe_2$ nanosheets: facile preparation and enhanced electrocatalytic activity towards the hydrogen evolution reaction *Phys. Chem. Chem. Phys.* **18** 70–4

[13] Tan C, Luo Z, Chaturvedi A, Cai Y, Du Y, Gong Y, Huang Y, Lai Z, Zhang X and Zheng L 2018 Preparation of high-percentage 1T-phase transition metal dichalcogenide nanodots for electrochemical hydrogen evolution *Adv. Mater.* **30** 1705509

[14] Xu S, Li D and Wu P 2015 One-pot, facile, and versatile synthesis of monolayer MoS_2/WS_2 quantum dots as bioimaging probes and efficient electrocatalysts for hydrogen evolution reaction *Adv. Funct. Mater.* **25** 1127–36

[15] Lin L, Xu Y, Zhang S, Ross I M, Ong A C and Allwood D A 2013 Fabrication of luminescent monolayered tungsten dichalcogenides quantum dots with giant spin-valley coupling *ACS Nano* **7** 8214–23

[16] Ouyang Y, Ling C, Chen Q, Wang Z, Shi L and Wang J 2016 Activating inert basal planes of MoS_2 for hydrogen evolution reaction through the formation of different intrinsic defects *Chem. Mater.* **28** 4390–6

[17] Raza A, Hassan J Z, Qumar U, Zaheer A, Babar Z U D, Iannotti V and Cassinese A 2024 Strategies for robust electrocatalytic activity of 2D materials: ORR, OER, HER, and CO2RR *Mater. Today Adv.* **22** 100488

[18] Hong J, Hu Z, Probert M, Li K, Lv D, Yang X, Gu L, Mao N, Feng Q and Xie L 2015 Exploring atomic defects in molybdenum disulphide monolayers *Nat. Commun.* **6** 1–8

[19] Li H, Tsai C, Koh A L, Cai L, Contryman A W, Fragapane A H, Zhao J, Han H S, Manoharan H C and Abild-Pedersen F 2016 Activating and optimizing MoS_2 basal planes for hydrogen evolution through the formation of strained sulphur vacancies *Nat. Mater.* **15** 48–53

[20] Voiry D, Salehi M, Silva R, Fujita T, Chen M, Asefa T, Shenoy V B, Eda G and Chhowalla M 2013 Conducting MoS_2 nanosheets as catalysts for hydrogen evolution reaction *Nano Lett.* **13** 6222–7

[21] Chang K, Hai X, Pang H, Zhang H, Shi L, Liu G, Liu H, Zhao G, Li M and Ye J 2016 Targeted synthesis of 2H-and 1T-phase MoS_2 monolayers for catalytic hydrogen evolution *Adv. Mater.* **28** 10033–41

[22] Tsai C, Chan K, Abild-Pedersen F and Nørskov J K 2014 Active edge sites in $MoSe_2$ and WSe_2 catalysts for the hydrogen evolution reaction: a density functional study *Phys. Chem. Chem. Phys.* **16** 13156–64

[23] Zhang Y *et al* 2015 Chemical vapor deposition of monolayer WS_2 nanosheets on Au foils toward direct application in hydrogen evolution *Nano Res.* **8** 2881–90

[24] Acerce M, Voiry D and Chhowalla M 2015 Metallic 1T phase MoS_2 nanosheets as supercapacitor electrode materials *Nat. Nanotechnol.* **10** 313–8

[25] Hinnemann B, Moses P G, Bonde J, Jørgensen K P, Nielsen J H, Horch S, Chorkendorff I and Nørskov J K 2005 Biomimetic hydrogen evolution: MoS_2 nanoparticles as catalyst for hydrogen evolution *J. Am. Chem. Soc.* **127** 5308–9

[26] Jawaid A, Che J, Drummy L F, Bultman J, Waite A, Hsiao M-S and Vaia R A 2017 Redox exfoliation of layered transition metal dichalcogenides *ACS Nano* **11** 635–46
[27] Xie J, Zhang H, Li S, Wang R, Sun X, Zhou M, Zhou J, Lou X W and Xie Y 2013 Defect-rich MoS_2 ultrathin nanosheets with additional active edge sites for enhanced electrocatalytic hydrogen evolution *Adv. Mater.* **25** 5807–13
[28] He Q, Wang L, Yin K and Luo S 2018 Vertically aligned ultrathin 1T-WS_2 nanosheets enhanced the electrocatalytic hydrogen evolution *Nanoscale Res. Lett.* **13** 1–9
[29] Voiry D *et al* 2013 Enhanced catalytic activity in strained chemically exfoliated WS_2 nanosheets for hydrogen evolution *Nat. Mater.* **12** 850–5
[30] Tan C *et al* 2018 Preparation of high-percentage 1T-phase transition metal dichalcogenide nanodots for electrochemical hydrogen evolution *Adv. Mater.* **30** 1705509
[31] Sokolikova M S, Sherrell P C, Palczynski P, Bemmer V L and Mattevi C 2019 Direct solution-phase synthesis of 1T' WSe_2 nanosheets *Nat. Commun.* **10** 712
[32] Novoselov K S, Mishchenko A, Carvalho A and Castro Neto A H 2016 2D materials and van der Waals heterostructures *Science* **353** aac9439
[33] Gong M *et al* 2014 Nanoscale nickel oxide/nickel heterostructures for active hydrogen evolution electrocatalysis *Nat. Commun.* **5** 4695
[34] Zheng Y, Jiao Y, Zhu Y, Li L H, Han Y, Chen Y, Du A, Jaroniec M and Qiao S Z 2014 Hydrogen evolution by a metal-free electrocatalyst *Nat. Commun.* **5** 3783
[35] Duan J, Chen S, Jaroniec M and Qiao S Z 2015 Porous C_3N_4 nanolayers@N-graphene films as catalyst electrodes for highly efficient hydrogen evolution *ACS Nano* **9** 931–40
[36] Chen S, Duan J, Jaroniec M and Qiao S-Z 2014 Nitrogen and oxygen dual-doped carbon hydrogel film as a substrate-free electrode for highly efficient oxygen evolution reaction *Adv. Mater.* **26** 2925–30
[37] Duan J, Chen S, Chambers B A, Andersson G G and Qiao S Z 2015 3D WS_2 nanolayers@heteroatom-doped graphene films as hydrogen evolution catalyst electrodes *Adv. Mater.* **27** 4234–41
[38] Gao M-R, Liang J-X, Zheng Y-R, Xu Y-F, Jiang J, Gao Q, Li J and Yu S-H 2015 An efficient molybdenum disulfide/cobalt diselenide hybrid catalyst for electrochemical hydrogen generation *Nat. Commun.* **6** 5982
[39] Staszak-Jirkovský J *et al* 2016 Design of active and stable Co–Mo–Sx chalcogels as pH-universal catalysts for the hydrogen evolution reaction *Nat. Mater.* **15** 197–203
[40] Yang J, Voiry D, Ahn S J, Kang D, Kim A Y, Chhowalla M and Shin H S 2013 Two-dimensional hybrid nanosheets of tungsten disulfide and reduced graphene oxide as catalysts for enhanced hydrogen evolution *Angew. Chem. Int. Ed.* **52** 13751–4
[41] Huang X, Zeng Z, Bao S, Wang M, Qi X, Fan Z and Zhang H 2013 Solution-phase epitaxial growth of noble metal nanostructures on dispersible single-layer molybdenum disulfide nanosheets *Nat. Commun.* **4** 1444
[42] Li Y, Wang H, Xie L, Liang Y, Hong G and Dai H 2011 MoS_2 nanoparticles grown on graphene: an advanced catalyst for the hydrogen evolution reaction *J. Am. Chem. Soc.* **133** 7296–9
[43] Du X *et al* 2019 Modulating electronic structures of inorganic nanomaterials for efficient electrocatalytic water splitting *Angew. Chem. Int. Ed.* **58** 4484–502
[44] Xiong Q, Wang Y, Liu P-F, Zheng L-R, Wang G, Yang H-G, Wong P-K, Zhang H and Zhao H 2018 Cobalt covalent doping in MoS_2 to induce bifunctionality of overall water splitting *Adv. Mater.* **30** 1801450

[45] Wang R *et al* 2019 Beyond 1T-phase? Synergistic electronic structure and defects engineering in 2H-$MoS_{2x}Se_{2(1-x)}$ nanosheets for enhanced hydrogen evolution reaction and sodium storage *ChemCatChem.* **11** 3200–11
[46] Fu Q, Yang L, Wang W, Han A, Huang J, Du P, Fan Z, Zhang J and Xiang B 2015 Synthesis and enhanced electrochemical catalytic performance of monolayer $WS_{2(1-x)}Se_{2x}$ with a tunable band gap *Adv. Mater.* **27** 4732–8
[47] Zhang K *et al* 2014 Unconventional pore and defect generation in molybdenum disulfide: application in high-rate lithium-ion batteries and the hydrogen evolution reaction *ChemSusChem* **7** 2489–95
[48] Sun X, Dai J, Guo Y, Wu C, Hu F, Zhao J, Zeng X and Xie Y 2014 Semimetallic molybdenum disulfide ultrathin nanosheets as an efficient electrocatalyst for hydrogen evolution *Nanoscale* **6** 8359–67
[49] Xiong Q, Zhang X, Wang H, Liu G, Wang G, Zhang H and Zhao H 2018 One-step synthesis of cobalt-doped MoS_2 nanosheets as bifunctional electrocatalysts for overall water splitting under both acidic and alkaline conditions *Chem. Commun.* **54** 3859–62
[50] Liu G, Cui Z, Han M, Zhang S, Zhao C, Chen C, Wang G and Zhang H 2019 Ambient electrosynthesis of ammonia on a core–shell-structured Au@CeO_2 catalyst: contribution of oxygen vacancies in CeO_2 *Chem. Eur. J.* **25** 5904–11
[51] Yang L, Fu Q, Wang W, Huang J, Huang J, Zhang J and Xiang B 2015 Large-area synthesis of monolayered $MoS_{2(1-x)}Se_{2x}$ with a tunable band gap and its enhanced electrochemical catalytic activity *Nanoscale* **7** 10490–7
[52] Xie J, Zhang J, Li S, Grote F, Zhang X, Zhang H, Wang R, Lei Y, Pan B and Xie Y 2013 Controllable disorder engineering in oxygen-incorporated MoS_2 ultrathin nanosheets for efficient hydrogen evolution *J. Am. Chem. Soc.* **135** 17881–8
[53] Ding Q *et al* 2015 Designing efficient solar-driven hydrogen evolution photocathodes using semitransparent MoQ_xCl_y (Q = S, Se) catalysts on Si micropyramids *Adv. Mater.* **27** 6511–8
[54] Zhang X, Meng F, Mao S, Ding Q, Shearer M J, Faber M S, Chen J, Hamers R J and Jin S 2015 Amorphous MoS_xCl_y electrocatalyst supported by vertical graphene for efficient electrochemical and photoelectrochemical hydrogen generation *Energy Environ. Sci.* **8** 862–8
[55] Sun Y, Alimohammadi F, Zhang D and Guo G 2017 Enabling colloidal synthesis of edge-oriented MoS_2 with expanded interlayer spacing for enhanced HER catalysis *Nano Lett.* **17** 1963–9
[56] Zhao Z *et al* 2017 Vertically aligned MoS_2/Mo_2C hybrid nanosheets grown on carbon paper for efficient electrocatalytic hydrogen evolution *ACS Catal.* **7** 7312–8
[57] Zhang J, Chen Y, Liu M, Du K, Zhou Y, Li Y, Wang Z and Zhang J 2018 1T@2H-$MoSe_2$ nanosheets directly arrayed on Ti plate: an efficient electrocatalytic electrode for hydrogen evolution reaction *Nano Res.* **11** 4587–98
[58] Cao J, Zhou J, Zhang Y, Wang Y and Liu X 2018 Dominating role of aligned MoS_2/Ni_3S_2 nanoarrays supported on three-dimensional Ni foam with hydrophilic interface for highly enhanced hydrogen evolution reaction *ACS Appl. Mater. Interfaces* **10** 1752–60
[59] Zhou H *et al* 2016 Efficient hydrogen evolution by ternary molybdenum sulfoselenide particles on self-standing porous nickel diselenide foam *Nat. Commun.* **7** 12765
[60] Yang Y *et al* 2019 Hierarchical nanoassembly of MoS_2/Co_9S_8/Ni_3S_2/Ni as a highly efficient electrocatalyst for overall water splitting in a wide pH range *J. Am. Chem. Soc.* **141** 10417–30
[61] Ding Q, Song B, Xu P and Jin S 2016 Efficient electrocatalytic and photoelectrochemical hydrogen generation using MoS_2 and related compounds *Chem.* **1** 699–726

[62] Seh Z W, Kibsgaard J, Dickens C F, Chorkendorff I, Nørskov J K and Jaramillo T F 2017 Combining theory and experiment in electrocatalysis: insights into materials design *Science* **355** eaad4998

[63] Zhao G, Rui K, Dou S X and Sun W 2018 Heterostructures for electrochemical hydrogen evolution reaction: a review *Adv. Funct. Mater.* **28** 1803291

[64] Quaino P, Juarez F, Santos E and Schmickler W 2014 Volcano plots in hydrogen electrocatalysis—uses and abuses *Beilstein J. Nanotechnol.* **5** 846–54

[65] Parsons R 1958 The rate of electrolytic hydrogen evolution and the heat of adsorption of hydrogen *Trans. Faraday Soc.* **54** 1053–63

[66] Skúlason E, Tripkovic V, Björketun M E, Gudmundsdóttir S, Karlberg G, Rossmeisl J, Bligaard T, Jónsson H and Nørskov J K 2010 Modeling the electrochemical hydrogen oxidation and evolution reactions on the basis of density functional theory calculations *J. Phys. Chem.* C **114** 18182–97

[67] Wei J, Zhou M, Long A, Xue Y, Liao H, Wei C and Xu Z J 2018 Heterostructured electrocatalysts for hydrogen evolution reaction under alkaline conditions *Nano-Micro Lett.* **10** 75

[68] Danilovic N, Subbaraman R, Strmcnik D, Chang K-C, Paulikas A P, Stamenkovic V R and Markovic N M 2012 Enhancing the alkaline hydrogen evolution reaction activity through the bifunctionality of $Ni(OH)_2$/metal catalysts *Angew. Chem. Int. Ed.* **51** 12495–8

[69] Higgins D, Zamani P, Yu A and Chen Z 2016 The application of graphene and its composites in oxygen reduction electrocatalysis: a perspective and review of recent progress *Energy Environ. Sci.* **9** 357–90

[70] Shao M, Chang Q, Dodelet J-P and Chenitz R 2016 Recent advances in electrocatalysts for oxygen reduction reaction *Chem. Rev.* **116** 3594–657

[71] Stacy J, Regmi Y N, Leonard B and Fan M 2017 The recent progress and future of oxygen reduction reaction catalysis: a review *Renew. Sustain. Energy Rev.* **69** 401–14

[72] Yang Z, Nie H, Chen X a, Chen X and Huang S 2013 Recent progress in doped carbon nanomaterials as effective cathode catalysts for fuel cell oxygen reduction reaction *J. Power Sources* **236** 238–49

[73] Zhu J, Xiao M, Zhao X, Liu C, Ge J and Xing W 2015 Strongly coupled Pt nanotubes/N-doped graphene as highly active and durable electrocatalysts for oxygen reduction reaction *Nano Energy* **13** 318–26

[74] Xie B, Zhang Y and Zhang R 2017 Coassembly and high ORR performance of monodisperse Pt nanocrystals with a mesopore-rich nitrogen-doped graphene aerogel *J. Mater. Chem.* A **5** 17544–8

[75] Hu W and Yang J 2017 Two-dimensional van der Waals heterojunctions for functional materials and devices *J. Mater. Chem.* C **5** 12289–97

[76] Wang J and Wang H 2018 One-step fabrication of coating-free mesh with underwater superoleophobicity for highly efficient oil/water separation *Surf. Coat. Technol.* **340** 1–7

[77] Wu M, Dou Z, Chang J and Cui L 2016 Nitrogen and sulfur co-doped graphene aerogels as an efficient metal-free catalyst for oxygen reduction reaction in an alkaline solution *RSC Adv.* **6** 22781–90

[78] Jaramillo T F, Jørgensen K P, Bonde J, Nielsen J H, Horch S and Chorkendorff I 2007 Identification of active edge sites for electrochemical H_2 evolution from MoS_2 nanocatalysts *Science* **317** 100–2

[79] Wang J, Zhang H, Wang C, Zhang Y, Wang J, Zhao H, Cheng M, Li A and Wang J 2018 Co-synthesis of atomic Fe and few-layer graphene towards superior ORR electrocatalyst *Energy Storage Mater.* **12** 1–7

[80] Tao L, Wang Q, Dou S, Ma Z, Huo J, Wang S and Dai L 2016 Edge-rich and dopant-free graphene as a highly efficient metal-free electrocatalyst for the oxygen reduction reaction *Chem. Commun.* **52** 2764–7

[81] Huang H, Feng X, Du C, Wu S and Song W 2015 Incorporated oxygen in MoS_2 ultrathin nanosheets for efficient ORR catalysis *J. Mater. Chem.* A **3** 16050–6

[82] Huang H, Feng X, Du C and Song W 2015 High-quality phosphorus-doped MoS_2 ultrathin nanosheets with amenable ORR catalytic activity *Chem. Commun.* **51** 7903–6

[83] Li L, Wei Z, Chen S, Qi X, Ding W, Xia M, Li R, Xiong K, Deng Z and Gao Y 2012 A comparative DFT study of the catalytic activity of MnO_2 (211) and (2-2-1) surfaces for an oxygen reduction reaction *Chem. Phys. Lett.* **539–40** 89–93

[84] Zhang H, Tian Y, Zhao J, Cai Q and Chen Z 2017 Small dopants make big differences: enhanced electrocatalytic performance of MoS_2 monolayer for oxygen reduction reaction (ORR) by N- and P-doping *Electrochim. Acta* **225** 543–50

[85] Liu C, Dong H, Ji Y, Hou T and Li Y 2018 Origin of the catalytic activity of phosphorus doped MoS_2 for oxygen reduction reaction (ORR) in alkaline solution: a theoretical study *Sci. Rep.* **8** 13292

[86] Xiao B B, Zhang P, Han L P and Wen Z 2015 Functional MoS_2 by the Co/Ni doping as the catalyst for oxygen reduction reaction *Appl. Surf. Sci.* **354** 221–8

[87] Urbanová V, Lazar P, Antonatos N, Sofer Z, Otyepka M and Pumera M 2020 Positive and negative effects of dopants toward electrocatalytic activity of MoS_2 and WS_2: experiments and theory *ACS Appl. Mater. Interfaces* **12** 20383–92

[88] Di J *et al* 2019 Defect-tailoring mediated electron–hole separation in single-unit-cell Bi_3O_4Br nanosheets for boosting photocatalytic hydrogen evolution and nitrogen fixation *Adv. Mater.* **31** 1807576

[89] Wang Z, Zhao J, Cai Q and Li F 2017 Computational screening for high-activity MoS_2 monolayer-based catalysts for the oxygen reduction reaction via substitutional doping with transition metal *J. Mater. Chem.* A **5** 9842–51

[90] Jovanov Z P, Hansen H A, Varela A S, Malacrida P, Peterson A A, Nørskov J K, Stephens I E L and Chorkendorff I 2016 Opportunities and challenges in the electrocatalysis of CO_2 and CO reduction using bifunctional surfaces: a theoretical and experimental study of Au–Cd alloys *J. Catal.* **343** 215–31

[91] Sun X, Zhu Q, Kang X, Liu H, Qian Q, Ma J, Zhang Z, Yang G and Han B 2017 Design of a Cu(i)/C-doped boron nitride electrocatalyst for efficient conversion of CO_2 into acetic acid *Green Chem.* **19** 2086–91

[92] Velez-Fort E, Mathieu C, Pallecchi E, Pigneur M, Silly M G, Belkhou R, Marangolo M, Shukla A, Sirotti F and Ouerghi A 2012 Epitaxial graphene on 4H-SiC(0001) grown under nitrogen flux: evidence of low nitrogen doping and high charge transfer *ACS Nano* **6** 10893–900

[93] Chang X, Wang T and Gong J 2016 CO_2 photo-reduction: insights into CO_2 activation and reaction on surfaces of photocatalysts *Energy Environ. Sci.* **9** 2177–96

[94] Sun Z, Ma T, Tao H, Fan Q and Han B 2017 Fundamentals and challenges of electrochemical CO_2 reduction using two-dimensional materials *Chem* **3** 560–87

[95] Gao S, Lin Y, Jiao X, Sun Y, Luo Q, Zhang W, Li D, Yang J and Xie Y 2016 Partially oxidized atomic cobalt layers for carbon dioxide electroreduction to liquid fuel *Nature* **529** 68–71

[96] Fan Q, Zhang M, Jia M, Liu S, Qiu J and Sun Z 2018 Electrochemical CO_2 reduction to C_2^+ species: heterogeneous electrocatalysts, reaction pathways, and optimization strategies *Mater. Today Energy* **10** 280–301

[97] Dinh C-T *et al* 2018 CO_2 electroreduction to ethylene via hydroxide-mediated copper catalysis at an abrupt interface *Science* **360** 783–7

[98] Kumar B, Asadi M, Pisasale D, Sinha-Ray S, Rosen B A, Haasch R, Abiade J, Yarin A L and Salehi-Khojin A 2013 Renewable and metal-free carbon nanofibre catalysts for carbon dioxide reduction *Nat. Commun.* **4** 2819

[99] Asadi M *et al* 2014 Robust carbon dioxide reduction on molybdenum disulphide edges *Nat. Commun.* **5** 4470

[100] Rosen B A, Salehi-Khojin A, Thorson M R, Zhu W, Whipple D T, Kenis P J A and Masel R I 2011 Ionic liquid-mediated selective conversion of CO_2 to CO at low overpotentials *Science* **334** 643–4

[101] Rosen B A and Hod I 2018 Tunable molecular-scale materials for catalyzing the low-overpotential electrochemical conversion of CO_2 *Adv. Mater.* **30** 1706238

[102] Jouny M, Luc W and Jiao F 2018 High-rate electroreduction of carbon monoxide to multi-carbon products *Nat. Catal.* **1** 748–55

www.ingramcontent.com/pod-product-compliance
Ingram Content Group UK Ltd.
Pitfield, Milton Keynes, MK11 3LW, UK
UKHW051854100926
8556IPUK00005B/621

9 780750 352895